Job #: 116032

Author Name: Karanfil

Title of Book: Disinfection By-Products ACS

ISBN #: 9780841269507

Disinfection By-Products in Drinking Water

ACS SYMPOSIUM SERIES **995**

Disinfection By-Products in Drinking Water

Occurrence, Formation, Health Effects, and Control

Tanju Karanfil, EDITOR
Clemson University

Stuart W. Krasner, EDITOR
Metropolitan Water District of Southern California

Paul Westerhoff, EDITOR
Arizona State University

Yuefeng Xie, EDITOR
The Pennsylvania State University

Sponsored by the ACS Division of Environmental Chemistry, Inc.

American Chemical Society, Washington, DC

ISBN: 978–0–8412–6950–7

The paper used in this publication meets the minimum requirements of American National Standard for Information Sciences—Permanence of Paper for Printed Library Materials, ANSI Z39.48–1984.

Distributed by Oxford University Press

Foreword

The ACS Symposium Series was first published in 1974 to provide a mechanism for publishing symposia quickly in book form. The purpose of the series is to publish timely, comprehensive books developed from ACS sponsored symposia based on current scientific research. Occasionally, books are developed from symposia sponsored by other organizations when the topic is of keen interest to the chemistry audience.

Before agreeing to publish a book, the proposed table of contents is reviewed for appropriate and comprehensive coverage and for interest to the audience. Some papers may be excluded to better focus the book; others may be added to provide comprehensiveness. When appropriate, overview or introductory chapters are added. Drafts of chapters are peer-reviewed prior to final acceptance or rejection, and manuscripts are prepared in camera-ready format.

As a rule, only original research papers and original review papers are included in the volumes. Verbatim reproductions of previously published papers are not accepted.

ACS Books Department

Contents

Preface .... **xi**

Introduction

1. **Recent Advances in Disinfection By-product Formation, Occurrence, Control, Health Effects, and Regulations** **2**
   Tanju Karanfil, Stuart W. Krasner, Paul Westerhoff, and Yuefeng Xie

Health and Regulatory Concerns of Disinfection By-Products

2. **The Evolution of Disinfection By-Products Regulations: Past, Present, and Future** **22**
   J. Alan Roberson

3. **Comparative Mammalian Cell Toxicity of Nitrogen-Containing Disinfection By-Products and Carbonaceous Disinfection By-Products** **36**
   Michael J. Plewa, Elizabeth D. Wagner, Mark G. Muellner, Kang-Mei Hsu, and Susan D. Richardson

4. **Use of Chemical Models and Structure–Activity Relationships to Identify Novel Disinfection By-Products of Potential Toxicological Concern** **51**
   Richard J. Bull and David. A. Reckhow

5. **Contribution of Organic Bromines to the Genotoxicity of Chlorinated Water: A Combination of Chromosomal Aberration Test and Total Organic Bromine Analysis** **65**
   Shinya Echigo, Sadahiko Itoh, and Ryo Ando

Reactivity of Natural Organic Matter toward Disinfection By-Product Formation

6. **Disinfection By-Product Precursor Content of Natural Organic Matter Extracts...80**
David A. Reckhow, Gladys Makdissy, and Paula S. Rees

7. **Parameters Affecting Haloacetic Acid and Trihalomethane Concentrations in Treated United Kingdom Drinking Waters............95**
Cynthia M. M. Bougeard, Imran H. S. Janmohamed, Emma H. Goslan, Bruce Jefferson, Jonathan S. Watson, Geraint H. Morgan, and Simon A. Parsons

8. **Relationship between Brominated Trihalomethanes, Haloacetic Acids, and Total Organic Bromine during Drinking Water Chlorination...108**
Guanghui Hua and David A. Reckhow

9. **Haloacetic Acids Formation and Speciation during Chloramination...123**
Tanju Karanfil, Ying Hong, and Hocheol Song

10. **Prechlorination Induced Dissolved Organic Carbon and Disinfection By-Products Formation from *Microcystis aeruginosa* in Treatment Processes...140**
Chao-An Chiu and Gen-Shuh Wang

Probing Formation of Disinfection By-Products

11. **Spectroscopic Studies of the Roles of Distinct Chromophores in Natural Organic Matter Chlorination and Disinfection By-Products Formation..158**
Gregory V. Korshin and Hyun-Shik Chang

12. **Formation of Volatile Disinfection By-Products from Chlorination of Organic-N Precursors in Recreational Water..........172**
Jing Li and Ernest R. Blatchley III

13. **Reactivity of Bromine-Substituted Haloamines in Forming Haloacetic Acids...182**
Phillip G. Pope and Gerald E. Speitel Jr.

14. **Comparison of the Performance of Spectroscopic Indices Developed to Quantify the Halogenation of Natural Organic Matter at Varying Chlorine Concentrations, Reaction Times, and Temperatures**......198
Paolo Roccaro, Federico G. A. Vagliasindi, and Gregory V. Korshin

Formation and Control of Disinfection By-Products

15. **Combined Treatments for Enhanced Reduction of Trihalomethane Precursors**......214
Rolando Fabris, Christopher W. K. Chow, and Mary Drikas

16. **Removal of Trihalomethane Precursors Using the MIEX® Dissolved Organic Carbon Process in Combination with Granular Activated Carbon**......227
Mary Drikas, Mike Dixon, and Jim Y. Morran

17. **Natural Dissolved Organic Matter Removal and Subsequent Disinfection By-Product Formation: A Comparison of Ion Exchange and Activated Carbon**......242
Yongrui Tan, James E. Kilduff, and Tanju Karanfil

18. **Fate and Transport of Wastewater-Derived Disinfection By-Products in Surface Waters**......257
Baiyang Chen, Paul Westerhoff, and Stuart W. Krasner

19. **Organic Nitrogen Occurrence and Characterization**......274
Aaron D. Dotson, Paul Westerhoff, Baiyang Chen, and Wontae Lee

20. **Analysis of *N*-Nitrosamines Formed in Various Source Waters Treated with Eleven Disinfection Processes**......289
Yuan Yuan Zhao, Jessica M. Boyd, Matthew Woodbeck, Robert Andrews, Steve Hrudey, and Xing-Fang Li

21. **Formation of Nitrosamines in Effluent-Impacted Drinking Waters**......304
Stuart W. Krasner, Michael J. Sclimenti, Chih Fen Tiffany Lee, and Jessica Schramm

22. Free Radical Chemistry of Advanced Oxidation Process Removal of Nitrosamines in Water.....319
Stephen P. Mezyk, Nicholas A. Landsman, Katy L. Swancutt, Christine N. Bradford, Casandra R. Cox, James J. Kiddle, and Tim J. Clore

23. Degradation of Halogenated Disinfection By-Products in Water Distribution Systems.....334
Raymond M. Hozalski, William A. Arnold, Chanlan Chun, Timothy M. LaPara, Jeong-Yub Lee, Carrie R. Pearson, and Ping Zhang

Indexes

Author Index.....351

Subject Index.....353

Preface

The formation, control, and health effects of disinfection by-products (DBPs) in drinking water are issues of international concern. Many countries, as well as the World Health Organization, have regulations and/or guidelines on acceptable concentrations of DBPs in water. Because potable water is typically treated with a chemical disinfectant, DBPs are contaminants to which many people are exposed. Because of the health effects (e.g., bladder cancer and certain adverse reproductive–development impacts) associated with exposure to chlorinated water and/or certain DBPs, the water industry has made major efforts to balance disinfection and DBP control.

In recent years, extensive research has been conducted on emerging DBPs of health and regulatory concern. Although the focus on DBPs in the past was on chlorine- and bromine-containing carbonaceous DBPs, emerging DBPs include iodine-containing species, as well as halogenated and non-halogenated nitrogenous DBPs. Moreover, new toxicity testing has suggested that some of the newer DBPs are of higher toxicity than some of the regulated chemicals. Thus, research has been conducted to better understand how to cost-effectively control a wide range of regulated and emerging DBPs. This includes the use of innovative treatment and disinfection strategies.

Moreover, exposure to DBPs is not limited to drinking water. Treated wastewater, which can impact drinking water supplies, is a source of DBPs and DBP precursors. Swimming pools provide another milieu for exposure to DBPs. Furthermore, the types of DBPs that can be produced in these other environments may be somewhat different than those formed in drinking water.

This book represents some of the latest research efforts at understanding these important DBP-related issues. The authors of the chapters in this book are a multidisciplinary mixture of scientists and engineers, who are conducting studies in many parts of the world. The book includes chapters on the health and regulatory concerns of DBPs, the reactivity of natural organic matter (NOM) (i.e., DBP precursors) toward DBP formation, probing the formation of DBPs, and the formation and control of DBPs. The chapters in this book address both regulated and emerging DBPs, as well as innovative treatment technologies and analytical methods. This book will be of interest to researchers, drinking water utility scientists and engineers, toxicologists, epidemiologists, and regulators interested in the formation and control of and exposure to DBPs.

This book is part of a continuing series of books based on chlorination conferences of the 1970s and 1980s and more recent American Chemical Society (ACS) symposia on NOM and DBPs. These books have reflected the state-of-the-art in DBP research and, more importantly, have continued to be cited for decades. This current book is based on a symposium entitled *Occurrence, Formation, Health Effects and Control of Disinfection By-Products in Drinking Water,* held at the 233rd National Meeting of the ACS Division of Environmental Chemistry, Inc., in Chicago, Illinois, on March 27–29, 2007.

We thank all of the people who participated in the symposium and those who provided chapters for this book. We also thank those who served as peer reviewers of chapter manuscripts: Gary Amy, Helene Baribeau, Mark Benjamin, Richard Bull, Stephen Booth, Zhuo Chen, Jean-Phillippe Croué, John Crittenden, Jean Debroux, David DeMarini, Vasil Diyamandoglu, Stephen Duirk, Gary Emmert, Raymond Hozalski, Guanghui Hua, Djanette Khiari, James Kilduff, Mehmet Kitis, Detlef Knappe, Gregory Korshin, Karl Linden, Benito Marinas, Michael McGuire, William Mitch, Simon Parsons, Michael Plewa, David Reckhow, Alexa Obolensky, Kenan Ozekin, Susan Richardson, David Sedlak, Chii Shang, Gerald Speitel, Windsor Sung, Richard Valentine, Peter Vikesland, Urs von Gunten, Gen-Shuh Wang, Martha Wells, Yintak Woo, and Christian Zweiner. Furthermore, we thank the following organizations for their financial assistance in supporting the symposium: American Water Works Association Research Foundation, Malcolm Pirnie, Inc., and Carollo Engineers. Finally, we acknowledge the Association of Environmental Engineering and Science Professors for helping disseminate information on the symposium.

Tanju Karanfil
Department of Environmental Engineering and Earth Sciences
Clemson University
Anderson, SC 29625

Stuart W. Krasner
Metropolitan Water District of Southern California
700 Moreno Avenue
La Verne, CA 91750–3399

Paul Westerhoff
Department of Civil and Environmental Engineering
Arizona State University
Tempe, AZ 85287–5306

Yuefeng Xie
Environmental Engineering Programs
Capital College
The Pennsylvania State University
Middletown, PA 17057

Introduction

Chapter 1

Recent Advances in Disinfection By-Product Formation, Occurrence, Control, Health Effects, and Regulations

Tanju Karanfil[1], Stuart W. Krasner[2], Paul Westerhoff[3], and Yuefeng Xie[4]

[1]Department of Environmental Engineering and Science, Clemson University, 342 Computer Center, Anderson, SC 29625–6510
[2]Water Quality Standards Branch, Metropolitan Water District of Southern California, 700 Moreno Avenue, La Verne, CA 91750–3303
[3]Department of Civil and Environmental Engineering, Arizona State University, Tempe, AZ 85287–5306
[4]Department of Environmental Engineering, Pennsylvania State University at Harrisburg, Middletown, PA 17057

Since the discovery of disinfection by-products (DBPs) in drinking water in the 1970s and with increasing understanding about their occurrence and health effects, the control of DBP formation has become one of the major issues for the drinking water industry. A significant amount of research has been conducted on trihalomethanes (THMs) and haloacetic acids (HAAs), the two regulated (in the United States [U.S.]) and largest classes of DBPs (on a weight basis). In addition to other measured DBPs, the sum of known DBPs has accounted for 30 to 60% of the total organic halogen (TOX) detected during chlorination, which is the oldest and the most commonly used disinfection technique in water treatment around the world. As a result, the formation and control of THMs and HAAs and other DBPs have been widely researched and discussed in the literature *(1-4)*. However, recent research has shown that there are many emerging DBPs of health concern in drinking waters. The objective of this chapter is to summarize the recent advances in understanding DBP formation, occurrence, control, health effects, and regulations that have occurred mainly within the past five years. It was noteworthy that inorganic DBPs (e.g., bromate, chlorate, chlorite) have received less attention over the past few years as compared to emerging organic DBPs that exhibit high geno- and cytotoxicity.

Occurrence

Two groups of DBPs, iodinated DBPs and nitrogenous DBPs, have attracted much attention from researchers in recent years. In addition, emerging carbonaceous DBPs have been studied. Another important area of research has been an evaluation of spatial and temporal variations in DBP levels in distribution systems. Furthermore, another group of emerging by-products are the formation of toxic chemicals through the disinfection (e.g., chlorination) of pharmaceuticals or other emerging contaminants that can be present in water (e.g., bisphenol A). However, the latter group of by-products are not discussed in this chapter.

Iodinated DBPs

Iodinated THMs, which are responsible for medicinal taste-and-odor problems in drinking water *(5)*, and iodinated acids have been detected in waters containing iodide. Because of their potential health risks *(6)*, studies have been conducted to investigate their occurrence and formation in drinking water. Krasner and colleagues *(7)* conducted a survey of DBP occurrence at 12 drinking water treatment plants in the U.S. This study found that the ratio of the iodinated THMs to the four regulated THMs was 2% on a median basis (maximum = 81%). The highest formation of iodinated THMs was at a plant that used chloramines only (i.e., no pre-chlorination). In addition, at the latter plant, five iodinated acids were detected. Richardson *(8)* conducted a follow-up survey to evaluate the formation of iodinated acids in other chloraminated waters.

Karpel Vel Leitner and colleagues *(9)* reported that ammonia and organic amines enhanced iodoform formation in chlorinated water containing iodide. Bichsel and von Gunten *(10)* investigated iodinated THM formation in chlorinated, chloraminated, and ozonated waters. Their study indicated that the oxidation of iodide to hypoiodous acid (HOI) increased in the order of ozone > chlorine (Cl_2) > chloramines. Moreover, ozone and chlorine, but not chloramines, could further rapidly oxidize HOI to iodate, which would be a sink for the iodide (I^-) and, thus, be unavailable for forming iodinated organic DBPs. Alternatively, in chloraminated waters, HOI could react with natural organic matter (NOM) to form iodinated DBPs. Hua and colleagues *(11)* investigated the effect of iodide on the formation of iodinated DBPs and found that the Cl_2/total organic carbon (TOC) or Cl_2/I^- ratio was critical to the formation of iodinated DBPs. These researchers proposed one possible strategy to control the formation of iodinated DBPs in drinking water, by utilizing an appropriate dose of chlorine to convert the majority of iodide to iodate and then adding ammonia to form chloramines for distribution residual maintenance. Finally, Hua and Reckhow *(12)* demonstrated that iodinated DBPs could form during chlorine dioxide treatment.

Nitrogenous DBPs

Nitrogenous DBPs have attracted much attention because chloramination favors the formation of certain nitrogenous DBPs (e.g., cyanogen chloride *[13]* and N-nitrosodimethylamine (NDMA) *[14]*) and many water utilities in the U.S. have been considering using chloramination to reduce the levels of regulated carbonaceous DBPs in their water *(15)*. Of the nitrogenous DBPs, NDMA has been extensively studied. In addition, halonitromethanes (HNMs) and haloacetamides are emerging nitrogenous DBPs for which new research has been conducted.

Yang and colleagues *(16)* investigated various water quality parameters and monochloramine application modes on the formation of various nitrogenous DBPs from Suwannee River NOM. Good linear relationships were observed between monochloramine dosage and dichloroacetonitrile and cyanogen chloride formation. Dosing with free chlorine and ammonium salts in different orders produced much higher chloroform than dosing preformed monochloramine did. However, the chloramination modes had little effect on the formation of dichloroacetonitrile and cyanogen chloride.

Chen and Valentine *(17)* developed a kinetic model to validate proposed reactions and to predict NDMA formation in chloraminated water. The results indicated that NDMA formation was limited by the oxidation of NOM by the protonated chlorine species, HOCl. Schreiber and Mitch *(18)* reported on the critical importance of dichloramine and dissolved oxygen in nitrosamine formation and proposed a new nitrosamine formation pathway. Lee and colleagues *(19)* reported the destruction or transformation of NDMA precursors using ozone and chlorine dioxide. Likewise, Charrois and Hrudey *(20)* found that prechlorination minimized NDMA formation during subsequent chloramination.

In addition to evaluating the formation and control of NDMA, various researchers have studied the source of NDMA precursors. Krasner and colleagues *(21)* found treated wastewater to be a major source of NDMA precursors. Moreover, these researchers found that wastewater treatment plants that achieved some nitrification resulted (on a central tendency basis) in reducing the level of precursors for NDMA. Schmidt and colleagues *(22)* found that certain tertiary dimethylamines (e.g., the pharmaceutical ranitidine) showed much higher conversion rates to NDMA than dimethylamine. Some cationic polymers contain NDMA precursors, therefore it is important to not overdose the polymers used in the coagulation process *(23-25)*.

In terms of other emerging nitrogenous DBPs, preozonation was found to increase the formation of certain HNMs during postdisinfection *(7)*. In addition, chloropicrin formation was greatly enhanced by medium-pressure UV, possibly due to photonitration *(26)*. In other research, nitrite was found to be a potential source of nitrogen in the nitro group of chloropicrin *(27)*. Haloacetamides, another nitrogenous DBP class, were found to be present at levels similar to other commonly measured DBPs *(7)*.

Similar to trihalohalocatones, trihaloacetaldehyes, and brominated trihaloacetic acids, some nitrogenous DBPs, such as haloacetonitriles and cyanogen chloride, can undergo base-catalyzed hydrolysis. Alkaline hydrolysis of haloacetonitriles can form haloacetamides (and ultimately HAAs).

Emerging Carbonaceous DBPs

Emerging carbonaceous DBPs include haloaldehydes and halogenated furanones. In a recent U.S. DBP occurrence survey, haloacetaldehydes represented the third major class (after THMs and HAAs) formed in many of the waters *(7)*. Ozonation (without biological filtration) and chloramination controlled the formation of chloral hydrate (trichloroacetaldehyde); however, this treatment process increased the formation of dihaloaldehydes *(7)*. It has been shown that acetaldehyde (an ozone DBP) can react with chlorine to form chloral hydrate, whereas in the former study acetaldehyde most likely reacted with chloramines and formed dihaloaldehydes. However, it should be possible to minimize the formation of haloaldehydes at ozone plants through the use of biological filtration.

In a recent U.S. DBP occurrence survey, for some waters with high TOC and or high bromide, relatively high levels of the chlorinated furanone 3-chloro-4-(dichloromethyl)-5-hydroxy-2-(*5H*)-furanone (MX) and its brominated analogues (BMXs) were detected *(7)*. The maximum occurrence of some of the halogenated furanones was high compared to previous limited surveys.

Spatial and Temporal Variations in DBP Levels

The spatial variation of DBP levels in distribution systems are well recognized. To address these variations and to provide equal protection to consumers in the entire distribution system, the newly promulgated Stage 2 Disinfectants and DBP Rule in the U.S. specified the use of locational running annual averages for DBP compliance *(28)*.

THM levels in distribution systems generally increase with increasing residence time. This is due to the combination of continuous reactions between NOM and chlorine to form THMs and to hydrolysis reactions of intermediate DBPs that break down to form THMs.

Pereira and colleagues *(29)* investigated temporal and spatial variations of DBPs in a chloraminated disitribution system. Considerable short-term temporal fluctuations in the levels of THMs, HAAs, and TOX were observed. The fluctuations appeared to be associated with operational variations in chloramination practices at the plant. The average values of THMs and HAAs at three various sampling locations were similar over five to six days of sampling. Speight and Singer *(30)* investigated temporal and spatial variations of HAAs in five distribution systems. The study demonstrated in full-scale distribution systems that the loss of residual chlorine was a necessary condition for the degradation of HAAs. The study also showed that HAA levels varied at various locations (including at various depths in storage tanks) and over time. This variation was also true for systems using chloramines as the secondary disinfectant. The seasonal variation and spatial fate of DBPs were investigated in chlorinated water distribution systems in Canada *(31)*. Variations in THMs and HAAs were observed not only on a seasonal basis but also on an intra-seasonal basis. The pattern of spatial variations was very different for THMs and HAAs. THMs increased and became stable in the distribution system extremities. However, HAAs, especially dichloroacetic acid, decreased approaching the extremities, a phenomenon probably related to biological degradation of HAAs.

The relative dominance of THMs and HAAs in finished water was also investigated *(32)*. The distribution of THMs and HAAs was affected by water pH, characteristics of the NOM, temperature, and bromide. Baribeau and colleagues *(33)* evaluated the biodegradation of HAAs in annular reactors. These researchers found that dihalogenated HAAs degraded when the disinfectant residual was absent and when the water was warm, whereas the trihalogenated HAAs were stable. Biological degradation of HAAs by bacterial enrichment cultures was investigated by different researchers *(34,35)*. Using radiolabeled HAAs, one study indicated that the ^{14}C was primarily converted to $^{14}CO_2$, with minor incorporation into cell biomass. Exploratory studies were also conducted to investigate community structures in enrichment cultures using molecular techniques. Identifying bacteria responsible for HAA degradation will help researchers to better understand HAA degradation in distribution systems.

Control

Insights to DBP Precursors

Dissolved organic matter (DOM) in fresh waters typically constitutes over 80% of the total organic matter in fresh waters *(36)* and are the organic precursors of DBPs. Particulate organic matter does not play a major role in DBP formation *(37)*. Colloidal DOM, which passes through a 1-μm filter and is captured on a 3,500-Da dialysis membrane, has much lower TOX, THM, and HAA yields, as compared to DOM *(38)*. However, colloids, which can be nitrogen-rich, may be a source of precursors for emerging nitrogenous DBPs *(39)*. Although hydrophobic acids (humic substances) are a major source of precursors for THMs and HAAs, other DOM fractions, which can also be nitrogen-rich, may be important sources of precursors for other classes of emerging DBPs.

The concentration and characteristics of DOM are site specific and influence the formation and speciation of DBPs. Karanfil and colleagues *(36)* examined the factors influencing the determination of dissolved organic carbon (DOC) and ultraviolet absorbance (UVA) in natural waters and provided guidelines for filter selection and cleaning and sample processing. There are a number of techniques available for characterizing DOM *(40)*, and one commonly used technique is to fractionate the DOM into hydrophobic (acids, bases, and neutrals) and hydrophilic/transphilic (acids, bases, and neutrals) fractions, depending on their affinities to different resins and their back-elution efficiencies. Emerging methods for characterizing DOM include the measurement of dissolved organic nitrogen (DON) using a dialysis pretreatment technique to remove dissolved inorganic nitrogen *(41)*, size exclusion chromatography (SEC) with on-line UVA and DOC detectors *(42-43)*, and fluorescence excitation-emission matrix (EEM) measurements *(44)*. SEC-DOC provides molecular weight distributions of the DOM, which separated the organic matter into three regions: (1) polysaccharides, proteins, and colloids, (2) humic substances, and (3) low-molecular-weight organic acids. An EEM is represented by a 3-dimensional spectrum showing fluorescence intensity as a function of excitation wavelength and emission wavelength, with "peak" location(s) indicative of DOM composition (e.g., humic, soluble microbial product [SMP]-like). Because drinking water treatment plants may be under the influence of treated wastewater discharges and because there has been an increase in water reuse in arid regions, recent research has examined the formation of DBPs in effluent organic matter (EfOM), which is composed of background DOM from drinking water and SMPs from the biological wastewater treatment processes *(21)*. SMPs consist of macromolecules and cellular debris with a protein (nitrogen-enriched) and polysaccharide signature. Besides drinking water treatment plants being under the influence of EfOM from wastewater treatment plants, a related problem is the eutrophication of a drinking water supply and the associated presence of algal organic matter (AOM), another microbially derived form of NOM similar to EfOM. The concentration of DON in the raw waters of 28 U.S. drinking water treatment plants averaged 0.19 mg N/L and, on average, the plants removed 20% of the DON during the use of conventional treatment processes *(45)*.

Improvements in DBP Precursor Removal

Coagulation practices have changed over the past decade. Among the most prominent changes is the necessity for many water utilities in the U.S. to practice enhanced coagulation *(46,47)*. Enhanced coagulation is a best available technology for removing DBP precursors and is mandated by the Stage 1 Disinfectants/DBP Rule for the treatment of most surface waters. As a result of this Rule, many water treatment plants have increased the coagulant dosage applied to their water and/or added acids to coagulate waters at a lower pH than that used for conventional coagulation for turbidity removal.

A recent study with 27 waters from across the U.S. showed that the alum dose required for enhanced coagulation removed precursors for regulated and unidentified halogenated DBPs, as measured by TOX formation potential (FP) tests *(48)*. Additionally, results of that study and a companion study of the U.S. Information Collection Rule (ICR) database demonstrated the ability of specific ultraviolet absorbance (SUVA) to reflect the amenability of the NOM to removal by coagulation and the reactivity of the NOM to chlorine as measured by TOXFP *(48-49)*.

Laboratory- and field-scale research has also shown that a significant percentage of the DBPs observed in distribution systems form rapidly and within water treatment plants *(50)*. Consquently, two classes of DBP precursors have been characterized as 1) faster-reacting and 2) slower-reacting with disinfectants *(51)*. Coagulation processes remove both fractions, but preferentially remove the faster-reacting class, which tend to have higher SUVA.

DBP precursors can also be removed by precipitative softening *(52)*. One study found that the use of high Mg-content lime improved the removal of NOM and reduced THM formation compared with dolomitic quick lime hydrated under atmospheric conditions, presumably due to formation of $Mg(OH)_2$-NOM complexes *(53)*.

Among the newest innovations for DBP control has been inclusion or sole use of magnetic ion exchange (MIEX) resins for DOC removal *(54-56)*. This process targets NOM acids. In addition, this process can remove bromide, but the presence of sulfate and other anions can interfere with the efficacy of this usage. MIEX was more effective than alum coagulation at removing UV-absorbing substances, DOC, and THM and HAA precursors. Coagulation of MIEX-treated water may not provide additional removal of DOC, but may still be required for turbidity removal.

Another innovation in coagulation has been its combination with micro- or ultrafiltration membranes *(57)*. Most of this research focused on removing TOC to limit membrane fouling *(58)*, but removing TOC will reduce DBP formation too. The use of alum coagulants and recirculating powder activated carbon (PAC) in an ultrafiltration system was also found effective to reduce membrane fouling and remove TOC *(59)*. In terms of membranes, nanofiltration (NF) has become the membrane of choice in NOM removal. Other combined processes have also emerged to meet a variety of water quality needs, including ozone/biofiltration, which can remove TOC and, consequently, reduce DBP formation *(60,61)*.

Most of the aforementioned DBP control practices have focused on the removal of NOM before chlorination, where DOC is a contaminant. In Central Europe, treatment has focused on the removal of biodegradable organic carbon (BDOC) (or assimilable organic carbon [AOC]) as a contaminant.

Changes in Oxidation/Disinfection Strategies

There are different disinfection practices globally, which lead to different DBP trends. In the United States, there has been an increased usage of alternative disinfectants (ozone, chlorine dioxide, and/or UV for primary disinfection; chloramines for secondary disinfection). In addition, many U.S. utilities have switched from using gaseous chlorine to hypochlorite solutions. In Southern Europe and the United Kingdom, chlorine has predominantly been used; however, chloramines have started to be used in portions of the United Kingdom. In Central Europe (e.g., Berlin, Amsterdam, Zurich, Vienna), treatment has focused on achieving biostability so that chlorine can be eliminated in the distribution system. In Paris, low chlorine levels have been used in the distribution system, where booster chlorination has been used where appropriate. In many of these Central European countries, ozone has been used for disinfection and micropollutant (e.g., atrazine) destruction.

Bromate has been found in hypochlorite solutions *(62)*. As a result, NSF International has set a limit on bromate concentrations in commercially prepared hypochlorite solutions. In terms of controlling bromate formation during ozonation, recent research has shown that the use of chlorine dioxide as a pre-oxidant can reduce bromate formation during subsequent ozonation *(63-64)*. Chlorine dioxide satisfied some of the initial oxidant demand of the water, reducing the ozone dose required to meet treatment goals. However, the chlorine dioxide by-product chlorite was oxidized to chlorate, an emerging DBP of concern. In other bromate-control research, chlorine was added to oxidize the bromide and then ammonia was added prior to the ozone contactor to form bromamines *(65,66)*. In the "chlorine-ammonia" process, bromate formation was highly minimized, but other regulated DBPs (e.g., THMs) were produced. Recently, the "ammonia-chlorine" process, where chloramines (with no free chlorine contact time) were added prior to ozonation, has been evaluated. This process, especially at pH 7, reduced bromated formation with minimal THM formation *(67)*.

Two significant changes in disinfection strategies have occurred over the past decade: 1) increased use of chloramines for secondary disinfection in distribution systems in the U.S. *(15)* and 2) use of UV irradiation for in-plant disinfection. Chloramines are often applied as a secondary disinfectant in order to minimize THM and HAA formation and to maintain disinfectant residuals in distributions systems with long detention times. Because chloramines are weak disinfectants, primary disinfection with chlorine, chlorine dioxide, ozone or UV is almost always required for surface water systems. THMs and HAAs still form in the presence of chloramines, but at a much slower rate than in the presence of free chlorine. However, dihalogenated HAAs can form in the presence of chloramines *(68)*. Overall, chloramines decrease THM, HAA, and TOX formation, but tend to increase the percentage of nonidentifiable TOX *(12,69)*. Additionally, certain nitrogen-containing DBPs preferentially form during chloramination, including nitrosamines and cyanogen halides (*14,16*), some of which may pose more serious potential health issues than non-nitrogen containing DBPs.

Although the use of ozone as a primary, in-plant, disinfectant has continued to grow, there has been a significant growth in the use of UV irradiation (low and medium pressure lamps) for primary disinfection, mainly due to the need for some systems to inactivate *Cryptosporidium (70,71)*. At UV dosages used for microbial inactivation, there has typically been no evidence that UV irradiation affects (increases or decreases) DBP formation *(72)*. One study did find that the order of chlorine and UV application can affect THM production, but had less effect on other DBP classes *(73)*. However, as noted above, medium pressure UV can increase the formation of chloropicrin and may also increase the formation of other HNMs of health concern (*74*). In addition, it is possible to form nitrite with medium pressure UV. Also, it is possible that haloaldehyde formation could increase with UV treatment *(26)*. Increased use of UV irradiation as a primary disinfectant is likely to continue, thus reducing reliance upon ozone or free chlorine and, hopefully, reduce the overall DBP formation in water delivered to the public.

DBP Removal after Formation

Removing DBP precursors and using alternative disinfectants have been the main strategies for water utilities to control DBPs in their finished water. However, several recent studies have indicated that removing certain DBPs after their formation using biologically active granular activated carbon (GAC) filters (filter adsorbers or filter caps) is another potential cost-effective DBP control strategy. The removal of HAAs inside biologically active GAC filters are greatly affected by water temperature and empty-bed-contact-time (EBCT). Shorter EBCTs can be used at higher water temperatures *(75)*.

A substantial removal of HAAs in GAC filters has been reported in pilot and field studies *(30,32,75-79)*. The removal of HAAs in GAC filters was mainly due to biological degradation, not physical adsorption. Baribeau and colleagues *(33)* and Xie and colleagues *(80)* reported that the removal of HAAs was greatest when water temperatures were high and residual chlorine concentrations were low. Dichloroacetic acid appeared to be removed to a greater degree than trichloroacetic acid. The results indicated that there was no substantial difference between new and old GAC filters, especially after being on-line for a few months. A period of one to three months was needed for GAC filters to establish biological activity and to exhibit an effective removal of HAAs *(77,79)*. The length of this period can be affected by water temperature and chlorine residual in the filter influent. For dichloroacetic acid, a GAC bed life of 14-22 days was reported for a full-scale GAC filter *(79)*. Removal of THMs (via adsorption) was only observed for two to three months after GAC media were changed. The effect of GAC filters on HAA and THM levels in finished water were also investigated using the ICR database *(78)*.

In addition to the biodegradation of DBPs, another DBP loss mechanism is hydrolysis. Many DBPs, including trihaloacetones, trihaloacetaldehydes, and brominated trihaloacetic acids, can undergo base-catalyzed hydrolysis. Trihalomethanes are the end products of these hydrolusis reactions.

Health Effects

An expert toxicology review of the ~500 DBPs reported in the literature was conducted with an in-depth mechanism-based structural activity relationship (SAR) analysis—supplemented by an extensive literature search for genotoxicity and other data *(81)*. Approximately 50 DBPs that received the highest ranking for potential toxicity, and that were not already included in the ICR, were selected for a new U.S. occurrence study *(7)*. Using *in vitro* mammalian cell assays, certain DBPs studied in the latter project (e.g., iodinated THMs, certain nitrogenous DBPs [e.g., HANs, HNMs, haloacetamides] were found to be more cytotoxic and genotoxic than many of the regulated DBPs *(6,74,82,83)*. In addition, iodinated and brominated DBPs were found to be more toxic than their chlorinated analogues. Moreover, emerging DBPs of health concern not addressed by the previous SAR effort (e.g., nitrosamines) are being identified in disinfected drinking water. A recently conducted project *(84)* used a quantitative SAR model and other health-effects information to examine the likely carcinogenic potential of a group of possible other DBPs. These researchers came up with a list of plausible candidates for chemicals that could be associated with bladder cancer from the consumption of chlorinated drinking water.

Bladder cancer risk has been associated with long-term exposure to THMs in chlorinated waters. The most recent epidemiology study to evaluate this was Villanueva and colleagues *(85)*. However, a major issue in epidemiology studies has been exposure assessment *(86)*. Amy and colleagues *(87)* redid the exposure assessment for two previously conducted cancer epidemiology studies. In one location, in Iowa, historical annual-average THM and HAA exposure were predicted to be as high as 156 and 100 μg/L, respectively. The strongest and most consistent increases in cancer risk for all three cancer sites studied (bladder, colon, rectum) were for males exposed to >96 μg/L of THMs or >53 μg/L of HAAs.

In addition, various epidemiology and toxicology studies have shown there to be adverse reproductive/developmental risks associated with high exposure to DBPs *(88)*. However, in a recent epidemiology study—with improved exposure assessment methods—pregnancy loss was not associated with high personal THM exposure *(89)*. Sporadic elevations in risk were found across DBPs, most notably for ingested TOX.

Another important area of health effects research is the study of mechanisms of action. For example, recent research suggests that bromate may be a practical threshold carcinogen *(90)*. If so, continued research on this issue could impact how this DBP is regulated in the future.

Regulations

There has been an active regulatory process for DBPs around the world within the past five to ten years. In the U.S., Stage 2 of the D/DBP Rule was published in January 2006 (*28*), requiring water utilities to comply with maximum contaminant levels (MCLs) of 80 μg/L of THMs and 60 μg/L of five haloacetic acids (HAA5) at each individual monitoring location in a distribution system, which will commence in the next six years. As part of this Rule, the MCL for bromate was kept at 10 μg/L. However, the bromate MCL will be reexamined as part of the 6-yr review of the Rule. Moreover, there is an MCL for chlorite (but not chlorate) at 1.0 mg/L. A detailed discussion on the evaluation of DBP regulations in the U.S. is discussed in the following chapter in this book *(91)*.

The World Health Organization has guideline values for DBPs in drinking water. They are 10 μg/L for bromate, 60 μg/L for bromodichloromethane, 100 μg/L for bromoform, 700 μg/L for chlorate, 700 μg/L for chlorite, 300 μg/L for chloroform, 70 μg/L for cyanogen chloride, 70 μg/L for dibromoacetonitrile, 100 μg/L for dibromochloromethane, 50 μg/L for dichloroacetate, 20 μg/L for dichloroacetonitrile, 20 μg/L for monochloroacetate, 200 μg/L for trichloroacetate, and 200 μg/L of or 2,4,6-trichlorophenol. In addition, for THMs, the sum of the ratio of the concentration of each to its respective guideline value should not exceed 1.

In the European Union, the European Council adopted Directive 98/83/EC, establishing standards for the member states. Total THMs and bromate were selected as representative by-products of chlorination and ozonation, respectively. The standard for THMs was set at 150 μg/L by 2003 and 100 μg/L by 2008. The bromate standard was 10 μg/L, with an interim value of 25 μg/L.

However, many member states have adopted more stringent standards for THMs (Table I). In addition, some countries have set standards for chlorite and chlorate and some other compounds, such as formaldehyde and 2,4,6-trichlorophenol. Currently, there are no regulatory limits for HAAs in Europe. Compliance with European Union regulations are based on absolute maximums, not running annual averages as used in the U.S.

In Japan, the new DBP regulations that became effective on April 1, 2004 set the standards for individual THM species at 60, 30, 100, and 90 μg/L for chloroform, bromodichloromethane, dibromochloromethane, and bromoform, respectively. The total THMs, the sum of the four species, was regulated at 100 μg/L. In addition, bromate, monochloroacetic acid, dichloroacetic acid, trichloroacetic acid, and formaldehyde were regulated at 10, 20, 40, 200, and 80 μg/L, respectively. A more active regulatory process is expected in the coming years, as more occurrence and toxicological data become available. Currently, occurrence and toxicology data are being collected for NDMA, bromoacetic acid, and MX by a national research project lead by the Ministry of Health, Labor and Welfare.

In China, the standards for Drinking Water Quality became effective on July 1, 2007. The new standards set limits for chloroform, bromodichloromethane, dibromochloromethane, bromoform, and the sum of all four THM

Table I. Total THM Regulatory Limits in Different Countries and Regions

Country and Region	*Limit (μg/L)*
Austria	30
Belgium	30
China	100
Czech Republic	100
Germany	50
Ireland	100
Italy	30
Japan	100
Luxembourg	50
Norway	100
Spain	100
Sweden	50
Scotland	100
Taiwan	80
United Kingdom	100
United States	80

species at 60, 60, 100, 100, and 100 μg/L, respectively. The sum of the ratios of the measured concentrations of each THM specie to its regulatory limit should be less than one. In addition, dichloroacetic acid, trichloroacetic acid, trichloroacetaldehyde, bromate, chlorite, and chlorate were also regulated at 50, 100, 10, 10, 70, and 70 μg/L, respectively. In Taiwan, the latest DBP regulation became effective on July 1, 2006 and set the limits for total THMs and bromate at 80 and 10 μg/L, respectively. Currently, there is no regulation for HAAs, however a rule of 60 μg/L has been proposed. In addition, there is an increase in the use of chlorine dioxide in Taiwan. Therefore, regulations for HAAs and chlorine dioxide (and its by-products) may become effective in the near future.

Australia does not have national DBP regulations, instead there are drinking water quality guidelines published by the Australian National Health and Medical Research Council. Each state can adopt regulations using these guidelines. To date, only Victoria adopted these guidelines as drinking water regulations. The current DBP guidelines are as follows: total THMs (250 μg/L), monochloroacetic acid (150 μg/L), dichloroacetic acid (100 μg/L), trichloroacetic acid (100 μg/L), chloral hydrate (20 μg/L), chlorite (300 μg/L), cyanogens chloride (80 μg/L), formaldehyde (500 μg/L), and bromate (20 μg/L).

Acknowledgments

Mary Drikas (Australia), Shinya Echigo (Japan), Wenjun Liu (China), Jinfeng Lu (China), Simon Parsons (U.K.), Paolo Roccaro (Italy), and Gen-Shuh Wang (Taiwan) provided information about the DBP regulations in their countries and regions.

References

1. *Disinfection By Products in Water Treatment: The Chemistry of their Formation and Control*; Minear, R. A.; Amy, G. L., Eds.; CRC Press Inc.: Boca Raton, FL, 1996.
2. *Formation and Control of Disinfection By-Products in Drinking Water*; Singer, P. C., Ed.; American Water Works Association (AWWA), Denver, CO, 1999.
3. *Natural Organic Matter and Disinfection By-Products: Characterization and Control in Drinking Water*; Barrett, S. E.; Krasner, S. W.; Amy, G. L., Eds.; American Chemical Society (ACS) Symposium Series No 761; ACS, Washington, DC, 2000.

4. Xie, Y. F. *Disinfection Byproducts in Drinking Water: Formation, Analysis, and Control*; Lewis Publishers: New York, NY, 2003.
5. Hansson, R. C.; Henderson, M. J.; Jack, P.; Taylor, R. D. *Wat. Res.* **1987**, *21*, 1265-1271.
6. Plewa, M. J.; Wagner, E. D.; Richardson, S.D.; Thruston, A.D., Jr.; Woo, Y-K.; McKague, A.B. *Environ. Sci. Technol.* **2004**, *38*, 4713-4722.
7. Krasner, S. W.; Weinberg. H. S.; Richardson, S. D.; Pastor, S. J.; Chinn, R.; Sclimenti, M. J.; Onstad, G. D.; Thurston, A. D., Jr. *Environ. Sci. Technol.* **2006**, *40*, 7175-7185.
8. Richardson, S. D. Presented at 2005 AWWA Ann. Conf., San Francisco, CA, 2005.
9. Karpel Vel Leitner, N.;Vessella, J.; Dore, M.; Legube, B. *Environ. Sci. Technol.* **1998**, *32*, 1680-1685.
10. Bichsel, Y.; von Gunten, U. *Environ. Sci. Technol.* **2000**, *34*, 2784-2791.
11. Hua, G.; Reckhow, D.; Kim, J. *Environ. Sci. Technol.* **2006**, *40*, 3050-3056.
12. Hua, G. H.; Reckhow, D. *Wat. Res.* **2007**, *41*, 1667-1678.
13. Hirose, Y.; Maeda, N.; Ohya, T.; Nojima, K.; Kanno, S. *Chemosphere* **1988**, *17*, 865-873.
14. Mitch, W. A.; Sharp, J. O.; Trussell, R. R.; Valentine, R. L.; Alvarez-Cohen, L.; Sedlak, D. L. *Environ.l Eng. Sci.* **2003**, *20*, 389-404.
15. Seidel, C.; McGuire, M. J.; Summers, R. S.; Via. S. *J. AWWA* **2005**, *97*(10), 87-97.
16. Yang, X.; Shang, C.; Westerhoff, P. *Wat. Res.* **2007**, *41*, 1193-1200.
17. Chen, Z.; Valentine, R. L. *Environ. Sci. Technol.* **2006**, *40*, 7290-7297.
18. Schreiber, I. M.; Mitch, W. A. *Environ. Sci. Technol.* **2005**, *39*, 3811-3818.
19. Lee, C.; Schmidt, C.; Yoon, J.; von Gunten, U. *Environ. Sci. Technol.* **2007**, *41*, 2056-2063.
20. Charrois, J. W. A.; Hrudey, S. E. *Wat. Res.* **2007**, *41*, 674-682.
21. Krasner, S. W.; Westerhoff, P.; Chen, B.; Amy, G.; Nam, S.-N.; Chowdhury, Z. K.; Sinha, S.; Rittmann, B. E. *Contribution of Wastewater to DBP Formation*; Awwa Research Foundation: Denver, CO, 2007 (in press).
22. Schmidt, C. K.; Sacher, F.; Brauch, H.-J. *Proc. of the 2006 AWWA Wat. Qual. Technol. Conf. (WQTC)*; AWWA: Denver, CO, 2006.
23. Najm, I.; Trussell, R. R *J. AWWA* **2001**, *93*(2), 92-99.
24. Kohut, K. D.; Andrews, S. A. *Wat. Qual. Res. J. Canada* **2003**, *38*(4), 719-735.
25. Bolto, B. *J. Wat. Supply Res. Technol.-Aqua* **2005**, *54*(8): 531-544.
26. Reckhow, D. U.S. Environmental Protection Agency (USEPA) Workshop on Optimizing the Design and Interpretation of Epidemiologic Studies to Consider Alternative Disinfection of Drinking Water; Raleigh, NC, 2005.

27. Choi, J.; Richardson, S. D. *Proc. of the 2004 AWWA WQTC*; AWWA, Denver, CO, 2004.
28. USEPA. National Primary Drinking Water Regulations: Stage 2 Disinfectants and Disinfection Byproducts Rule; Final Rule, *Fed. Reg.* Part II, 40 CFR Part 9, 141 and 142, 71: 2: 388, 2006.
29. Pereira, V. J.; Weinberg, H. S.; Singer, P. C. *J. AWWA* **2004**, *96*(11), 91-102.
30. Speight, V. L.; Singer, P. C. *J. AWWA* **2005**, *97*(2), 82-91.
31. Rodriguez, M. J.; Serodes, J. B.; Levallois, P. *Wat. Res.* **2004**, *38*, 4367-4382.
32. Singer, P. C.; Weinberg, H. S.; Brophy, K.; Liang, L., Roberts, M.; Grisstede, I.; Krasner, S. W.; Baribeau, H., Arora, H.; Najm, I. *Relative Dominance of Haloacetic Acids and Trihalomethanes in Treated Drinking Water.* Awwa Research Foundation: Denver, CO, 2002.
33. Baribeau, H.; Krasner, S. W.; Chinn, R.; Singer, P. C. *J. AWWA* **2005**, *97*(2), 69-81.
34. McRae, B. M.; LaPara, T. M.; Hozalski, R. M. *Chemosphere* **2004**, *55*, 915-925.
35. Tung, H.; Regan, J. M.; Unz, R. F.; Xie, Y. F. *Wat. Sci. Technol.: Wat. Supply* **2006**, *6*(2), 267-271.
36. Karanfil, T.; Erdogan, I.; Schlautman, M. A. *J. AWWA* **2005**, *95*(5), 125-136.
37. Najm, I.; Marcinko, J.; Oppenheimer, J. *J. AWWA* **2000**, *92*(8), 84-92.
38. Hwang, C. J.; Krasner, S. W.; Sclimenti, M. J.; Amy, G. L.; Dickenson, E.; Bruchet, A., Prompsy, C.; Flippi, G.; Croué, J.-P.; Violleau, D.; Leenheer, J. A. *Polar NOM: Characterization, DBPs, Treatment.* Awwa Research Foundation, Denver, CO, 2001.
39. Leenheer, J. A.; Dotson, A., Westerhoff, P. *Annals Environ. Sci.* **2007**, *1*, 45-56.
40. Leenheer, J. A.; Croué, J-P. *Environ. Sci. Technol.* **2003**, *37*, 18A-26A.
41. Lee, W.; Westerhoff, P. *Environ. Sci. Technol.* **2005**, *39*, 879-884.
42. Her, N.; Amy, G., Foss, D.; Cho, J.; Yoon, Y.; Kosenka, P. *Environ. Sci. Technol.* **2002**, *36*, 1069-1076.
43. Her, N.; Amy, G., Foss, D.; Cho, J. *Environ. Sci. Technol.* **2002**, *36*, 3393-3399.
44. McKnight, D. M.; Boyer, E. W.; Westerhoff, P. K.; Doran, P. T.; Kulbe, T.; Andersen, D. T. *Limnol. Ocean.* **2001**, *46*, 38-48.
45. Lee, W.; Westerhoff, P.; Esparza-Soto, M. *J. AWWA* **2006**, *98*(10), 102-110.
46. Edzwald, J. K.; Tobiason, J. E. *Wat. Sci. Technol.* **1999**, *40*(9), 63-70.
47. USEPA; *Enhanced Coagulation and Enhanced Precipitative Softening Guidance Manual*; EPA 815-R-99-012, 1999.

48. Archer, A. D.; Singer, P. C. *J. AWWA* **2006**, *98*(7), 110-123.
49. Archer, A. D.; Singer, P. C. *J. AWWA* **2006**, *98*(8), 97-107.
50. Adams, C.; Timmons, T.; Seitz, T.; Lane, J.; Levotch, S. *J Environ. Eng.-ASCE* **2005**, *131*(4), 526-534.
51. Westerhoff, P.; Reckhow D.; USEPA Grant # R 826831-01-0, 2002.
52. Kalscheur, K. N.; Gerwe, C. E.; Kweon, J.; Speitel, G. E., Jr.; Lawler, D. F. *J. AWWA* **2006**, *98*(11), 93-106.
53. Bob, M.; Walker, H. W. *J. Environ. Eng.-ASCE* **2006**, *132*(2), 158-165.
54. Boyer, T. H; Singer, P. C. *Wat. Res.* **2005**, *39*(7), 1265-1276.
55. Humbert, H.; Gallard, H.; Suty, H.; Croué, J.-P. *Wat. Res.* **2005**, *39*, 1699-1708.
56. Singer, P. C.; Schneider, M.; Edwards-Brandt, J., Budd, G. C. *J. AWWA* **2007**, *99*(4), 128-139.
57. Vickers, J. C.; Thompson, M. A.; Kelkar, U. G. *Desalination* **1995**, *102*, 57-61.
58. Howe, K. J., Clark, M. M. *J. AWWA* **2006**, *98*(4), 133-146.
59. Campos, C.; Schimmoller, L.; Marinas, B. J.; Snoeyink, V. L.; Baudin, I.; Laine, J. M. *J. AWWA* **2000**, *92*(8), 69-83.
60. Chaiket, T.; Singer, P. C.; Miles, A.; Moran, M.; Pallotta, C. *J. AWWA* **2002**, *94*(12), 81-95.
61. Yavich, A. A.; Masten, S. J. *J. AWWA* **2003**, *95*(4), 159-171.
62. Weinberg, H. S.; Delcomyn, C. A.; Unnam, V. *Environ. Sci. Technol.* **2003**, *37*, 3104-3110.
63. Zhou, P.: Neeman, P.; *Use of Chlorine Dioxide and Ozone for Control of Disinfection By-Products*; Awwa Research Foundation: Denver, CO, 2004.
64. Krasner, S. W.; Yun, T.; Yates, R.; Mofidi, A.; Liang, S.; Coffey, B.; *Proc. of the 2004 AWWA WQTC*; AWWA: Denver, CO, 2004
65. Neeman J.; Hulsey, R.; Rexing, D.; Wert, E. *J. AWWA* **2004**, *96*(2), 26-29,
66. Buffle, M. O.;, Calli, S.; von Gunten, U.; *Environ. Sci. Technol.***2004**, *38*(19), 5187-5195.
67. Krasner, S. W.; Yates, R.; Gabelich, C. J.; Liang, S. *Proc. of the 2007 AWWA Ann. Conf.*; AWWA: Denver, CO, 2007.
68. Karanfil, T.; Hong, Y.; Song, H.; Orr, O.; *Exploring the Pathways of HAA Formation during Chloramination*; Awwa Research Foundation: Denver, CO, 2007.
69. Richardson, S. D., Thruston, A. D.; Caughran, T. V.; Chen, P. H.; Collette, T. W.; Schenck, K. M.; Lykins, B. W., Jr.; Rav-Acha, C.; Glezer, V. *Wat. Air Soil Poll.* **2000**, *123*(1-4), 95-102.
70. Clancy, J. L.; Hargy, T. M.; Marshall, M. M.; Dyksen, J. E. *J. AWWA* **1998**, *90*(9), 92-102.
71. McGuire, M. J. *J. AWWA* **2006**, *98*(3), 123-126.

72. Kashinkunti, R. D. *J. AWWA* **2004,** *96*(6), 114-127.
73. Liu, W.; Cheung, L. M.; Yang, X.; Shang, C. *Wat. Res.* **2006,** *40*, 2033-2043.
74. Plewa, M. J.; Wagner, E. D.; Richardson, S. D.; Chen, P. H.; McKague, A. B. *Environ. Sci. Technol.* **2004,** *38*(1), 62-68.
75. Wu, H. W.; Xie, Y. F. *J. AWWA* **2005,** *97*(11), 94-101.
76. Wobma, P.; Pernitsky, D.; Bellamy, B.; Kjartanson, K.; Sears, K. *Ozone-Sci. Eng.* **2000,** *22*, 393-413.
77. Xie, Y. F.; Zhou, H. J. *J. AWWA* **2002,** *94*(5), 126-134.
78. Tung, H.-H.; Xie, Y. F. *Preprints of Extended Abstracts of Spring 2007 ACS Nat'l Mtg*; ACS, Div. Environ. Chem., Chicago, IL, 2007.
79. Tung, H.-H.; Unz, R. F.; Xie, Y. F. *J. AWWA* **2006,** *98*(6), 107-112.
80. Xie, Y. F.; Wu, H.; Tung, H.-H.; *Haloacetic Acid Removal by Granular Activated Carbon*; Awwa Research Foundation, Denver, CO, 2004.
81. Woo, Y.-T.; Lai, D.; McLain, J. L.; Manibusan, M. K.; Dellarco, V.; *Environ. Health Perspect.* **2002**; *110*(suppl. 1):75-87.
82. Plewa, M. J.; Wagner, E. D.; Muellner, M. G.; Hsu, K. M.; Richardson, S. D. *Preprints of Extended Abstracts of Spring 2007 ACS Nat'l Mtg*; ACS, Div. Environ. Chem., Chicago, IL, 2007.
83. Muellner, M. G.; Wagner, E. D.; McCalla, K.; Richardson, S. D.; Woo, T.; Plewa, M. J. *Environ. Sci. Technol.* **2007**, *41*, 645-651.
84. Bull, R. J; Reckhow, D. A.; Rotello, V.; Bull, O. M.; Kim, J. *Use of Toxicological and Chemical Models to Prioritize DBP*; Awwa Research Foundation: Denver, CO, 2006.
85. Villanueva, C. M.; Cantor, K. P.; Grimalt, J. O.; Malats, N.; Silverman, D.; Tardon, A.; Garcia-Closas, R.; Serra, C.; Carrato, A.; Castano-Vinyals, G.; Marcos, R.; Rothman, N.; Real, F. X.; Dosemeci, M.; Kogevinas, M. *Amer. J. Epidem.* **2007**, *165*, 148-156.
86. Arbuckle, T. E.; Hrudey, S. E.; Krasner, S. W.; Nuckols, J. R.; Richardson, S. D.; Singer, P.; Mendola, P.; Dodds, L.; Weisel, C.; Ashley, D. L.; Froese, K. L.; Pegram, R. A.; Schultz, I. R.; Reif, J.; Bachand, A. M.; Benoit, F. M.; Lynberg, M.; Poole, C.; Waller, K. *Environ. Health Perspect.* **2002,** *110*, 53-60.
87. Amy, G.; Graziano, N.; Craun, G.; Krasner, S.; Cantor, K.; Hildesheim, M.; Weyer, P.; King, W.; *Improved Exposure Assessment on Existing Cancer Studies*; Awwa Research Foundation: Denver, CO, 2005.
88. Waller, K.; Swan, S. H.; DeLorenze, G.; Hopkins, B.; *Epidem.* **1998,** *9*(2), 134-140.
89. Savitz, D. A,;Singer, P. C.; Herring, A. H.; Hartmann, K. E.; Weinberg, H. S.; Makarushka, C.; *Amer. J. Epidem.* **2006,** *164*(11), 1043-1051.

90. Bull, R. J.; Cottruvo, J. A. *Toxicology* **2006**, *221*(2-3), 135-144.
91. Roberson, A. In *Disinfection By-Products in Drinking Water: Occurrence, Formation, Health Effects, and Control.* Karanfil, T.; Krasner, S. W.; Westerhoff, P.; Xie, Y., Eds.; ACS Symposium Series 995. American Chemical Society: Washington, DC, 2008, chapter 2, pp 22-35.

Health and Regulatory Concerns of Disinfection By-Products

Chapter 2

The Evolution of Disinfection By-Products Regulations: Past, Present, and Future

J. Alan Roberson

American Water Works Association, 1300 Eye Street, N.W., Washington, DC 20005

The regulation of disinfection by-products (DBPs) has evolved significantly in the last thirty years since the initial discovery of trihalomethanes (THMs) in drinking water in 1974. DBP regulations are important from a public health perspective due to the potential widespread exposure. These changing regulations have both significantly reduced DBP exposure and challenged water utilities to maintain compliance. Compliance challenges will continue as recently finalized regulations will impact more water utilities and potential treatment technologies for compliance become more complex. Many questions still remain about unknown and unregulated DBPs that might result in future additional DBP regulations.

Introduction

The regulation of DBPs continues to be important for water utilities due to the widespread practice of disinfection and the potential for exposure to a large fraction of the population. The majority of water utilities disinfect, primarily for the prevention of waterborne disease outbreaks. Disinfection of water supplies started in the early 1900s and resulted in a significant decrease in the annual death rate from typhoid fever (*1*). Yet, even in those early times before the actual scientific discovery of DBPs, utilities recognized that chlorine disinfection had to be balanced with potential taste and odor problems from chlorine itself and certain chlorination reaction products (e.g., chlorophenols). As the science evolved with the initial discovery of DBPs, the risk balancing between the formation of DBPs and adequate disinfection became more significant.

Going beyond the valid and significant taste and odor concerns, utilities and their customers continue to be legitimately concerned about potential adverse health effects from DBPs. More health effects research on DBPs is needed, as the body of current health effects data still generates controversy. From the federal perspective, DBPs are regulated on a bladder cancer endpoint assuming a lifetime exposure of 70 years, i.e., a chronic health endpoint. Some limited and mixed health effects data points towards potential adverse reproductive and developmental endpoints. Addressing more acute health endpoints would constitute a major regulatory shift, leading to substantial changes in the compliance framework, which would pose compliance challenges for utilities.

The universe of knowledge is still not complete on DBPs, as the initial discovery of trihalomethanes (THMs) in drinking water was only a little over thirty years ago (*2, 3*). As analytical methods have improved, more and more DBPs are being identified and quantified (*4*). Many research questions remain on the health effects from these new DBPs, the potential exposure, and how treatment can prevent their formation or remove them.

Historical DBP Regulations

The Safe Drinking Water Act (SDWA) was signed into law on December 16, 1974 (PL 93-523). In 1975, using existing Public Health Service guidelines as the foundation, EPA finalized the National Interim Primary Drinking Water Regulations (NIPDWRs) for 22 well-known chemical and microbial contaminants (*5*). Since DBPs had just been recently discovered, they were not included in this initial rulemaking.

The Total Trihalomethane (TTHM) Rule

The Total Trihlalomethane (TTHM) Rule was finalized in 1979 and was the first federal regulation addressing DBPs (*6*). The TTHM Rule established a Maximum Contaminant Level (MCL) for TTHMs at 0.10 mg/l with compliance based on a running annual average of quarterly samples taken at locations throughout the distribution system. The TTHM Rule applied only to systems serving >10,000 people that added a disinfectant, so small systems were not impacted by this first DBP regulation. From the national perspective, the TTHM Rule was significant as it was the first federal regulation that followed a broad range of new SDWA regulatory requirements for USEPA, including toxicological and health risk assessments, occurrence and exposure assessments, specified analytical techniques, options for treatment technologies, and a detailed cost-benefit analysis (*7*).

The TTHM Rule was successful in significantly reducing THM exposure and most utilities complied by shifting the point of chlorination to later in the treatment plant, reducing the amount of chlorine added, or switching to chloramination for secondary disinfection in the distribution system (*8*). Soon thereafter, USEPA finalized the Surface Water Treatment Rule (SWTR) in 1989, mandating filtration and specific inactivation (disinfection) requirements for viruses and *Giardia lamblia* that varied slightly based on pH and temperature and had to be achieved on essentially a continuous basis (*9*). The risk balancing between disinfecting for microbial protection and minizing DBP exposure was now instituted in a regulatory framework, and EPA had additional concerns with utilities simply increasing the amount of disinfectant to meet their inactivation requirements.

Evolving from the TTHM Rule

To address ongoing concerns with potential adverse health effects from DBPs, USEPA informally released a "straw man" rule in 1989 that broadly framed future DBP regulations. At that time, USEPA was considering a future TTHM MCL of 25 or 50 ug/l, and listed other DBPs that it would consider regulating. This was a substantial reduction in the MCL, and many utilities would have faced major compliance challenges. EPA followed up this "straw man" with a status report in 1991.

Due to the significant concerns with the "straw man" rule that were raised by the water utilities and the recognition of the need for appropriate risk balancing, USEPA elected to use a negotiated rulemaking (Reg-Neg) process for future DBP and microbial regulations and to continue to pair these regulations in the future. USEPA constituted a Reg-Neg Committee that included representatives of State and local health and regulatory agencies, utilities, elected officials, and consumer and environmental advocates. The Reg-Neg Committee met from November 1992 through June 1993. The Reg-Neg Committee recommended the development of three categories of proposed regulations:

- A two-stage approach for DBPs, i.e., Stage 1 and Stage 2 Disinfection By-Products Rules (*10*);
- An "interim" Enhanced Surface Water Treatment Rule (ESWTR) (*11*); and
- An Information Collection Rule (ICR) (*12*).

The approach used by the Reg-Neg Committee considered the constraints of simultaneously treating the water to control for both DBPs and microbial contaminants. The Reg-Neg Committee further stipulated that the compliance schedules for the paired rules should be linked to assure simultaneous compliance and the appropriate risk-risk balancing.

The Stage 1 Disinfection By-Products Rule (DBPR)

USEPA published the final Stage 1 Disinfection By-Products Rule (DBPR) on December 16, 1998, along with the final IESWTR (13,14). The Stage 1 DBPR lowered the TTHM MCL to 0.080 mg/l and established a new MCL of 0.060 mg/l for the sum of five of the haloacetic acids (HAA5). The Stage 1 DBPR also established a MCL for bromate at 0.010 mg/l for plants that use ozone and a MCL for chlorite of 1.0 mg/l for plants that use chlorine dioxide. The Stage 1 DBPR set new standards for disinfectants, with the Maximum Residual Disinfectant Levels (MRDLs) for chlorine and chloramines of 4.0 mg/l and a MRDL for chlorine dioxide of 0.8 mg/l. The Stage 1 DBPR applied to all systems that added a disinfectant, with phased compliance deadlines for large and small systems. The compliance impacts of pulling small systems into the DBP regulatory framework will be discussed later. Systems complied with the Stage 1 DBPR with the typical treatment strategies of shifting the point of chlorination to later in the treatment plant, reducing the amount of chlorine added, switching to ozone or chlorine dioxide for primary disinfection, implementing enhanced coagulation or enhanced softening, and/or switching to chloramination for secondary disinfection in the distribution system.

In an attempt to address unknown DBPs, a Treatment Technique (TT) was established using enhanced coagulation or enhanced softening to improve removal of DBP precursors. A 3X3 matrix mandated a percentage removal of Total Organic Carbon (TOC) based on source water alkalinity and source water TOC and alternative compliance criteria were also developed for "difficult-to-treat" waters. Many utilities were able to comply with the DBP MCLs by taking advantage of the TOC removal from enhanced coagulation/softening.

The intent of the ICR was to collect raw water quality, treatment informaion, and DBP occurrence data for 18 months from the large systems serving >100,000 to be used in the next round of negotiations for the Stage 2 DBPR. The results of the ICR monitoring data have been summarized elsewhere (*15, 16*).

The Stage 2 Disinfection By-Products Rule (DBPR)

A Federal Advisory Committee (FACA) met from September 1999 to July 2000 to work on the Stage 2 DBPR and Long-Term 2 Enhanced Surface Water

Treatment Rule (LT2ESWTR). The Advisory Committee developed an Agreement in Principle that laid out their recommendations on how to further control DBPs (*17*). This Agreement was the basis for the final Stage 2 DBPR that was published on January 4, 2006 (*18*).

The Stage 2 DBPR maintained the same numeric MCLs from the Stage 1 DBPR but added three significant modifications to compliance monitoring design and compliance determinations. First, as part of the Stage 2 DBPR, systems would be required to conduct an Initial Distribution System Evaluation (IDSE). The IDSE is a monitoring program (or a hydraulic study) to locate potential future compliance monitoring locations that would be more representative of higher DBP concentrations in the distribution system. Second, future compliance at these locations would also be based on the Locational Running Annual Average (LRAA), i.e., based on the running annual average of each sampling location as opposed to averaging across all compliance locations in the distribution system. Finally, the concept of operational evaluation levels was established to curtail peak DBP levels by providing utilities with a structured approach in the regulation to remain in compliance. Compliance with the Stage 2 DBPR was predicted to drive the many of the surface water systems still using free chlorine to switch to chloramination for secondary disinfection in the distribution system.

To address groundwater systems that could potentially need disinfection (comparable to the SWTR), USEPA published the final Ground Water Rule (GWR) on November 8, 2006 (*19*). The GWR established a risk-based approach for ground water systems to collect data to determine if their wells were susceptible to fecal contamination, submit that data to the state, and then the state would make the determination if disinfection was required (or not).

DBP Compliance Data

USEPA tracks compliance with the federal regulations with the Safe Drinking Water Information System/Federal (SDWIS/FED), and this compliance data is available on the EPA website. The author downloaded the FY 1998-2005 data for DBP MCL, MRDL, and TT (enhanced coagulation or softening) violations.

As previously mentioned, the Stage 1 DPBR had phased compliance deadlines for large and small systems. Systems serving $\geq$10,000 people (that were already impacted by the TTHM Rule) had to comply with the Stage 1 DBPR by January 1, 2002. Smaller systems serving <10,000 people had an additional two years and had to comply by January 1, 2004. Additional time could be granted for compliance if capital improvements were required.

As shown in Figure 1, DBP MCL violations were relatively minor in 1998-2002, and began to increase slightly in 2003 and 2004 as the initial compliance deadline for the Stage 1 DBPR for large systems passed. The DBP violations jumped significantly in 2005 as the smaller systems faced compliance with DBP regulations for the first time and did not have the operational experience with the TTHM Rule, which the larger systems did. Compliance challenges will likely continue for the small systems as they work through compliance with the Stage 1 DBPR and prepare for compliance with the Stage 2 DBPR. Compliance with the Stage 2 LRAA may not be as problematic for small systems due to the smaller geographical size of their distribution systems.

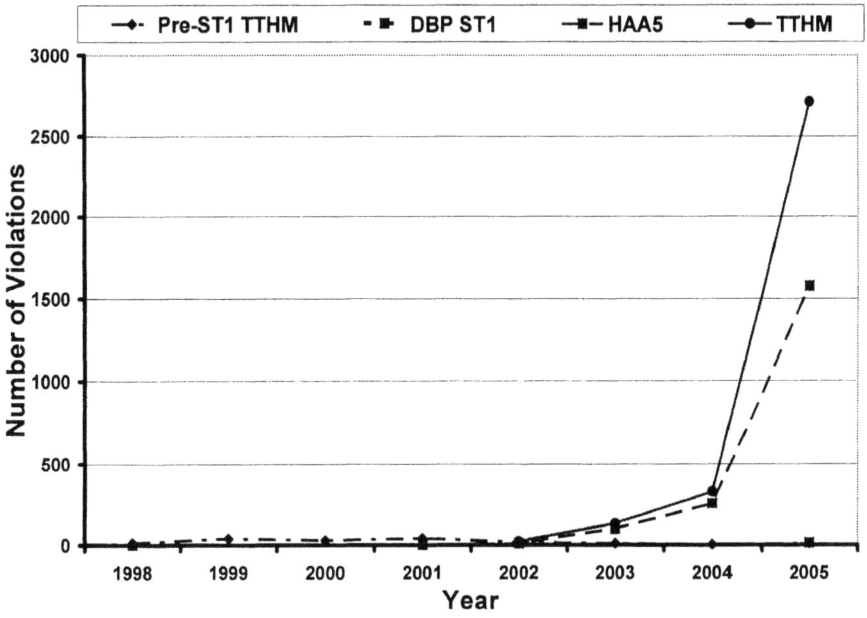

Figure 1. DBP Violations, 1998-2005

How do DBP violations compare with the violations from all of the other drinking water regulations? As shown on Figure 2, DBP violations are now the second largest number of violations, trailing only the Total Coliform Rule (TCR). The TCR has consistently had the largest number of violations of any of the drinking water regulations, with approximately 10,000 MCL violations annually (*20*). Prior to the Stage 1 DBPR, the Lead Copper Rule (LCR) was second for the total number of violations, but now is in third place by a significant margin. So, the Stage 1 DBPR is posing significant compliance challenges in comparison to other chemical contaminants.

The percentage compliance with regulations is important, as this is a performance and budgetary benchmark used by USEPA and other agencies such

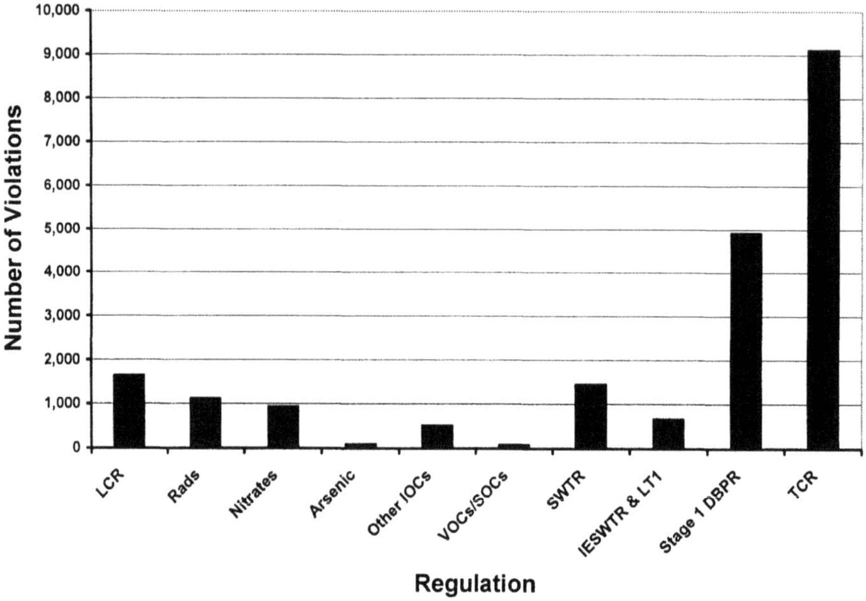

Figure 2. 2005 Violations by Regulation (IOC = inorganic contaminant, VOC = volatile organic contaminant, SOC = synthetic organic contaminant)

as the Office of Management and Budget (OMB). In the *2006-2011 EPA Strategic Plan,* under Goal 2 (Clean and Safe Water), Objective 2.1 (Protect Human Health), EPA establishes a target that, by 2011, 90 percent of community water systems will provide water that meets all applicable health-based standards. This target is up from 89 percent in 2005 (*21*). While one percent does not seem like much of an increase, the significant increases in DBP violations in 2004 and 2005 as shown in Figure 1 may make that target difficult to meet.

DBPs in Future Regulations

Two statutory regulatory processes form the foundation of the development of future national drinking water regulations. The first is the Contaminant Candidate List (CCL) that identifies potential contaminants of concern. At a later date, regulatory determinations are made from the CCL to determine if a contaminant should or should not be regulated, requires guidance development, or needs more research. If USEPA decides that a contaminant warrants regulation under the SDWA criteria, then, a proposed rule is published at a later date, followed by the publication of a final rule after public comment on the proposed rule. The second regulatory process is a review every six years of existing drinking water regulations mandated by the 1996 SDWA Amendments.

The Contaminant Candidate List (CCL)

USEPA published the first CCL (CCL1) in 1998 (*22*). This list of 60 contaminants was developed through expert input and nine contaminants were listed as priorities for potential future regulations. DBPs were not included on CCL1 because the Stage 1 and Stage 2 FACA processes were ongoing. In 2003, USEPA decided not to regulate any of the nine priority contaminants listed on CCL1 (*23*).

In 2005, USEPA published the second CCL (CCL2), and CCL2 listed the leftover 51 contaminants from CCL1 (*24*). The CCL2 did not list any priority contaminants like CCL1. USEPA recently published the preliminary second round of regulatory determinations from CCL2, and has preliminarily decided not to regulate eleven contaminants from CCL2 (*25*).

Future CCLs need a more robust process than what was used in the first two CCLs and USEPA requested expert input from both the National Research Council (NRC) and the National Drinking Water Advisory Council (*26*). In 2006, USEPA requested nominations for the third CCL (CCL3) as part of the new and improved process for developing CCL3 (*27*). The American Water Works Association (AWWA) nominated 21 chemicals and 12 microbes, including the following DBPs:

- Bromochloroaceonitrile;
- Aggregated nitrosamines including N-nitrosomorpholine (NMOR);
- Haloacetaldehydes; and
- Halonitromethanes/Halopicrins

Building on the nominations previously received, USEPA anticipates publishing the proposed CCL3 in early 2008, and finalizing it in mid-2009.

Six-Year Review of Existing Regulations

As part of the 1996 SDWA Amendments, on a six-year cycle, USEPA is required to go back and review all existing drinking water regulations and determine if a regulation needs to be revised (or not) due to new health effects or occurrence data, or the development of a more sensitive analytical method. In 2003, USEPA published the results of their first six-year review, and determined only to revise the Total Coliform Rule (TCR) (*28*). As previously shown in

Figure 2, the TCR has the largest number of violations, so revision is probably appropriate. Due to the complexities of a typical distribution system and the challenges in maintaining water quality, USEPA recently announced the formation of a Federal Advisory Committee (FACA) to address potential TCR revisions and other distribution system issues for future research. (*29*).

USEPA anticipates publishing the preliminary second six-year review in early to mid-2008, and publishing the final second six-year review in mid-2009. The third six-year review would then be scheduled for 2015, and so on. At any point during these six-year review cycles, if new health effects or occurrence data became available, USEPA could review the Stage 1 and Stage 2 DBPRs MCLs, and MRDLs and reevaluate whether those levels were appropriate.

Potential Future DBP Issues

Currently Regulated DBPs

The current regulations focus on TTHM and HAA5, and compliance is based on the sum of the species that make up these classes. Some have argued that the standard for haloacetic acids (HAAs) should include all nine HAAs, i.e., the standard should be for HAA9 (*30*). The analytical methods for the balance of the HAAs were still evolving during the formulation of the Stage 1 and Stage 2 DBPRs, and the Federal Advisory Committee for the Stage 2 DBPR did not want to change the HAA5 standard during their negotiations.

From another perspective, all species are not alike from a toxicity perspective. Both trichloroacetic acid (TCAA) and chloroform have relatively high Maximum Contaminant Level Goals (MCLGs) as compared to other species in their respective classes, and some have argued that these two species should not be regulated at all. For both TTHM and HAA5, the brominated species are generally considered to be more toxic (*13, 18*). Therefore, one could ask, should we be regulating each species individually?

Regulating each species individually could lead to MCLs being raised for some species as compard to current class sums and/or potentially being dropped for chloroform and/or TCAA. However, the 1996 SDWA Amendments included an "anti-backsliding" provision such that any revised regulation "maintain equivalent public health protection". That statutory language is a bit vague, and the impact of this provision in the six-year review process is not completely clear. In the intial six-year review in 2003, USEPA could have potentially lowered a handful of existing MCLs based on new data, but elected not to change these MCLs, as changing the MCLs would not provide "a meaningful opportunity for risk reduction" as mandated by the SDWA. Litigation could be one potential avenue to question USEPA's decision to revise or lower (or not revise) a MCL through the six-year review process.

Some are still concerned with the bromate MCL potentially being lowered, as the MCL of 0.010 mg/l was set to maintain the feasibility of ozone disinfection for *Cryptosporidium*. The Milwaukee outbreak occurred in the middle of the Reg-Neg Committee's negotiations for the Stage 1 DBPR and Interim and Long-Term 1 Enhanced Surface Water Treatment Rules. The Committee wanted to preserve ozone disinfection, even though at that time, the health effects data could potentially have drive the bromate standard lower. However, the current standard may be more than protective based on new health effects data hypothesizing that bromate is a threshold carcinogen based on bromate's mode of action and the chemical reactions of bromate in the stomach (*31*).

Unregulated and Newly Discovered DBPs

Many known and unknown DBPs are not regulated at this time. Nitrosamines are an example of a class of DBPs that are currently unregulated. There are many data gaps for nitrosamines, such as health effects, formation, removal by conventional and/or advanced treatment, and relative source contribution between food and drinking water. Several water reclamation plants in California and other parts of the country have installed advanced treatment for the removal of nitrosamines. Health effects research on nitrosamines is ongoing. To assist in developing a national occurrence profile, USEPA included six nitrosamines on the screening survey portion of the second Unregulated Contaminant Monitoring Rule (UCMR2) (32). The inclusion of these six nitrosamines on UCMR2 is a bit out of step with the CCL process as no nitrosamines are currently listed on CCL2, so the limited number of utilities required to conduct the UCMR2 screening survey will be generating occurrence data ahead of any decision to list nitrosamines on a CCL.

The current DBP regulations focus on TTHMs and HAA5, and these two compounds typically make up approximately 30-40% of the Total Organic Halide (TOX) found in drinking water (*15*). More research is needed to identify the compounds that make up the balance of the TOX. A recent survey of DBP occurrence was conducted at 12 drinking water plants to determine the occurrence of more than 50 unregulated DBPs (*4*). Many were found at low concentrations, and this survey also found 28 previously unidentified DBPs.

Clearly, more research is needed to determine the human health significance of these unregulated DBPs and newly discovered DBPs. Occurrence is not necessarily a cause for concern, as assessing the relevance of occurrence data in the absence of health information is difficult. The shifts in treatment, such as the general shift to chloramination for secondary disinfection, could be creating a shift to more toxic (or less toxic) DBPs. A better understanding is needed of the relative concentrations and the relative toxicities of these unregulated DBPs and newly discovered DBPs in order to determine if the regulations need to be revised at some point in the future.

The Big Health Effects Question

The current regulations focus on TTHMs and HAA5, and compliance is based on a running annual average of quarterly samples. The adverse health endpoint of concern in the current regulations is bladder cancer based on 70 years of exposure. The current compliance averaging appropriately reflects this long-term chronic exposure.

Conflicting health effects data currently surrounds potential adverse reproductive and developmental endpoints from DBPs. For these types of endpoints, a much shorter exposure window might be relevant, possibly a single exposure (acute), similar to microbials. The first large-scale epidemiological study showed some positive associations between DBPs and spontaneous abortions, and that study had some impact on the first round of FACA negotiations (33). However, additional research has provided mixed results, with some studies showing positive associations, and others not showing such associations. USEPA summarized the relevant health effects data on potential adverse reproductive and developmental endpoints in the final Stage 2 DBPR, and based on its overall assessment of the science, maintained the averaging component of the compliance determination (18). However, the new location-based compliance and improved monitoring designs are expected to significantly lower peak exposure.

More health effects research is needed due to the widespread exposure to DBPs and to address the remaining uncertainties surrounding adverse health impacts to children and/or fetuses. More analytical method research is needed to identify the unknown DBPs in the balance of the TOX. More treatment research is needed to better understand the formation of these unknown DBPs and their fate through treatment and distribution. More research is also needed on the impacts of the dynamics of the distribution system on DBP occurrence and on the spatial and temporal variability of DBPs in the distribution system. Some limited information on DBP variability is being collected through the Stage 2 IDSEs, but the IDSEs focus on seasonal variability at a limited number of locations. More research is needed on the hour-to-hour variability of DBPs in the distribution system due to the typical diurnal water demand and the influence of multiple treatment plants with varying water qualities.

Any shift in the compliance determination from a locational running annual average based on quarterly samples would significantly impact many water utilities. If new health effects data show acute reproductive and development endpoints from DBPs, then more frequent monitoring would be needed based on the timeframe of concern, which could possibly range from a thirty-day or ninety-day window to a single acute event. Reliable online monitoring would be needed so that utilities could optimize their treatment with greater certainty of staying below potential acute DBP MCLs at all times.

Conclusions

The regulations for DBPs have evolved significantly from the initial TTHM Rule in 1979 to the recently promulgated Stage 2 DBPR, which becomes effective in 2012. The initial TTHM MCL of 0.10 mg/l has been lowered, and MCLs have been established for HAA5, bromate, and chlorite. In an attempt to address unknown DBPs, a treatment technique has been established using enhanced coagulation or enhanced softening to improve removal of DBP precursors. All systems that disinfect are now covered by DBP regulations and all systems will eventually be required to meet the MCLs at each compliance monitoring location through the Stage 2 DBPR LRAA.

Current DBP compliance data shows relatively minor non-compliance in 1998-2002, with a slight increase in 2003 and 2004 as the initial compliance deadline for the Stage 1 DBPR for large systems passed. The DBP violations jumped significantly in 2005 as the smaller systems faced compliance with DBP regulations for the first time. Small systems did not have the operational experience with the TTHM Rule, which the larger systems did. In 2005, the Stage 1 DPBR had the second largest number of violations, trailing only the TCR, the perennial leader in the number of violations from all of the regulations. It is difficult to predict what future DBP compliance rates will look like. Compliance challenges will likely continue for the small systems as they continue to adjust to the Stage 1 DBPR and ultimately face compliance with the Stage 2 DBPR LRAA.

USEPA could potentially develop future DBP regulations through the CCL process for unregulated DBPs and newly discovered DBPs. The existing MCLs could be revised through the six-year review process if new health effects data or occurrence data warranted such a revision. It is not completely clear at this time whether USEPA will revise any of the existing MCLs or decide to regulate any unregulated DBPs or newly discovered DBPs. A better understanding is needed of the relative concentrations and the relative toxicities of these unregulated DBPs and newly discovered DBPs in order to determine if the regulations need to be revised at some point in the future.

The big health effects question is whether potential adverse reproductive and development endpoints will fundamentally shift the focus of DBP regulations from a chronic endpoint with corresponding annual average compliance determinations to a potential acute endpoint with more frequent monitoring and shorter-term compliance determinations. Such a shift would be significant for all systems that disinfect, which is the majority of the systems.

More health effects research is needed on the unanswered questions of potential adverse reproductive and development endpoints from DBP exposure.

References

1. McGuire, M.J. Eight Revolutions in the History of US Drinking Water Disinfection. *Jour. AWWA* **2006**, 98:3:123.
2. Bellar, T.A., et al. Occurrence of Organohalides in Chlorinated Drinking Water. *Jour. AWWA* **1974**, 66:12:703.
3. Rook, J.J. Formation of Haloforms During Chlorination of Natural Water. *Water Treatment & Examination* **1974**, 23:3:234.
4. Krasner, S.W., et al. Occurrence of a New Generation of Disinfection By-Products. *Environ. Sci. Technol.* **2006**, *40(23)*, 7175-7185.
5. USEPA (US Environmental Protection Agency). National Interim Primary Drinking Water Regulations. *Fed. Reg.* **1975**, 40:248:59566.
6. USEPA. Total Trihalomethanes. National Interim Primary Drinking Water Regulations. *Fed. Reg.* **1979**, 44:228:68624.
7. Roberson, J.A. From Common Cup to *Cryptosporidium*: A Regulatory Evolution. *Jour. AWWA* **2006**, 98:3:198.
8. McGuire, M.J. and Meadow, R.G. AwwaRF Trihalomethane Survey. *Jour. AWWA* **1988**, 80:1:61.
9. USEPA. Filtration, Disinfection, Turbidity, *Giardia lamblia*, Viruses, *Legionella*, and Heterotrophic Bacteria. Final Rule. *Fed. Reg.* **1989**, 54:124:27486.
10. USEPA. Disinfectants/Disinfection By-Products. Proposed Rule. *Fed. Reg.* **1994**, 59:145:38668.
11. USEPA. Enhanced Surface Water Treatment Rule Requirements. Proposed Rule. *Fed. Reg.* **1994**, 59:145:38832
12. USEPA. Monitoring Requirements for Public Drinking Water Supplies. Proposed Rule. *Fed. Reg.* **1994**, 59:28:6332
13. USEPA. Disinfectants/Disinfection By-Products. Final Rule. *Fed. Reg.* **1998**, 63:241:69390.
14. USEPA. Interim Enhanced Surface Water Treatment Rule Requirements. Final Rule. *Fed. Reg.* **1998**, 63:241:69477.
15. *Information Collection Rule Data Analysis*; McGuire, M.J.; McLain, J.L.; Obolensky, A. (eds.); Awwa Research Foundation: Denver, CO; 2002.
16. Obolensky, A., Singer, P.C., and Shukairy, H.M. Information Collection Rule Data Evaluation and Analysis to Support Impacts on Disinfection By-Product Formation. *Jour. of Environmental Engineering,* **2007**, 133:1:53.
17. USEPA. Stage 2 Microbial and Disinfection By-Products Federal Advisory Committee Agreement in Principle. *Fed. Reg.* **2000**, 65:255:83015.

18. USEPA. Stage 2 Disinfectants/Disinfection By-Products Rule. Final Rule. *Fed. Reg.* **2006**, 71:2:388.
19. USEPA. Ground Water Rule. Final Rule. *Fed. Reg.* **2006**, 71:216:65574.
20. Roberson, J.A. Complexities of the New Drinking Water Regulations-Everything You Wanted to Know but Were Afraid to Ask. *Jour. AWWA*, **2003**, 95:3:48.
21. USEPA. 2006-2011 EPA Strategic Plan: Charting Our Course. September 30, **2006**.
22. USEPA. Announcement of the Drinking Water Contaminant Candidate List. Notice. *Fed. Reg.* **1998**, 63:40:10273.
23. USEPA. Announcement of the Regulatory Determinations for the Priority Contaminants on the Drinking Water Contaminant Candidate List. Notice. *Fed. Reg.* **2003**, 68:138:42897.
24. USEPA. Drinking Water Contaminant Candidate List 2. Final Notice. *Fed. Reg.* **2005**, 70:36:9071.
25. USEPA. Regulatory Determinations Regarding Contaminants on the Second Contaminant Candidate List. Preliminary Determinations. *Fed. Reg.* **2007**, 72:83:24016.
26. *Classifying Drinking Water Contaminants for Regulatory Consideration*; National Academy of Sciences; Washington, DC; 2001.
27. USEPA. Request for Nominations for Drinking Water Contaminants for the Contaminant Candidate List. *Fed. Reg.* **2006**, 71:199:60704.
28. USEPA. Announcement of Completion of EPA's Review of Existing Drinking Water Regulations. Notice. *Fed. Reg.* **2003**, 68:138:42907.
29. USEPA. Establishment of the Total Coliform Rule Distribution System Advisory Committee. Notice; Establishment of Federal Advisory Committee. *Fed. Reg.* **2007**, 72:125:35870.
30. Singer, P.C. DBPs in Drinking Water: Additional Scientific and Policy Considerations for Public Health Protection. *Jour. AWWA*, **2006**, 98:10:73.
31. Bull, R.J. and Cotruvo, J.A. Research Strategy for Developing Key Information on Bromate's Mode of Action. *Epidemiology,* **2006**, 221:2-3:135.
32. USEPA. Unregulated Contaminant Monitoring Rule for Public Water System Revisions. Final Rule. *Fed. Reg.* **2007**, 72:2:367.
33. Swan, S.H. et al. Is Drinking Water Related to Spontaneous Abortion? Reviewing Evidence from the California Department of Health Services. *Epidemiology,* **1992**, 3:83.

Chapter 3

Comparative Mammalian Cell Toxicity of N-DBPs and C-DBPs

Michael J. Plewa[1], Elizabeth D. Wagner[1], Mark G. Muellner[1], Kang-Mei Hsu[1], and Susan D. Richardson[2]

[1]Department of Crop Sciences, University of Illinois at Urbana-Champaign, Urbana, IL 61801
[2]U.S. Environmental Protection Agency, 960 College Station Road, Athens, GA 30605

In order to generate a quantitative, direct comparison amongst classes of drinking water disinfection by-products (DBPs), we developed and calibrated *in vitro* mammalian cell cytotoxicity and genotoxicity assays to integrate the analytical biology with the analytical chemistry of these important environmental contaminants. The generated database demonstrates the universality of the comparative toxicity of iodo- > bromo- >> chloro-DBPs across different structural DBP classes and the substantially greater toxicity of nitrogen-containing DBPs (N-DBPs) compared to carbonaceous DBPs (C-DBPs). These results are important in light of the generation of iodinated-DBPs and N-DBPs that may result from the use of alternative disinfectants.

Introduction

The drinking water community provides an important public health service for the nation by its generation of high quality, safe and palatable tap water. Drinking water disinfection by-products (DBPs) are an unintended consequence and were discovered over 30 years ago; since then over 600 DBPs have been identified (*1*). The vast majority of DBPs have not been chemically or biologically characterized (*1, 2*). Due to the lack of quantitative biological data, concerns exist over which DBPs pose the greatest health and environmental risks, and, therefore, which DBPs should be the focus of heightened federal regulation. There is generally less total organic halogen (TOX) formed in water treated with alternative disinfectants as compared to chlorine. However, there is a predominance of unknown DBPs as a percentage of TOX for these alternative disinfectants (*3*). As some water utilities move from chlorine to alternative disinfectants to meet the new U.S. EPA Stage 2 DBP Rule (*4*), they may generate greater amounts of emerging DBPs for which we have little data (*5, 6*).

The objective of this research was to develop a database on the mammalian cell chronic cytotoxicity and acute genotoxicity of DBPs and structurally related agents. Using this database we compared the impact of the halogenated species of DBP analogues as well as the toxicity of specific DBP chemical classes. This database may aid in identifying highly toxic emerging DBPs that warrant concern and further attention (*1*).

CHO Cell Cytotoxicity and Genotoxicity Assays

We developed and calibrated two *in vitro* cellular toxicological assays based on Chinese hamster ovary (CHO) cells. CHO cells are widely used in toxicology research; we used line AS52 clone 11-4-8 (*7*). This clone expresses a stable chromosome complement as well as functional p53 protein and normal DNA repair kinetics (*7-9*). CHO cells were maintained in Ham's F12 medium +5% fetal bovine serum (FBS) at 37°C in a humidified atmosphere of 5% CO_2. The cells exhibit normal morphology, express cell contact inhibition and grow as a monolayer without expression of neoplastic foci.

CHO Cell Chronic Cytotoxicity Assay

The CHO cell microplate chronic cytotoxicity assay measures the reduction in cell density as a function of the DBP concentration over a 72 h period (*10-12*). A 96-well flat-bottomed microplate is used to evaluate a series of DBP concentrations. One column served as the blank control consisting of 200 μL of F12 +FBS medium only. The concurrent negative control column consisted of 3×10^3 CHO cells plus F12 +FBS medium. The wells of the remaining columns contained 3×10^3 CHO cells, F12 +FBS and a known DBP concentration in a total of 200 μL. The wells were covered with a sheet of sterile AlumnaSeal™ and the cells were incubated for 72 h at 37°C at 5% CO_2. After the treatment time, the medium from each well was aspirated, the cells fixed in methanol for 10 min and stained for 10 min with a 1% crystal violet solution in 50% methanol. The microplate was washed, 50 μL of dimethyl sulfoxide/methanol

(3:1 v/v) was added to each well, and the plate was incubated at room temperature for 10 min. The microplate was analyzed at 595 nm with a BioRad microplate reader; the absorbancy of each well was recorded and stored on a spreadsheet file. The averaged absorbancy of the blank wells was subtracted from the absorbancy data from each well. The mean blank-corrected absorbancy value of the negative control was set at 100%. The absorbancy for each treatment group well was converted into a percentage of the negative control. For each DBP concentration, 8 replicate wells were analyzed per experiment, and the experiments were repeated 2-4×. These data were used to generate a concentration-response curve for each DBP (Figure 1). Regression analysis was applied to each DBP concentration-response curve, which was used to calculate the %C½ value which is analogous to a LC_{50} value. The %C½ value is the calculated DBP concentration that induced a cell density that was 50% of the negative control (Figure 1).

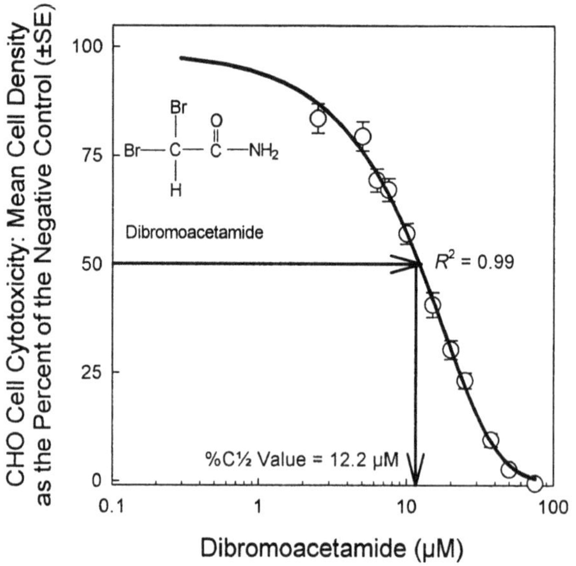

Figure 1. CHO cell chronic cytotoxicity concentration-response curve of dibromoacetamide and the derivation of the %C½ (LC_{50}) value.

A ranking of the %C½ values for all DBPs analyzed with this assay is presented in Figure 2. Abbreviations are listed at the end of the text. Note that the concentration range of the %C½ values exceeds 4 orders of magnitude. The cytotoxicity of over 55 DBPs was compared; their concentration-response curves were based on molar concentrations which is essential for structure activity relationship analysis.

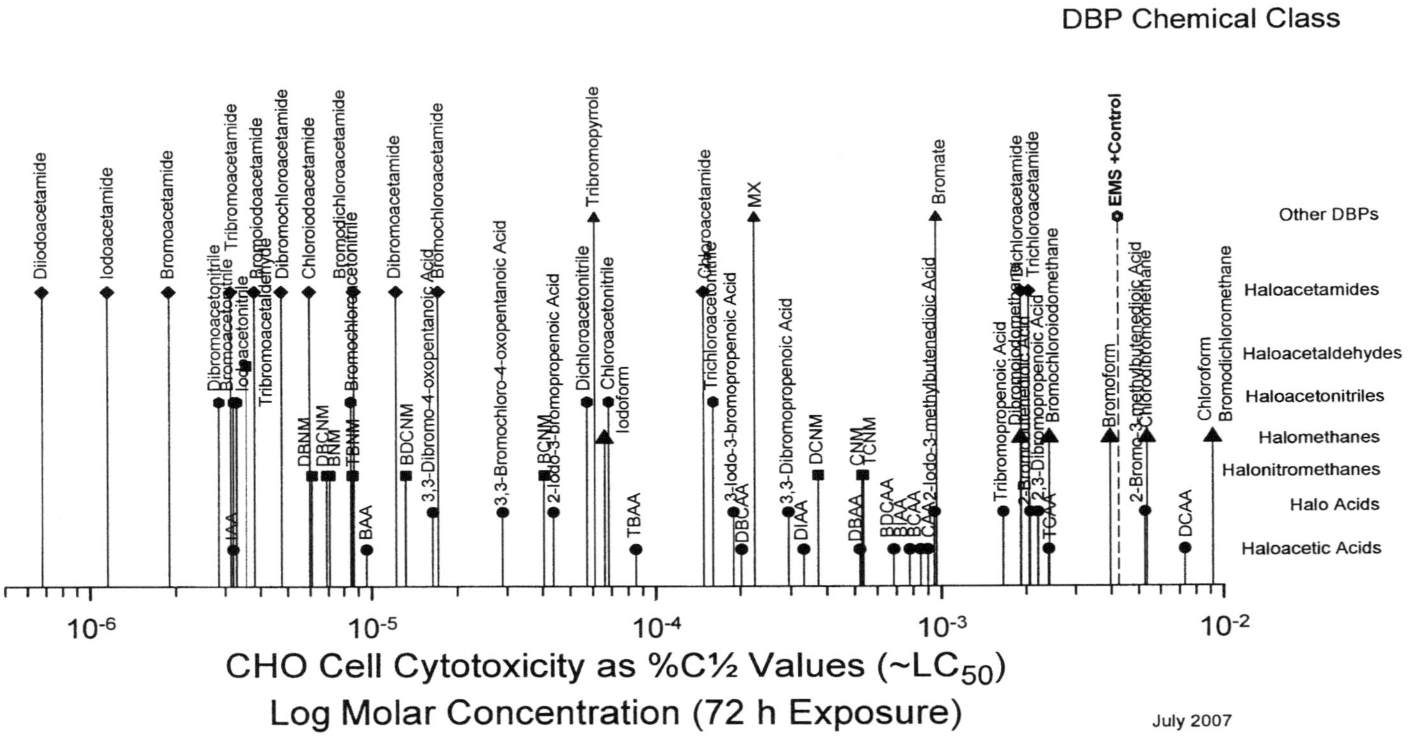

Figure 2. A distribution of DBPs ordered by their CHO cell chronic cytotoxicity as the %C½ value (M) derived from the individual DBP concentration response curves. Each %C½ value was derived from between 160 and 240 individual cell cultures. The DBP chemical classes are listed to the right of the figure.

Figure 2 presents the relative cytotoxic potency of DBPs within a chemical class and the cytotoxic potencies among DBP chemical classes. The cytotoxicity data were analyzed using the statistical and graphical functions of SigmaPlot 8.02, SigmaStat 3.1 and Table Curve 4.03 (SPSS Inc., Chicago, IL). For the mammalian cell chronic cytotoxicity experiments, the data were normalized as the percent of the concurrent negative control. The data from the independent experiments were combined. To determine if a significant toxic response occurred within the concentration-response curve, a one-way analysis of variance (ANOVA) test was conducted. If a significant F value of $P \leq 0.05$ was obtained, a Holm-Sidak multiple comparison versus the control group analysis was conducted. The power of the test statistic $(1-\beta)$ was ≥ 0.8 at $\alpha = 0.05$.

CHO Cell Acute Genotoxicity Assay

We employed single cell gel electrophoresis (SCGE) as the genotoxicity assay for this research. Single cell gel electrophoresis (or the comet assay) is a molecular genetic assay that quantitatively measures the level of genomic DNA damage induced in individual nuclei of treated cells (*9, 13*). The day before treatment, 4×10^4 CHO cells were added to each microplate well in 200 μL of F12 +FBS and incubated. The next day, the cells were washed with Hank's balanced salt solution (HBSS) and treated with a series of concentrations of a specific DBP in F12 medium without FBS in a total volume of 25 μL for 4 h at 37°C, 5% CO_2. The wells were covered with sterile AlumnaSeal™. After incubation, the cells were washed 2× with HBSS and harvested with 50 μL of 0.01% trypsin + 53 μM EDTA. The trypsin was inactivated with 70 μL of F12 +FBS. Acute cytotoxicity was measured using the trypan blue vital dye assay. SCGE data were not used if the acute cytotoxicity exceeded 30%. The remaining cell suspension from each well was embedded in a layer of low melting point agarose prepared with PBS upon clear microscope slides that were previously coated with a layer of 1% normal melting point agarose prepared with deionized water and dried overnight. The cellular membranes were removed by an overnight immersion in lysing solution (2.5 M NaCl, 100 mM Na_2EDTA, 10 mM Tris, 1% sodium sarcosinate, 1% Triton X-100, and 10% DMSO) at 4°C. The slides were placed in an alkaline buffer (pH 13.5) in an electrophoresis tank, and the DNA was denatured for 20 min. The microgels were electrophoresed at 25 V, 300 mA (0.72 V/cm) for 40 min at 4°C. The microgels were neutralized with Tris buffer (pH 7.5), rinsed in cold water, dehydrated in cold methanol, dried at 50°C, and stored at room temperature in a covered slide box. For analysis, the microgels were hydrated in cold water for 30 min and stained with 65 μL of ethidium bromide (20 μg/mL) for 3 min. The microgels were rinsed in cold water and analyzed with a Zeiss fluorescence microscope with an excitation filter of 546/10 nm and a barrier filter of 590 nm. For each experiment, 2 microgels were prepared per treatment group. Randomly chosen nuclei (25/microgel) were analyzed using a digital camera. A computerized image analysis system (Komet version 3.1, Kinetic Imaging Ltd., Liverpool, UK) was employed to determine the SCGE tail moment value (integrated value of migrated DNA density multiplied by the migration distance) of the nucleus as an index of DNA damage (Figure 3). The digitalized data were automatically transferred to a computer-based spreadsheet for subsequent statistical analysis. Within each experiment, a negative control, a positive control (3.8 mM ethylmethanesulfonate) (EMS) and 9 concentrations of a specific DBP were analyzed concurrently. The experiments were repeated 3 times for each DBP. Within the DBP concentration range with 70% or greater viable cells, a concentration-response curve was generated, and a regression analysis of the data was conducted. The SCGE genotoxic potency was calculated for each DBP from the concentration-response curve. It represents the midpoint of the curve within the concentration range that expressed >70% cell viability of the treated cells (Figure 3). The tail moment endpoint of the SCGE assay that measures genomic DNA damage is not normally distributed and violates the requirements for analysis by parametric statistics. The median tail moment value for each microgel was determined and the data were averaged. Averaged median values express a normal distribution according to the Central Limit Theorem (*14*). The averaged median tail moment values obtained from repeated experiments were used with a one-way analysis of variance test. If a significant F value of $P \leq 0.05$ was obtained, a Holm-Sidak multiple comparison versus the control group analysis was conducted ($1\text{-}\beta \geq 0.8$ at $\alpha = 0.05$).

The genotoxicity concentration-response curve for the newly discovered DBP bromoiodoacetamide is presented in Figure 4. For each DBP analyzed, the concentration-response data were regressed, and the concentration at the midpoint of the curve in the range that expressed >70% cell viability was calculated as the SCGE genotoxic potency. For bromoiodoacetamide the genotoxic potency was 72.1 μM. The SCGE genotoxic potency provides a single value that represents a molar concentration of the genotoxicity of the DBP. This allows a direct and sensitive comparison of the relative genotoxicities of all of the DBPs in our database.

A ranking of the SCGE genotoxic potency values for all DBPs analyzed with this assay is presented in Figure 5. Note that the concentration range of the SCGE genotoxic potency values exceeds 3 orders of magnitude. As with the cytotoxicity data, the SCGE genotoxic potency values are expressed on a molar basis. This characteristic permits sensitive structure activity relationship analysis. Figure 5 presents the relative SCGE genotoxic potency of DBPs within a chemical class and a comparison of the SCGE genotoxic potencies among DBP chemical classes.

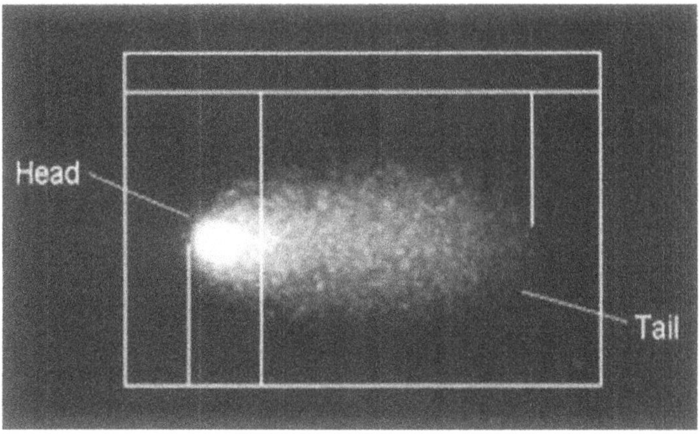

Figure 3. A digitalized SCGE image of a nucleus expressing DNA damage using Komet 3.1 software. The top rectangle of the image records the background fluorescence of the image. The head calipers measure the fluorescence intensity of the nucleus, while the tail calipers measure the fluorescence intensity and the migration distance of the DNA fragments within the agarose gel.

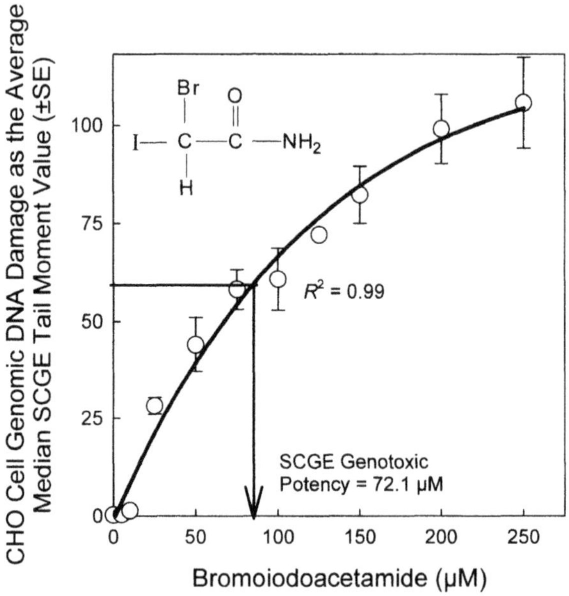

Figure 4. CHO cell genotoxicity concentration-response curve of bromoiodoacetamide and the derivation of the SCGE genotoxic potency.

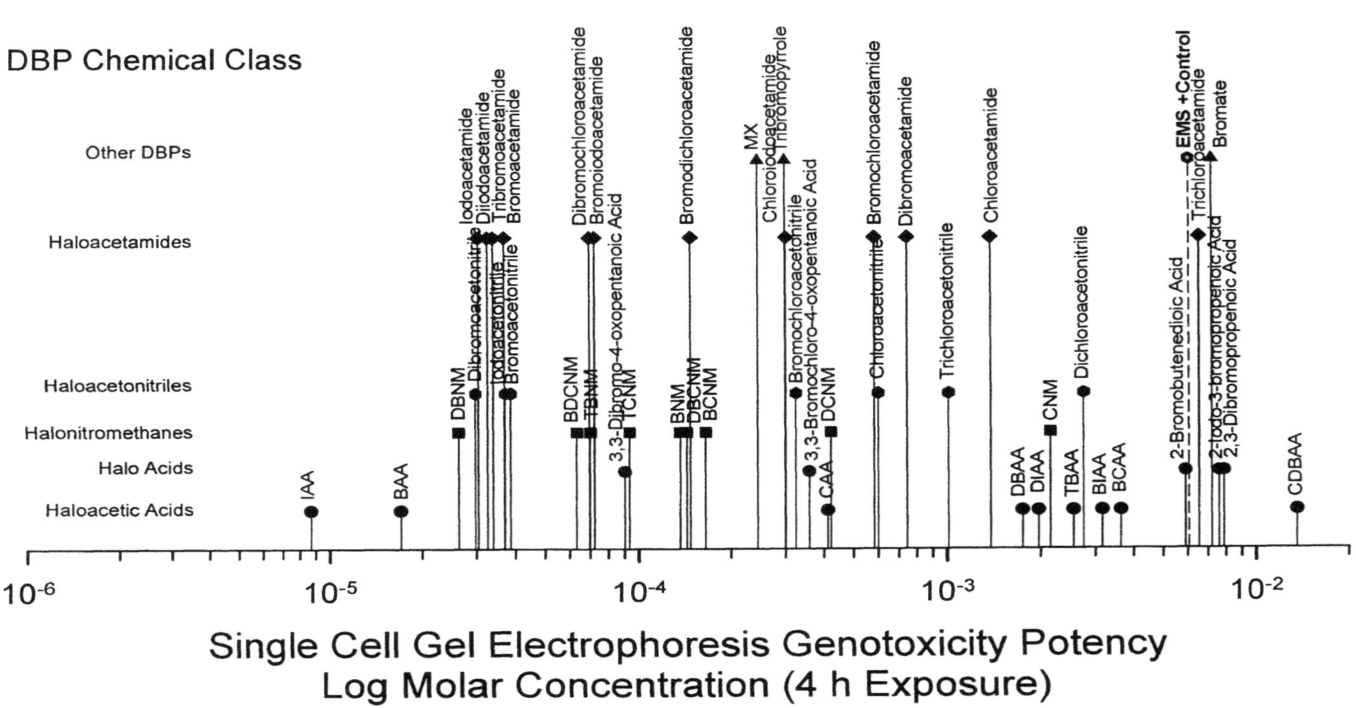

Not Genotoxic: DCAA, TCAA, BDCAA, Dichloroacetamide, Chloroform, Chlorodibromomethane, Bromoform, Iodoform, Bromochloroiodomethane, Dibromoiodomethane, Bromodichloromethane, 3,3-Dibromopropenoic Acid, 3-Iodo-3-bromopropenoic Acid, 2,3,3,Tribromopropenoic Acid

March 2007

Figure 5. A distribution of DBPs ordered by their CHO cell SCGE genotoxic potency (M) derived from the individual DBP concentration-response curves. The DBP chemical classes are listed to the left of the figure.

Cytotoxicity and Genotoxicity Indices

To compare among DBP chemical classes, we calculated CHO cell chronic cytotoxicity and acute genotoxicity indices. These indices allow a uniform comparison among groups of DBPs and one can ask global questions about their relative cytotoxicity or genotoxicity. The cytotoxicity index was determined by calculating the median %C½ value of all of the individual members of a single class of DBPs. The reciprocal was taken of this number so that a larger value was equated with that of higher cytotoxic potency. The genotoxicity index was determined by calculating the median SCGE genotoxicity potency value from the individual members within a single class of DBPs. The reciprocal was taken of this number so that a larger value was equated with that of higher genotoxicity. The cytotoxicity and genotoxicity indices for the classes of DBPs analyzed by our laboratory are presented in Figure 6.

These broader comparisons amongst DBP classes are possible because these agents were analyzed using identical assays, and all of the concentrations are in molar units. From these values it is clear that the halomethanes and haloacetic acids (which include the majority of U.S. EPA-regulated DBPs) are the least toxic chemical classes, while the emerging N-DBPs are the most toxic. Using the database generated by the CHO cell assays, we compared the cytotoxicity and genotoxicity indices for C-DBPs versus N-DBPs (Figure 7). The difference in their relative toxicities argues that additional attention should be focused on the N-DBPs and their possible adverse impacts on the public health and the environment (*12, 15*).

The species of the halogen atom of specific DBPs generated a difference in cytotoxicity and genotoxicity in CHO cells. The rank order of decreasing DBP cytotoxicity and genotoxicity was iodo- > bromo- >> chloro-. The cytotoxicity and genotoxicity indices were individually calculated for bromo-, dibromo-, chloro-, dichloro, iodo- and diiodo acetic acids and acetamides, bromo-, chloro-, and iodo-acetonitrile and chloroform, bromoform and iodoform. As illustrated in Figure 8, when a balanced design of these representative DBPs was analyzed, the iodinated DBPs were substantially more toxic than their brominated and chlorinated analogs. These data are important in that there are increasing numbers of treatment plants using chloramines to meet the EPA Stage 1 and Stage 2 D/DBP Rules (*4*), and there may be increased occurrence of iodo-DBPs in those drinking waters (*16-18*).

Future Research

Identification of New DBPs

This research is continuing with the analysis of emerging DBPs such as the haloaldehydes and nitrosamines. The application of this work is to also direct the

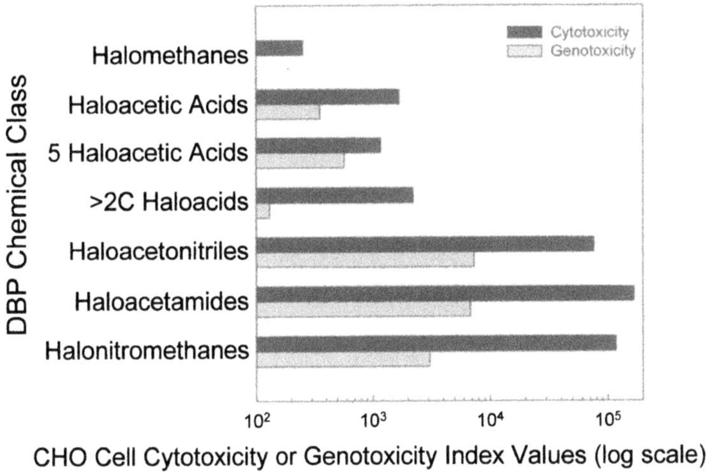

Figure 6. Cytotoxicity and genotoxicity indices for the major DBP classes.

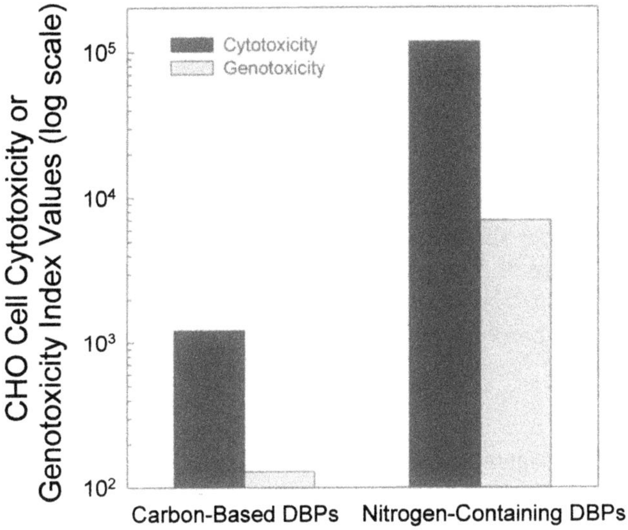

Figure 7. Cytotoxicity and genotoxicity indices for carbonaceous-DBPs (C-DBPs) versus nitrogen-containing DBPs (N-DBPs).

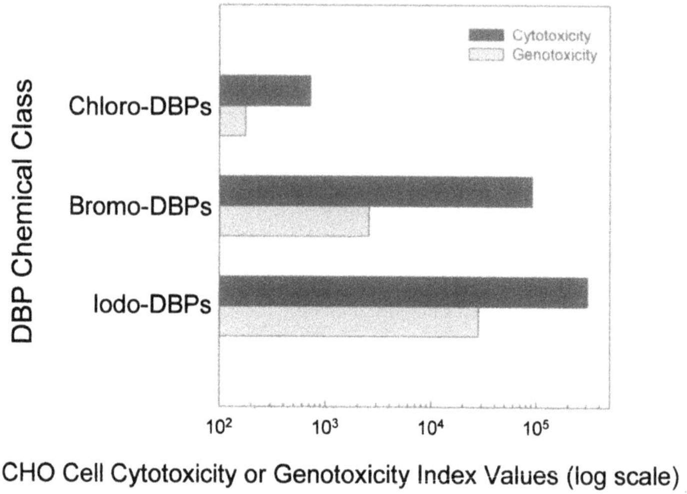

Figure 8. Cytotoxicity and genotoxicity indices of mono-, di- and trihalogenated DBPs from the CHO cell database (6 DBPs in each group).

discovery of new DBPs in drinking waters such as tribromopyrrole (*19*) and iodoacids (*16, 17*). Recently, we discovered bromoiodoacetamide in finished drinking water (*15*). Bromoiodoacetamide was isolated in drinking water from 12 treatment plants located in 6 U.S. states. One plant used chlorine for disinfection; 11 plants used chloramination. These plants had source waters with relatively high natural bromide (67 to 699 μg/L) and iodide levels (up to 65 μg/L). Using gas chromatography/electron ionization-mass spectrometry with selected ion monitoring of 5 key ions, this compound was identified in drinking water extracts (Figure 9) and confirmed by a match of the GC retention time. All of the haloamides showed a prominent peak at m/z 44, which represents the amide group (Figure 9). The presence of bromine and iodine was evident in the mass spectrum with 1-bromine doublets present at m/z 263/265, 220/222, and 136/138, and the presence of iodine at m/z 127 and loss of iodine at m/z 136/138.

Human Cell Toxicogenomic Analysis of DBPs

We are currently using non-transformed human embryonic cells for qRT-PCR gene array analysis with bromoacetic acid (*20*). In 6-well plates, human cells were treated with 60 μM bromoacetic acid for 4 h with concurrent negative

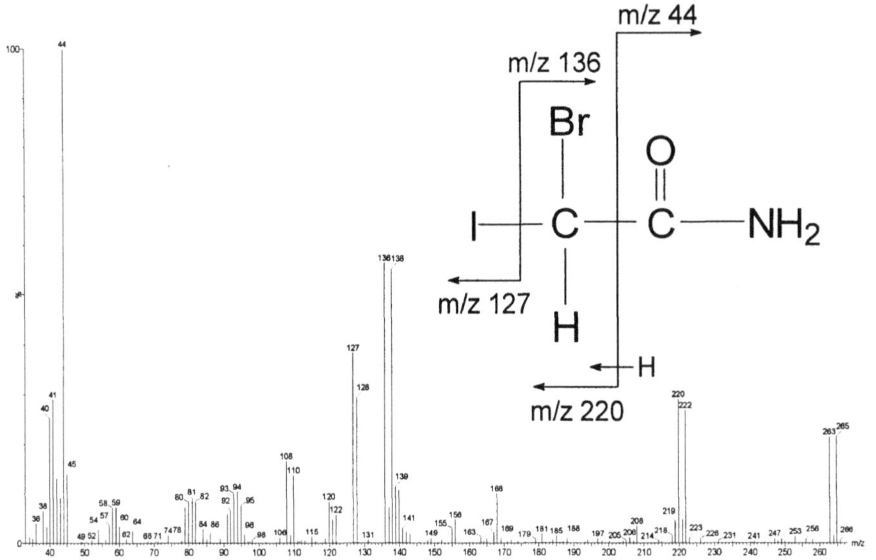

Figure 9. EI mass spectrum of bromoiodoacetamide.

controls. After the treatment time, acute cytotoxicity was examined with trypan blue vital dye; there was no cytotoxicity. The total RNA was isolated, and the purity and concentration was analyzed spectrophotometrically. Single-strand cDNA was prepared from the mRNAs for each group, and we analyzed for gene expression effects using the human DNA damage/repair PCR arrays. The data presented in Figure 10 represent the first non-transformed human cell gene array data for a regulated DBP. These data clearly demonstrate that significant differences in gene expression in human DNA damage/repair pathways are induced by bromoacetic acid. The expression of two genes was significantly up-regulated, and the expression of 4 genes was significantly down-regulated. If we can define functional gene groups and pathways whose expressions are impacted by DBPs, we may identify biomarkers that would be suitable for human monitoring. Such biomarkers may contribute to resolving epidemiological issues such as the induction of cancer and birth defects in human populations that consume disinfected water. This alone could be an exceedingly useful outcome for the drinking water community and the public health. Toxicogenomic analysis may form the foundation of a new analytical tool that will more closely integrate laboratory toxicological data with the experience of real people drinking real disinfected tap waters. This would be an approach to improve the high quality tap water that we currently enjoy.

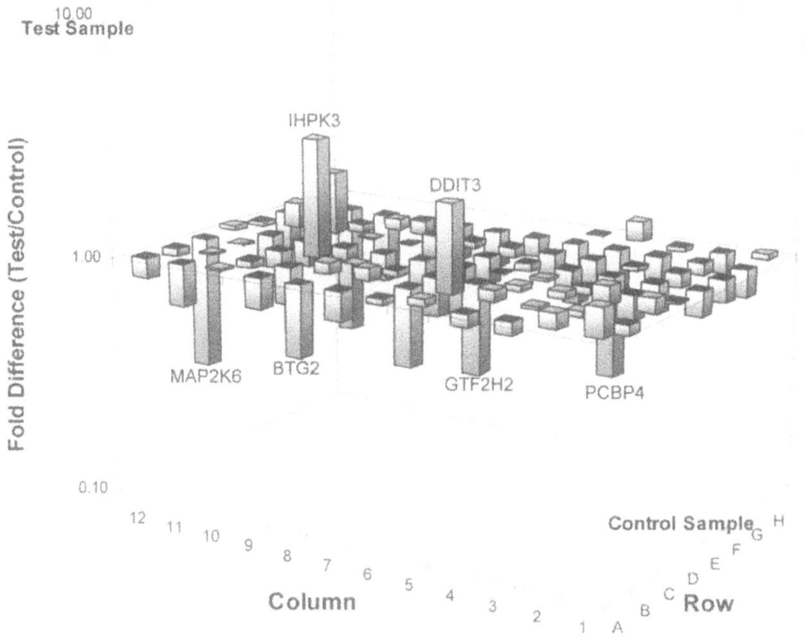

Figure 10. Human DNA damage/repair gene expression fold-differences between bromoacetic acid treated FHs cells and concurrent controls.

Abbreviations

The abbreviations for Figures 2 and 5 are AA = acetic acid, B = bromo, C = chloro, D = di, I = iodo, NM = nitromethane, T = tri, MX = 3-chloro-4-(dichloromethyl)-5-hydroxy-2(*5H*)-furanone. Abbreviations in the text are PBS = phosphate buffered saline.

Acknowledgements

This research was funded in part by AwwaRF 3089, EPA CR83069501 and Sea Grant R/WF-09-06 (M.J.P and E.D.W). M. Muellner was supported by T32 ES07326 (NIEHS). The authors thank AwwaRF for its assistance in funding and managing the project through which some of this information was discovered. The comments and views detailed herein may not necessarily reflect the views of the Awwa Research Foundation, its officers, directors, affiliates or agents. We acknowledge the partial support (MJP) of this work by the Center of Advanced

Materials for the Purification of Water with Systems, a National Science Foundation Science and Technology Center, under Award № CTS-0120978. This paper has been reviewed in accordance with the EPA's peer and administrative review policies and approved for publication. Mention of trade names or commercial products does not constitute endorsement or recommendation for use by the U.S. EPA.

References

1. Richardson, S.D.; Plewa, M.J.; Wagner, E.D.; Schoeny, R.; DeMarini, D.M. Mutat. Res. **2007**, *636*(1-3), 178-242.
2. Boorman, G.A.; Dellarco, V.; Dunnick, J.K.; Chapin, R.E.; Hunter, S.; Hauchman, F.; Gardner, H.; Cox, M.; Sills, R.C. *Environ. Health Perspect.* **1999**, *107*, 207-217.
3. Zhang, X.; Echigo, S.; Minear, R.A.; Plewa, M.J. In *Natural Organic Matter and Disinfection By-Products: Characterization and Control in Drinking Water,* Barrett, S.E., Krasner, S.W. and Amy, G.L., Eds.; American Chemical Society: Washington, D.C., 2000; pp. 299-314.
4. U.S. Environmental Protection Agency National Primary Drinking Water Regulations: Stage 2 Disinfectants and Disinfection Byproducts Rule. *Federal Register* **2006**, *71*, 387-493.
5. Woo, Y.T.; Lai, D.; McLain, J.L.; Manibusan, M.K.; Dellarco, V. *Environ. Health Perspect.* **2002**, *110 Suppl 1,* 75-87.
6. Krasner, S.W.; Weinberg, H.S.; Richardson, S.D.; Pastor, S.J.; Chinn, R.; Sclimenti, M.J.; Onstad, G.D.; Thruston, A.D., Jr. *Environ. Sci. Technol.* **2006**, *40*, 7175-7185.
7. Wagner, E.D.; Rayburn, A.L.; Anderson, D.; Plewa, M.J. *Environ. Mol. Mutagen.* **1998**, *32*, 360-368.
8. Tzang, B.S.; Lai, Y.C.; Hsu, M.; Chang, H.W.; Chang, C.C.; Huang, P.C.; Liu, Y.C. *DNA Cell Biol.* **1999**, *18*, 315-321.
9. Rundell, M.S.; Wagner, E.D.; Plewa, M.J. *Environ. Mol. Mutagen.* **2003**, *42*, 61-67.
10. Plewa, M.J.; Kargalioglu, Y.; Vankerk, D.; Minear, R.A.; Wagner, E.D. *Environ. Mol. Mutagen.* **2002**, *40*, 134-142.
11. Plewa, M.J.; Wagner, E.D.; Jazwierska, P.; Richardson, S.D.; Chen, P.H.; McKague, A.B. *Environ. Sci. Technol.* **2004**, *38*, 62-68.
12. Muellner, M.G.; Wagner, E.D.; McCalla, K.; Richardson, S.D.; Woo, Y.T.; Plewa, M.J. *Environ. Sci. Technol.* **2007**, *41*, 645-651.
13. Tice, R.R.; Agurell, E.; Anderson, D.; Burlinson, B.; Hartmann, A.; Kobayashi, H.; Miyamae, Y.; Rojas, E.; Ryu, J.C.; Sasaki, Y.F. *Environ. Mol. Mutagen.* **2000**, *35*, 206-221.

14. Box, G.E.P.; Hunter, W.G.; Hunter, J.S. *Statistics for experimenters: An introduction to design, data analysis, and model building.* Wiley & Sons, Inc.: New York, NY., 1978;
15. Plewa, M.J.; Muellner, M.G.; Richardson, S.D.; Fasano, F.; Buettner, K.M.; Woo, Y.T.; McKague, A.B.; Wagner, E.D. *Environ. Sci. Technol.* **2007** in press.
16. Plewa, M.J.; Wagner, E.D.; Richardson, S.D.; Thruston, A.D., Jr; Woo, Y.T.; McKague, A.B. *Environ. Sci. Technol.* **2004**, *38*, 4713-4722.
17. Richardson, S.D.; Crumley, F.G.; Ellington, J.J.; Evans, J.J.; Blount, B.C.; Silva, L.K.; Cardinali, F.L.; Plewa, M.J.; Wagner, E.D. *Proceedings of Symposium on Safe Drinking Water: Where Science Meets Policy, Carolina Environmental Program;* 2006.
18. Cemeli, E.; Wagner, E.D.; Anderson, D.; Richardson, S.D.; Plewa, M.J. *Environ. Sci. Technol.* **2006**, *40*, 1878-1883.
19. Richardson, S.D.; Thruston, A.D., Jr; Rav-Acha, C.; Groisman, L.; Popilevsky, I.; Juraev, O.; Glezer, V.; McKague, A.B.; Plewa, M.J.; Wagner, E.D. *Environ. Sci. Technol.* **2003**, *37*, 3782-3793.
20. Quellhorst, G.; Prabhakar, S.; Han, Y.; Yang, J. PCR Array: A simple and quantitative method for gene expression profiling. **2006**, *White Paper 1.* http://www.superarray.com/support_whitepapers.php

Chapter 4

Use of Chemical Models and Structure–Activity Relationships to Identify Novel Disinfection By-Products of Potential Toxicological Concern

Richard J. Bull[1] and David. A. Reckhow[2]

[1]MoBull Conculting, 1928 Meadows Drive North, Richland, WA 99352
[2]University of Massachusetts at Amherst, 16C Marston Hall, Box 35205, Amherst, MA 01003–5305

Chlorinated drinking water has been epidemiologically associated with increased risk of cancer, especially in the bladder, and adverse reproductive outcomes. A large number of chemicals have been identified as disinfection by-products (DBPs). Several by-product classes, such as the trihalomethanes and haloacetic acids, have received substantial toxicological study. Although many of these have been found carcinogenic in animals, the combination of low potency as carcinogens and/or low concentrations in drinking water make them unlikely causes of the cancer risk of the magnitude suggested by epidemiology studies. To provide some direction for the resolution of this discrepancy, a strategy was developed to predict formation of novel by-products through reactions of chlorine with substructures found in natural organic matter (NOM). These potential disinfection by-products (DBPs) were examined using a quantitative structure toxicity relationship (QSTR) program, TOPKAT®, to identify those chemicals of sufficient toxicologic potency to account for epidemiologic findings, if present. A "weight of evidence" strategy was utilized to rectify the predictions of the multiple cancer models within TOPKAT® to identify those compounds that were most probably potent carcinogens. Potential developmental toxicants were identified with the single model provided in TOPKAT®. A substantial effort was made to verify/deny predictions by literature search; examining both descriptive and mechanistic data to evaluate the prediction. Several haloquinones and related chemicals appear to be the most interesting. It is of note that formation of haloquinones would be favored with chloramine relative to chlorine disinfection.

Introduction

Chlorinated drinking water has been epidemiologically associated with increased risk of cancer, especially of the bladder, and with a range of adverse reproductive outcomes (*1*). The USEPA (*1*) estimated the population attributable risks (PARs) from a series of published meta analyses. The PARs translate into lifetime risks that range from 2×10^{-4} to as high as 6×10^{-3} in the U.S. population (Table I). A large number of chemicals have been identified as by-products of the disinfection of drinking water. Several by-product classes, such as the trihalomethanes (THMs) and haloacetic acids (HAAs), have received substantial toxicological study. Many of these have been found carcinogenic in animals, but the combination of low potency as carcinogens and/or low concentrations in drinking water make them unlikely causes of risks of the magnitude suggested by epidemiology study (*2*).

There are several possible explanations for the discrepancy between the levels of risk projected from toxicological studies of the major chlorination by-products and epidemiological findings:

Table I. Risks of bladder cancer attributed to chlorinated drinking water

	Males	*Females*
Bladder Cancer Incidence	39/100,000†	10.1/100,000
Lifetime probability for developing bladder cancer	0.0356‡	0.0113‡
Population Attributable Risks to Cl_2 H_2O§	Cancer risk attributable to Cl_2 H_2O	
2%	0.0007	0.0002
17%	0.006	0.002

† Age adjusted incidence for years 1997-2001 (*3*)
‡ Years 1999-2001 (*3*)
§ (*1*)

1. Lack of occurrence and/or toxicological data for potent carcinogens that are produced by reaction of disinfectant(s) with natural organic matter.
2. Humans are innately more sensitive to chlorination by-products than experimental animals.
3. Chlorinated drinking water is a complex mixture of by-products that act synergistically to produce cancer or other adverse effects.
4. The epidemiological data available overestimate the actual risk.

These same issues may explain similar apparent descrepancies with respect to reproductive effects that have been associated with chlorinated drinking water. These explanations are not exclusive of one another. Therefore, it is necessary to develop an approach for each.

The present project was undertaken to evaluate the first potential explanation. The first question addressed was whether identified by-products might be predicted to have sufficient toxicological potency to contribute to some of the adverse health outcomes that have been associated with chlorinated drinking water. A second question was whether by-products may be formed that have yet to be identified that could contribute to such outcomes. Quantitative structure toxicity relationships (QSTR) were used to evaluate the probable general toxicity, carcinogenic and developmental toxicity of DBPs. The formation of novel DBPs was based upon predicted reactions of disinfectants with reactive centers in substructures found in natural organic matter (NOM) using principles of organic chemistry.

Methods

A schematic of the study design is provided in Figure 1. The study included all established chlorination by-products [detailed references in (5)]. In addition, formation of novel by-products were predicted based upon standard reactions of chlorine or chloramines with substructures that are found in natural organic matter (NOM). Several dozen probable DBPs were predicted as likely reaction products. The precursor structures under consideration included lignin monomers, nucleic acids, amino acids, terpenoids, simple aromatics and many others. General pathways for the formation and degradation of chlorination byproducts were proposed for each precursor type. These were derived from published studies of similar compounds, and known reaction pathways and products from the attack of chlorine on a wide range of reactive centers. The nature of the precursors is described in detail in Bull et al. (*5*). These probable disinfection by-products (DBPs) were examined using a QSTR program, TOPKAT® (*4*), supplemented by literature review, to identify those chemicals with a high probability of being carcinogens or developmental toxicants of sufficient toxicological potency to be candidate causes for epidemiologic findings, if they are present. By-products that were obviously of no toxicological importance (e.g. fatty acids) were not evaluated by QSTR. A total of 489 chemicals were evaluated by QSTR. A complete explanation of the desposition of each DBP evaluated is provided in Bull et al. (*5*).

QSTR relationships were first evaluated through the use of TOPKAT® (Accelrys). Second, the available toxicological literature on related compounds was evaluated. This was particularly important when the training set of the models employed were inadequate to support a prediction. QSTR programs, and TOPKAT® in particular, have limitations. TOPKAT® makes predictions based on the electrotopological characteristics of the molecule. Such predictions

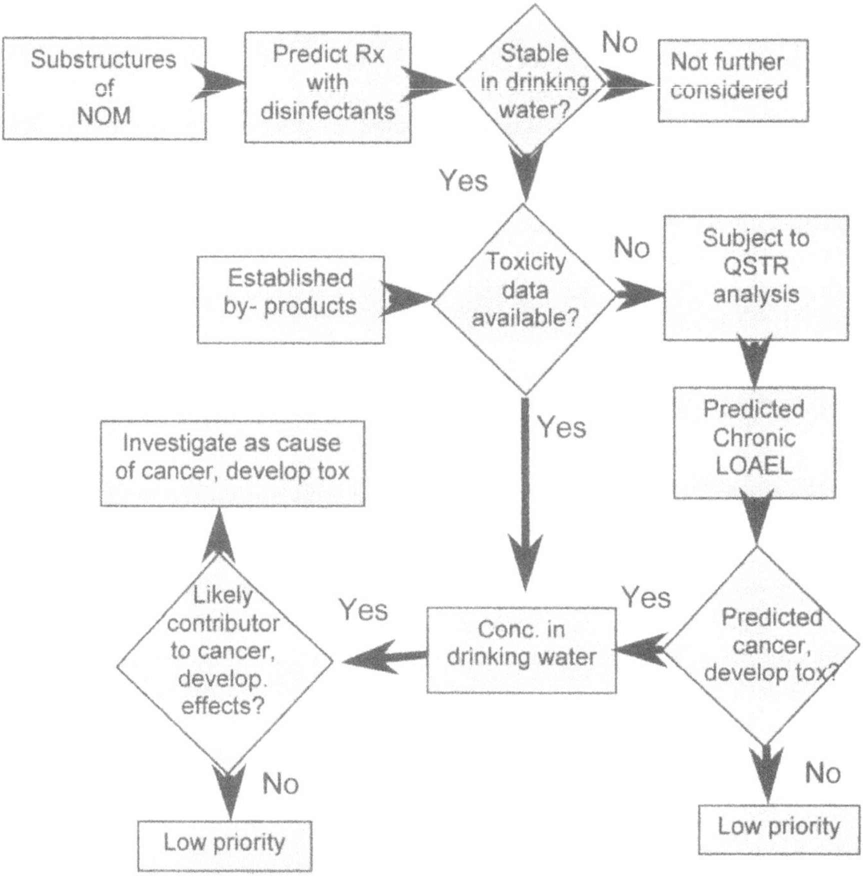

Figure 1. Outline of the prioritization scheme used.

may be within the optimum prediction space (OPS) of the model even though the training set does not include a close structural analog. Each model within TOPKAT® provides a listing of the chemicals in its training set, indicates which compounds were used in making the prediction, the structure of each chemical in the training set, identifies the weight assigned to each functional groups within the molecule, and an indication (defined here as the similarity index) of how closely related the chemical in the training set is related to the query structure. With the chemical diversity of DBPs, the simplest summary statistic that described the adequacy of the training set was the similarity index.

An effort was made to verify or deny predictions by literature search. Both descriptive and mechanistic data for related compounds were examined to determine if an independent rationale could be developed for the prediction. As with model predictions, the ability to perform this analysis depends very heavily on the availability of data in the open scientific literature. For example, predictions of carcinogenicity depend upon there being at least one related compound for which *in vivo* carcinogenic activity has been assessed.

The first criteria applied was to identify DBPs which had predicted chronic lowest observed adverse effect levels (LOAELs) < 1 mg/kg day^{-1}. This criterion was applied because chemicals of this toxicological potency would be of interest regardless of the actual health outcome. It was reasoned that chemicals of lesser potency would not contribute significantly to risk of the magnitude suggested by epidemiological studies. The regulated DBPs typically have chronic LOAELs in a range of 10 - 100 mg/kg day^{-1}.

For chemicals that were predicted to be carcinogens or developmental toxicants, the potency criteria was raised to 10 mg/kg day^{-1}. This adjustment was made to account for the finding that the chronic LOAEL was approximately an order of magnitude higher relative to the dose that produces a 50% increase in tumor incidence in a lifetime (TD_{50}) for potent carcinogens (Figure 2). The relationship between actual LOAELs in the TOPKAT® training set with TD_{50}s for the same compounds from the Cancer Potency Data Base (6) was based upon the correlations that are presented in Figure 2.

Chemicals were considered potential carcinogens only if a positive prediction was obtained for two species that were both within the acceptable limits of the optimum predictive space (OPS) of the respective model. The chemicals also had to be sufficiently similar to chemicals within the training set of the model from which the prediction was derived. This was judged if the similarity for at least one compound in the training set fell below 0.2 (0.00 = identity, 1.0 = no similarity). Within TOPKAT®, similarity is judged based on the electrotopological relationships amoung pairs of atoms within the molecule.

The ability of each of the 8 TOPKAT® carcinogen models to predict whether a chemical was a carcinogen was very dependent upon the chemicals in their respective training sets. The training sets were independently derived from

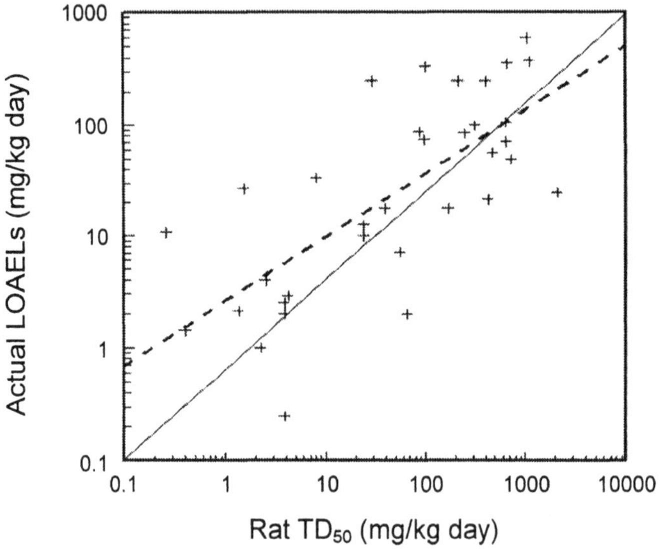

Figure 2. Relationship of the $TD_{50}s$ in rats to the actual chronic LOAELs for the same compounds within the TOPKAT® chronic LOAEL model training set. $TD_{50}s$ obtained from Gold et al. (6).

male and female rat and mouse studies in the National Toxicology Program (NTP) or the Food and Drug Administration (FDA) databases. Therefore, a chemical was judged a probable carcinogen if it met the criteria in the prior paragraph in at least one sex of two species in either the NTP or FDA databases. Negative predictions are frequently observed for well known carcinogens in one or the other of these models if there were no suitable structural analogs in the training set. While the carcinogen models were used to identify carcinogens, the ranking of carcinogens for potency depended upon the predicted chronic LOAEL and prediction of Ames' mutagenic activity as indicated above. In these cases, the similarity index was provided, but the chemical was not removed from consideration unless the value was outside of the OPS of the model or if there data in the literature that indicated that the prediction erred $\geq$ 10X.

In the prediction of potential developmental toxicity (DTP), the criteria of a probability of > 0.7 and a similarity index of < 0.2 was required. In this case there is only one model in TOPKAT®.

Results

DBPs and putative DBPs that met the criteria for predicted lowest observed adverse effect levels (LOAELs) of less than 1 mg/kg day^{-1} are presented in

Table II. Predictions of the chronic LOAEL in the halonitrile group of compounds were retained as being reasonable based upon the range of LOAELs identified with haloacetonitriles that have been studied in some detail in subchronic and/or reproductive toxicology studies (5).

Table II. DBPs and putative DBPs with predicted chronic LOAELs < 1 mg/kg day^{-1} listed in order of potency.

Chemical	*Pred. Ames' mut.*	*Simil. Index*	*LOAEL (mg/kg day^{-1})*	*Simil. Index*
2,3-Dibromopropionitrile	+	0.121	0.15	0.183
Dichloroacetonitrile	+	0.000	0.29$^{\dagger\#}$	0.170
2,3-Dichloropropamide	+	0.209	0.51	0.183
3,3-Dichloropropionitrile	+	0.109	0.77†	0.205
3-Bromopropionitrile	+	0.200	0.89†	0.149
2,3-Dichloropropenal	+	0.223	0.97	0.164

† Outside the optimal predictive space of the model and exceeds permissible limits
# This table corrects a typographical error in reference *5*.

Table III adds chemicals to the list of concern that were predicted to be carcinogenic and whose predicted chronic LOAELs fell below 10 mg/kg day^{-1}. For perspective, the predicted chronic LOAELs of chloroform and the regulated HAAs are 30 and in a range of 6-60 mg/kg day^{-1}, respectively. All chemicals in this list had a similarity < 0.2 in the carcinogenesis models in which they were predicted as probable carcinogens.

The similarity index in the chronic LOAEL model for the first five compounds are > 0.2 indicating that they were not well represented in the trainings set and a few compounds had predicted LOAELs outside the OPS of the model. In the case of haloquinone derivatives, a substantial literature on mechanism of quinone action lead credibility to the predictions. In the case of the furanone derivative, 3-chloro-4-(chloromethyl)-5-hydroxy-2-(5H)furanone (CMCF), the predicted LOAEL of MX was also outside the OPS, but approximated the value established experimentally in the study of MX (*7*), so it was retained.

Several halonitriles were identified as likely carcinogens. Dichloro-acetonitrile is well recongnized as a DBP and occurs in the low µg/L range. Two additional halonitriles were identified, 3-bromopropionitrile and 2,3-dibromopropionitrile, but no evidence of their occurrence was found. Dichloroacetaldehyde and 2,3-dichloropropenal were also identified as probable carcinogens. The former has been identified in drinking water in the low µg/L range (*8*).

Table III. Additional DBPs predicted to be carcinogens that were predicted to have chronic LOAELs < 10 mg/kg day^{-1}

Chemical	*Pred. Ames' Mut.*	*Simil. index*	*LOAEL (mg/kg day^{-1})*	*Simil. index*
2,3,6-Trichloro-1,4-benzoquinone	-	0.333	0.033	0.342
3-Chloro-4-(chloromethyl)-5-hydroxy-2(5H)furanone	-	0.318	0.029†	0.298
2,3-Dichloro-4,5-diketobenzoprop-2-enol	+	0.407	0.064*	0.280
2,3,6-Trichloro-1,4-benzoquinone-4-(N-chloro)imine	-	0.590	0.073‡	0.328
2,6-Dichloro-3-methyl-1,4-benzoquinone	-	0.398	0.079	0.341
2,3-Dibromoproprionitrile	+	0.120	0.15	0.183
Dichloroacetonitrile	+	0.000	0.29†	0.170
3-Bromopropionitrile	+	0.200	0.89†	0.149
2,3-Dichloropropenal	+	0.223	1.0	0.164
3-Chloro-4-(dichloromethyl)-2-furanone	+	0.324	1.7†‡	0.154
Trichloroacetonitrile	+†	0.494	4.5†	0.166
1,2,3-Trichloropropane	+	0.156	4.7	0.000
Dichloroacetaldehdye	+	0.192	5.8	0.047
3,3,3-Trichloropropionitrile	-	0.212	3.9†	0.169
2-Bromo-3,3-dichloropropionic acid	-	0.271	5.8	0.068

* Prediction is outside the optimal predictive space of the model, but within permissible limits

† Prediction is outside the optimal predictive space of the model and exceeds permissible limits

‡ Prediction in Bull et al. (*5*) was based on the default acyclic sub-model of TOPKAT®, this prediction is based upon the heteroaromatic sub-model.

Some chemicals came very close to meeting the criteria established. For example, both 3,3-dichloro- and 3,3-dibromopropenoic acid had similiarity indices >0.2 but <0.3 on one of two positive predictions, but met all other criteria. The 3,3-dichloropropenoic acid has been observed as a DBP in the low µg/L range (*8*). Dihaloacetamides also just missed the carcinogenicity criteria and have been reported to occur in drinking water (*8*). The predicted chronic LOAELs for these compounds approximate those of the haloacetic acids. However, the prediction of carcinogenic activity was more dependent upon the acetamide structure than the relationship to the haloacetic acids.

It is notable that MX is not on the list as it did not meet the criteria established to identify carcinogens. This was because the Komulainen study *(7)* results were not in the training sets of the carcinogenesis models. In addition there were no chemicals in the chronic LOAEL training set that were closely related to MX.

3-Chloro-4-(dichloromethyl)-2-furanone (reduced MX) was not included because it fell slightly outside the criteria established for identifying carcinogens, with three positive predictions all outside the OPS, but within acceptable limits. There were a series of compounds based on furan-like backbones (e.g. 3,5-dichloro-1-hydroxy-4-ketocyclopent-2-enoic acid) that had extremely low predicted LOAELs that were outside the OPS. They also did not meet the criteria applied for predicting carcinogens or developmental toxicants and are not included in the tables.

Figure 3 provides an abbreviated pathway for the formation of halogenated quinones by reaction of free chlorine or chloramines to phenols, known substructures within NOM. Also noted is the formation of 2,6-dichloro-1,4-dibenzoquinone 4-(N-chloro) imine. The addition of nitrogen from chloramines to the intermediate product, 2,4,6-trichlorophenol, was demonstrated to occur in high yield by Heasley et al. (*9*). Also noted in the figure is that in the presence of excess free chlorine, the quinone rings are cleaved to give rise to more familiar DBPs such as the HAAs. Haloquinones produced from cresol are destroyed with excess chlorine and results in the formation of THMs and other aliphatic products (*5*). It should be noted that tetrachloro-1,4-dibenzoquinone was not predicted to be carcinogenic, but the di- and tri-substituted compounds were, as was the related N-chlorimine derivative discussed above. The negative prediction of carcinogenicity by the tetrachloroquinone can be rationalized based on probable mechanism of action (*5*).

The functional group of the organic N-chloramines was not recognized by any of the TOPKAT® models. As discussed in Bull et al. *(5)*, the only biomedical data on these chemicals is related to their *in vivo* formation in inflammatory processes. Because of their well known occurrence in chlorinated and chloraminated water, this class of recognized DBPs should be of concern.

Several chemicals were predicted to be developmental toxicants that had predicted chronic LOAELs in the < 10 mg/kg day^{-1} (Table IV). They largely

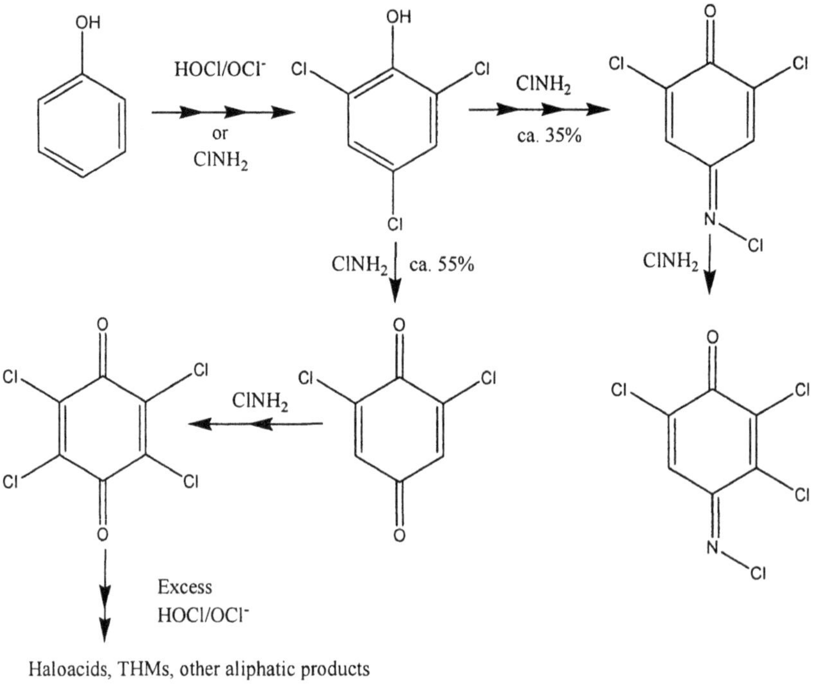

Figure 3. Formation of halogenated quinones and quinone 4-(N-chloro) imines by reaction of chlorine or chloramines with phenols (adapted from reference 9).

belonged to the halonitrile class of DBPs, but were distinct from those halonitriles predicted to be carcinogens. As noted for the halonitriles that appeared in Table III, the predictions of the chronic LOAEL for the halonitriles fell outside the permissible limits of the OPS, but were retained for reasons stated above. The developmental toxicity prediction can be rationalized on the cyanogenic halonitrile group, a precursor to isocyanate, a recognized anti-thyroid compound.

Discussion

The group of potential DBPs identified in the QSTR analysis of most interest is the haloquinones. These predictions of a probable high degree of toxicity were supported by a review of the toxicology of quinones in general and limited data that are available on the carcinogenic activity of members of haloquinone group, itself. The haloquinones were of particular interest because halogen substitution would enhance formation of glutathione conjugates in the

Table IV. Disinfection by-products predicted to be developmental toxicants by TOPKAT® with predicted chronic LOAELs < 10 mg/kg day^{-1}

Chemical	*Probability of DTP*	*Similarity index*	*LOAEL (mg/kg day^{-1})*
3-Chloro-2-butenoic acid	0.854	0.154	2.8
Tribromoacetonitrile	1	0.000	4.5[†]
Trichloroacetonitrile	1	0.000	4.5[†]
Bromodichloroacetonitrile	0.998	0.091	1.5[†]
Dibromochloroacetonitrile	0.998	0.091	1.5[†]
3-Bromobutanenitrile	1	0.205	2.5[†]
3,3,3-Trichloropropanenitrile	1	0.029	3.9[†]
3,3-Dichloropropenenitrile	0.975	0.127	0.77[†]
2,3,3-Trichloropropenenitrile	0.923	0.165	1[†]
Cyanogen chloride	0.869	0.157	0.23[†]
Cyanogen bromide	0.869	0.157	0.25[†]

[†] Outside the optimal predictive space of the model and exceeds permissible limits

liver relative to 1,4-dibenzoquinone. These conjugates are known to be further processed into forms that react with nucleophiles in target tissues *(10)*. The polypeptide and amino acid conjugates so formed are transported and concentrated in tissues such as the kidney and bladder (*10*). For this reason they would be a logical putative cause of bladder cancer if they occurred in drinking water. However, the net formation of these haloquinones are more likely to be favored with chloramines than with chlorine disinfection (*9*). Thus, this finding does not provide a simple explanation of epidemiological associations with bladder cancer, as bladder cancer risk appears to be decreased with chloramination vs. chlorination.

The prediction that 2,6-dichloro-1,4-benzoquinone-(4-N-chloro) imine is carcinogenic (also predicted to be quite toxic, but the similarity index in the chronic LOAEL model, makes this prediction less certain) is of interest. First, Heasley et al. (*9*) demonstrated that the nitrogen from chloramines can add to 2,4,6-trichlorophenol in a manner that is analogous to nitrosamine formation with monochloramine (*11*). Second, this compound could interact with tissue macromolecules in diverse ways. Quinone conjugates are known to participate in redox cycling to produce oxidative stress and the cysteine conjugates are processed into a form that alkylates macromolecules. The extensive and complex interorgan metabolic processing of quinones *in vivo* makes it very unlikely that they can be properly prioritized using simple *in vitro* mammalian cell assays or microbial mutagenesis systems. The N-chloroimine group adds another mechanism of interaction with macromolecules via alklylation or nucleophilic substitution of chlorine (*12*). We were unable to identify structural analogs of this potential chloramine by-product in our search of the toxicological literature.

Dichloroacetonitrile was the only commonly recognized DBP that arose in the QSTR study as of potential importance based on a predicted chronic LOAEL below 1 mg/kg day^{-1}. This compound, and its brominated analogs, have only received cursory toxicological study. No chronic bioassays have been conducted to estimate their carcinogenic potency. The prediction of carcinogenicity by TOPKAT® depended largely upon studies by Bull and coworkers (*13,14*) in which the activity of these compounds in mouse skin initiation-promotion and strain A/J lung ademona bioassays was examined. Results of such studies have generally not been accepted as conclusive proof of carcinogenic activity and are not generally employed in risk assessment.

Halonitriles were identified as the most important group studied as potential developmental toxicants. However, many of these predictions pointed to the trihaloacetonitriles, which tend not to be stable in water. Priopionitriles substituted in the 3-position were also predicted to be developmental toxicants. The identification of halonitriles as potential developmental toxicants was largely due to a series of studies conducted on the haloacetonitriles (*5*). Halogen substitutions in the 2-position are much more likely to occur than in the 3-position in reactions of chlorine with precursors in NOM and as a result should be considered of higher priority for toxicological characterization.

It should be noted that there are many DBPs for which there are very limited relevant toxicological data. This lack of data makes it difficult to evaluate related compounds in models or through literature search coupled with appropriate modeling. This project was conceived as a effort to identify those compounds for which the best cases of risk could be assembled. This does not mean that compounds not identified by this effort are harmless.

These data indicate the need for a more systematic study of DBP occurrence that includes low levels of potentially quite potent carcinogens. If these compounds are found to occur, it will be very important to experimentally confirm the QSTR predictions made in this process in vivo.

Acknowledgments

Supported by Awwa Research Foundation #2867 & EPA Cooperative Agreement R-82940901.

References

1. USEPA. *Health Risks to Fetuses, Infants, and Children: Proposed Stage 2 Disinfectant/Disinfection Byproducts: A Review.* **2003**, EPA-822-R-03-010.
2. Bull, R.J., S.W. Krasner, P.A. Daniel, and R.D. Bull. *Health Effects and Occurrence of Disinfection By-Products.* **2001**, Denver, Colo.: AWWA Research Foundation.
3. Cancer Facts. **2005**, American Cancer Society (http://www.cancer.org)
4. Accelrys. *TOPKAT® User Guide. Version 6.1.* **2001**, Accelrys, Inc., San Diego, Calif.
5. Bull, R.J., D.A. Reckhow, V. Rotello, O.M. Bull, and J. Kim. *Use of Toxicological and Chemical Models to Prioritize DBP Research.* **2006**, Awwa Research Foundation, Denver CO.
6. Gold, L. S. *Carcinogen Potency Database (CPDB)* [Online]. **2006**, Available: <http://potency.berkeley.edu/pub/tables/chemicalsummary>, [cited June 5, 2006].
7. Komulainen, H., V.-M. Kosma, S.-L. Vaittinen, T. Vartiainen, E. Kaliste-Korhonen, S. Lotjonen, R.K. Tuominen, and J. Tuomisto. Carcinogenicity of the drinking water mutagen 3-chloro-4-(dichloromethyl)-5-hydroxy-2(5H)-furanone. *J. Natl. Cancer Inst.* **1997**, *89*, 848-856.
8. Krasner, S.W., H.S. Weinberg, S.D. Richardson, S.J. Pastor, R. Chinn, M.J. Sclimenti, G.D. Onstad, and A.D. Thruston, Jr. Occurrence of a new generation of disinfection byproducts. *Environ. Sci. Technol.* **2006**, *40*(23), 7175-7185.
9. Heasley, V.L., A.M. Fisher, E.E. Herman, F.E. Jacobsen, E.W. Miller, A.M. Ramirez, N.R. Royer, J.M. Whisenand, D.L. Zoetewey, and D.F. Shellhamer. Investigations of the reactions of monochloramine and dichloramine with selected phenols: Examination of humic acid models and water contaminants. *Environ. Sci. Technol.* **2004**, *38*, 5022-5029.
10. Bolton, J.L., M.A. Trush, T.M. Penning, G. Dryhurst, and T.J. Monks. Role of quinones in toxicology. *Chem. Res. Toxicol.* **2000**, *13*, 135-159.
11. Choi, J. and R. L. Valentine. Formation of N-nitrosodimethylamine (NDMA) from reaction of monochloramine: A new disinfection by-product. *Water Res.*, **2002**, *36*(4), 817-824.
12. Calvo, P., J. Crugeiras, A. Rios, and M.A. Rios. Nucleophilic substitution reactions of N-chloramines: Evidence for a change in mechanism with increasing nucleophilic reactivity. *J. Org. Chem.* **2006**, *72*, 3171-3178.
13. Bull, R.J., Meier, J.R., Robinson, M., Ringhand, H.P., Laurie, R.D. and Stober, J.A. Evaluation of the mutagenic and carcinogenic properties of

brominated and chlorinated acetonitriles: By-products of chlorination. *Fundam. Appl. Toxicol.* **1985**, *5*, 1065-1074.

14. Bull, R.J. and M. Robinson. Carcinogenic activity of haloacetonitrile and haloacetone derivatives in the mouse skin and lung. In: Jolley, R.L et al. eds. Water Chlorination: Chemistry, Environmental Impact and Health Effects. **1985**, vol. 5, pp. 221-227.

Chapter 5

Contribution of Organic Bromines to the Genotoxicity of Chlorinated Water: A Combination of Chromosomal Aberration Test and Total Organic Bromine Analysis

Shinya Echigo[1], Sadahiko Itoh[1], and Ryo Ando[2]

[1]Department of Urban Management, Graduate School of Engineering, Kyoto University, Nishikyo, Kyoto 617-8540, Japan
[2]Sumitomo Heavy Industries, Ltd., 9-11, Kita-Shinagawa 5-chome, Shinagawa-ku, Tokyo 141-8686, Japan

The contribution of organic bromines to the genotoxicity of a chlorinated lake water was evaluated by chromosomal aberration test at various chlorine doses and bromide concentrations. The number of chromosomal aberrations increased as a function of initial bromide ion concentration, indicating that organic bromines are major contributors to the genotoxicity. This study highlights the importance of monitoring and controlling brominated disinfection by-products.

Bromide ion is a trace but ubiquitous anion in aquatic environment. Its origin in source waters can be both natural (salt water intrusion and dissolution from geological formation) and anthropogenic (*1*). Although bromide ion is quite stable in natural water, it can be involved in various chemical reactions once it enters a drinking water treatment process. Ozone (O_3) and hypochlorous acid (HOCl) used in drinking water treatment easily oxidize bromide ion to hypobromous acid (HOBr) (*2*). This intermediate, HOBr, rapidly reacts with natural organic matter (NOM) in source water to form organic bromines both during chlorination and ozonation (*3*). While the toxicity of bromide ion itself is considered negligible, some of these organic bromines are known to be toxic.

Several recent studies have suggested the toxicological importance of brominated DBPs. For example, Plewa *et al.* has shown that bromoacetic acids are more mutagenic than chloroacetic acids with an *in vitro* bioassay using mammalian cells (*4*). In particular, monobromoacetic acid is much more mutagenic than other chloro- or bromo- acetic acids (*4*). In addition, brominated nitromethanes and brominated nitriles are more cytotoxic and genotoxic than their chlorinated analogues (*5, 6*) Also, Nobukawa and Sanukida (*7*) showed that the mutagenicity of chlorinated water increased with increasing bromide ion concentration using Ames test. Kargalioglu *et al.* demonstrated that the rank order of the mutagenic potency in Salmonella typhimurium for the haloacetic acids was bromoacetic > dibromoacetic > chloroacetic > dichloroacetic acids (*8*). Further more, it is known that the products of the reaction between hypobromous acid and humic acid are several times more genotoxic than the ones between hypochlorous acid and humic acid on TOX basis (*9*). These results indicated that brominated compounds formed during chlorination could be major contributors to the toxicity of drinking water.

Also, the importance of bromination during chlorination has been highlighted by kinetic studies on the reactions between HOBr and organic compounds. HOBr is much more reactive to phenolic compounds, one of the most reactive components in NOM, than HOCl (*10, 11*). Westerhoff *et al.* observed the same trend for NOM isolates from a real source water: HOBr was more efficient at halogen substitution than HOCl (*12*). Therefore, bromination of NOM by HOBr during chlorination should not be neglected even if bromide ion concentration is much lower than HOCl dose.

Given the toxicological and kinetic importance of bromination during drinking water treatment, it is of practical importance to quantitatively identify raw water characteristics and treatment conditions under which the contribution of brominated DBPs to the toxicity of drinking water is not negligible. However, the information on the relationship between chlorination condition (*e.g.*, bromide ion concentration and chlorine dose) and the toxicity of chlorination by-products to mammalian cells in the presence of bromide ion is still very limited. We have reported a preliminary result on the effect of bromide ion on the toxicity of chlorinated water (*13*), but the chlorine dose was fixed in this evaluation. This is not sufficient to fully evaluate the importance of controlling brominated DBPs.

The present study sheds the light on this problem. More specifically, the relationship between the activity inducing chromosomal aberrations and the formation characteristics of total organic bromines (TOBr) from the chlorination of a real source water was explored with varying bromide ion concentration and chlorine dose. For this evaluation, we employed the combination of chromosomal aberration test and the differentiation technique of between TOBr and total organic chlorine (TOCl) (*14*).

Materials and Methods

Materials

Chemicals

All the chemical reagents used in this study were of reagent grade or better (mostly analytical grade), and were purchased from Wako pure chemical (Japan). All the aqueous solutions were prepared with ultra pure water treated by a Millipore Elix20 system.

Raw Water

The raw water used in this study was collected from Lake Biwa, Japan on September 15, 2004. This lake is the largest lake in Japan, and serves 14 million people as water source. The total sample volume was approximately 320 L. The ambient bromide concentration of the lake water was 50 μg/L. Also, DOC, pH, UV absorbance at 260 nm (E_{260}), and the chlorine demand for 24 hours were 1.6 mg/L, 7.6, 0.03 absorbance unit (AU)/cm and 1.71 mg Cl_2/L, respectively. The water was filtered through 0.45 μm membrane (ADVANTEC) to removal particulate matters right after collection, and chlorinated on the same day.

Chlorination and Sample Concentration

Chlorination of the Lake Biwa water was performed at five different bromide ion concentrations (50, 120, 240, 400, and 600 μg/L) and three different chorine doses (Cl_2/DOC=1.0, 1.5, and 2.0 mg Cl_2/L). Hence, 15 samples were prepared for chromosomal aberration test and TOBr analysis. The samples were chlorinated in the dark at 20 °C for 24 hours. The chlorinated samples were concentrated by solid phase extraction before chromosomal aberration test in the following procedure. After pH adjustment to 2 with HCl, each sample (20 L) was passed through a pair of two Sep-Pak® Plus (Long) CSP-800 solid phase extraction cartridges (Waters) in series at a flow rate of 50 mL/min. The materials extracted onto the solid phase were eluted with dimethylsulfoxide (DMSO) at a flow rate of 0.2 mL/min. The final volume of the extract was 2 mL. That is, the concentration factor of this process was 10^4.

Chromosomal Aberration Test

Chromosomal aberration test using Chinese hamster lung cells (CHL/IU, Dainihon Pharmaceutical) was performed to evaluate the initiating activity in the carcinogenesis process of the products produced from the chlorination of the Lake Biwa water under various chlorination conditions. The detail of the test procedure is described elsewhere (*15*). The dose of the extract to the cell culture media was fixed to 30 μL, and the sample-to-media ratio was 0.5% (v/v). The incubation time after contacting with sample solutions was 24 hours. For each sample, 50 metaphases were analyzed. The number of chromosomal aberrations was counted by visual observation under microscope.

Analytical Methods

Differentiation of TOBr and TOCl

The detail of the procedure is described elsewhere (*14*). But, briefly, the off gas from a TOX furnace that contains HCl and HBr corresponding to TOCl and TOBr was trapped into distilled water, and the corresponding chloride and bromide ions were separately quantified by ion chromatography (see the next paragraph for detail). A TOX analyzer (TOX-10Σ, Mitsubishi Chemical) was used as a furnace.

Anion Concentrations

Bromide and chloride concentrations were determined by ion chromatography (LC-VP, Shimadzu) with a Shim-pack IC-A3 analytical column (Shimadzu) protected by a Shim-pack IC-GA3 guard column (Shimadzu). The mobile phase was 50 mM of boric acid/ 8 mM of *p*-hydroxybenzoic acid/ 3.2 mM bistris.

Other Water Quality Parameters

For DOC and UV absorbance measurements, a TOC analyzer (TOC-5000A, Shimadzu) and a UV/Visible spectrophotometer (Multispec-1500, Shimadzu) were used respectively. Also, solution pH was measured by a D-51 pH meter (Horiba).

Results and Discussion

TOBr and TOCl Formation

The effects of bromide ion concentration and chlorine dose on TOCl and TOBr are shown in Figures 1 and 2, respectively. While TOCl was slightly higher at Cl_2/TOC=2.0 mg/mg than other conditions, no distinct difference in TOCl was observed with respect to chlorine dose. Similarly, TOBr was almost independent of chlorine dose in the range of Cl_2/TOC=1.0-2.0 mg/mg. On the other hand, bromide ion concentration had strong impact on both TOCl and TOBr. TOBr increased with increasing bromide ion, but only slightly for a bromide ion concentration higher than 200 μg/L. At the same time, TOCl decreased with increasing bromide ion concentration. TOX (=TOBr+TOCl) decreased slightly, but was not as sensitive to bromide level as TOCl and TOBr (Figure 3).

The reason for these constant TOX values with different chlorine dose is not clear at this point, but one should note that the chlorine doses used in this study were approximately at or above the chlorine demand. Thus, almost all the reaction sites in NOM may have been used up even at the lowest chlorine dose, and no further halogenation occurred with additional chlorine.

The reason for the constant TOBr values with different chlorine doses seems even more profound. However, this result makes sense if we assume that bromination is much faster than chlorination. That is, if we assume that bromination was so fast that bromination was completed before a competitive situation between chlorination and bromination occurs, it is reasonable to observe a constant TOBr level regardless HOCl concentration. These assumptions are supported by the high conversion of bromide ion to organic bromines (Figure 4) and higher TOBr-to-TOX ratios than corresponding initial bromide-to-chlorine ratios (Figure 5).

Chromosomal Aberration Test

Figure 6 shows the result of chromosomal aberration test. The trend was similar to that of TOBr. That is, the activity inducing chromosomal aberrations

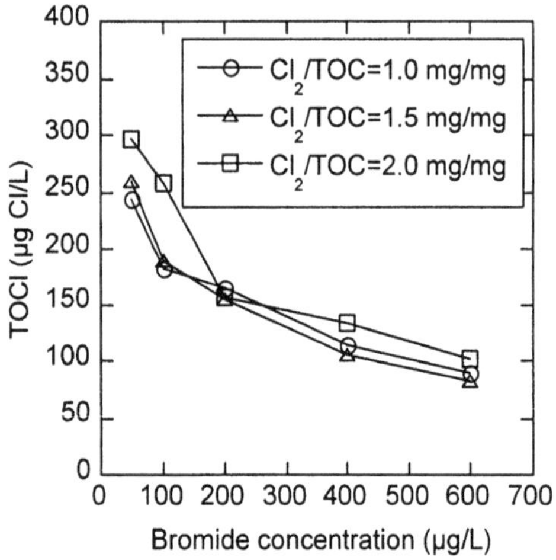

Figure 1. Effect of bromide concentration and chlorine dose on TOCl formation.

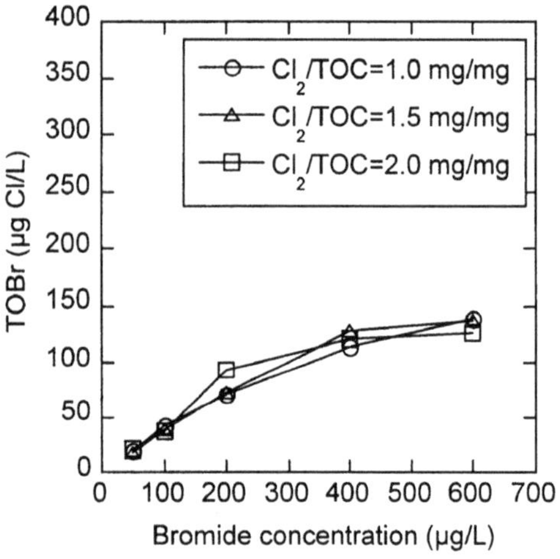

Figure 2. Effect of bromide concentration and chlorine dose on TOBr formation.

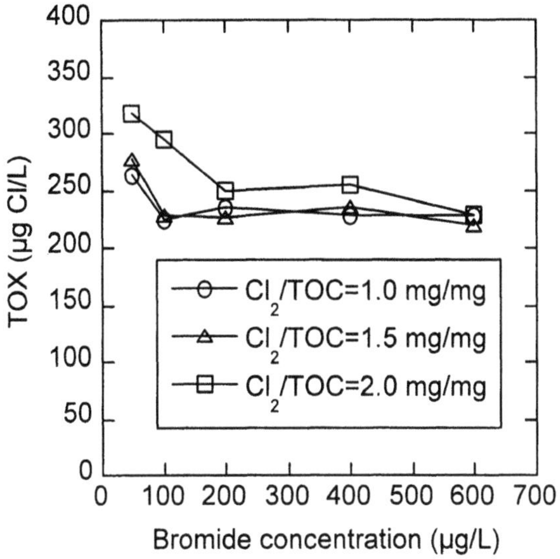

Figure 3. Effect of bromide concentration and chlorine dose on TOX formation.

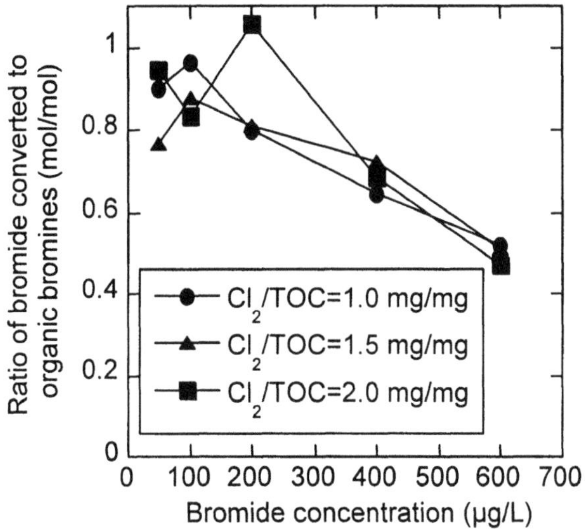

Figure 4. Bromide conversion to organic bromines.

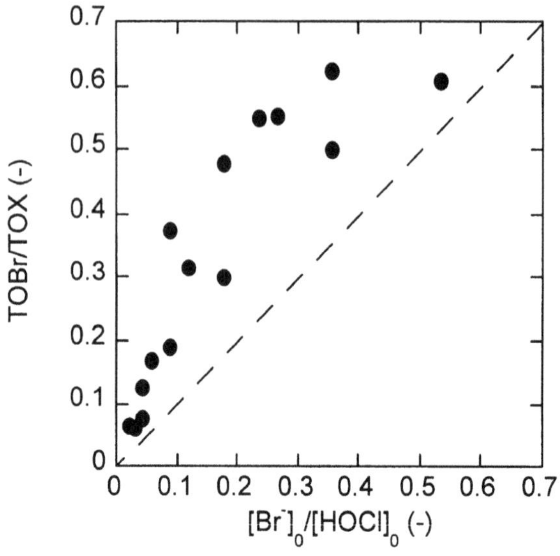

Figure 5. Competition between chlorination and bromination. Dashed line represents 1:1 TOBr/TOX: $[Br^-]_0/[HOCl]_0$ *line.*

increased with increasing bromide ion, but no clear change was observed with varying chlorine dose. This result suggests that organic bromines are major contributors to the activity inducing chromosomal aberrations of chlorinated drinking water.

Now, with the results of TOBr, TOCl, and chromosomal aberration test, the contribution of TOBr can be determined under the following assumptions.

- The numbers of chromosomal aberrations induced by a unit TOBr and TOCl on TOX basis does not change respectively, for the experimental conditions used in this study.
- The contributions of TOBr and TOCl to the number of chromosomal aberrations are additive.

With these assumptions, the contribution of TOBr is defined as below:

$$\text{Contribution of TOBr(\%)} = \frac{\text{Number of chromosomal aberrations from TOBr}}{\text{Total number of chromosomal aberrations}} \times 100 \quad (1)$$
$$= \frac{A_{\text{TOBr}} \bullet \text{TOBr}}{A_{\text{TOCl}} \bullet \text{TOCl} + A_{\text{TOBr}} \bullet \text{TOBr}} \times 100$$

where A_{TOCl} (L•counts/(100 cells•mg)) and A_{TOBr} (L•counts/(100 cells•mg)) are the toxicity per unit TOCl and TOBr, respectively. By multiple linear regression,

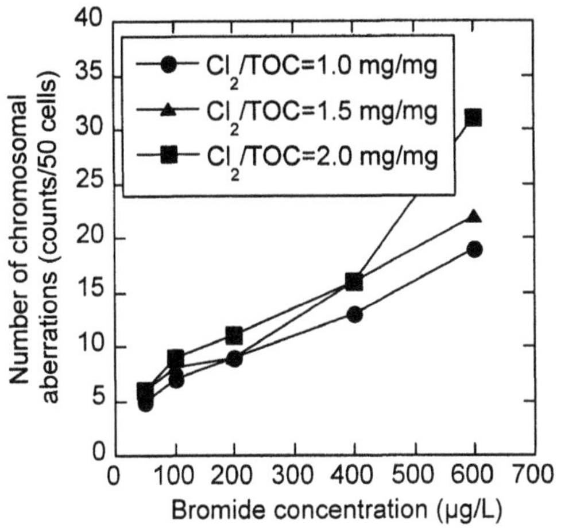

Figure 6. Effect of bromide concentration and chlorine dose on the number of chromosomal aberrations.

A_{TOCl} was found to be 0.71 L•counts/(100 cells•mg) and A_{TOBr} was 5.24 L•counts/(100 cells•mg). The R^2 value for the regression was 0.70.

The ratio of A_{TOBr} to A_{TOCl} was 7.38. Thus, on average, brominated DBPs formed from the chlorination of the lake water were 7.38 times more genotoxic than chlorinated DBPs on TOX basis (*i.e.*, when normalized by TOX concentration). This value (A_{TOBr}/A_{TOCl}=7.38) obtained form the chlorination of a lake water was comparable to that from the chlorination of Aldrich humic acid (A_{TOBr}/A_{TOCl}=4.8) (*13*). Also, the A_{TOBr} value was nearly one order magnitude higher than the number of chromosomal aberrations induced by a unit concentration of haloacetic acids (HAAs), except that of monobromoacetic acid, while the A_{TOCl} value was comparable to those of HAAs (Figure 7).

The contribution of TOBr calculated by eq. 1 at each condition is listed in Table I. As expected, the contribution of brominated DBPs tends to be high for higher bromide ion concentration. Also, it is of note that even at the lowest bromide concentration (ambient bromide concentration, 50 µg/L), the contribution of TOBr to the total number of chromosomal aberrations was above 30%, indicating that the contribution of brominated DBPs is not negligible even at relatively low bromide ion concentration.

For a bromide concentration above 100 µg/L, the contribution of TOBr was above 50%. Considering that the average and median bromide concentrations in US source waters are 60-100 µg/L and 30-40 µg/L, respectively (*1,17*), it would not be unusual that the contribution of TOBr exceeds 50% in actual drinking water practice. That is, brominated DBPs can be a dominant chemical class with respect to the genotoxicity of drinking water for many water utilities.

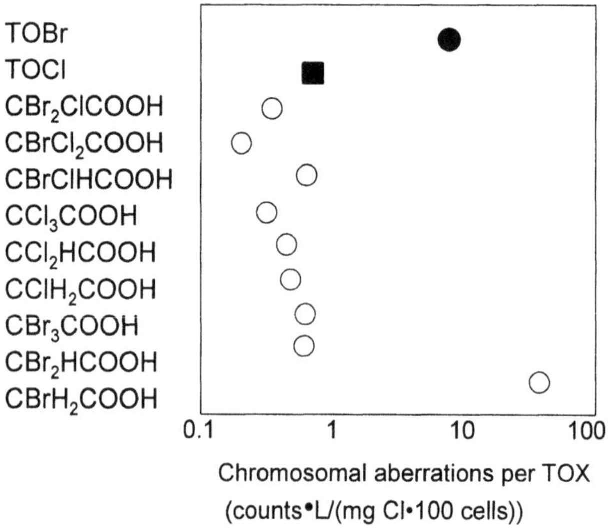

Figure 7. Comparison of chromosomal aberrations induced by a unit TOX. Data for individual HAAs cited from Tsujimura et al. (16)

Table I. Contribution of TOBr.

Cl_2/DOC	*Bromide concentration (µg/L)*				
(mg Cl_2/mg C)	*50*	*100*	*200*	*400*	*600*
1.0	37.3	63.3	75.8	88.1	92.0
1.5	32.9	59.9	77.3	89.9	92.3
2.0	33.7	51.5	81.5	87.0	90.0

Unit: %.

The discussion above suggests that bromide removal technologies would reduce the toxicity of drinking water significantly especially for a source water with high bromide. The technologies applicable to bromide removal are ion exchange (*18, 19*), reverse osmosis (RO) (*20*), selective electrochemical oxidation (*21*), and selective adsorption (*22*). In addition to the current efforts to remove organic precursors of DBPs (i.e., dissolved organic matter), more serious attention has to be paid to these technologies for improving the safety of drinking water.

Summary

Organic bromines formed during chlorination are found to be major contributors to the activity inducing chromosomal aberrations of chlorinated drinking water for bromide levels commonly found in source waters. Thus, TOBr would be a good indicator of the toxicity of chlorinated water, especially for a source water with varying bromide level.

Acknowledgements

This study was a part of a research project funded by Ministry of Health, Welfare and Labour, Japan.

References

1. Siddiqui, M. S.; Amy, G. L.; Rice, R. G. Bromate ion formation: a critical review. *J. Am. Water Works Assoc.* **1995**, *87*(10), 58-70.
2. Larson, R. A.; Weber, E. J. *Reaction Mechanisms in Environmental Organic Chemistry;* CRC Press: Boca Raton, FL, 1994.
3. Westerhoff, P.; Song, R.; Amy, G.; Minear, R. NOM's role in bromine and bromate formation during ozonation. *J. Water Works Assoc.* **1998**, *89*(11), 82-94.
4. Plewa, M. J.; Kargalioglu, Y.; Vakerk, D.; Minear, R. A.; Wagner, E. D. Mammalian cell cytotoxicity and genotoxicity analysis of drinking water disinfection by-products. *Environ. Mol. Mutagen.* **2002**, *40*, 134-142.
5. Plewa, M.J.; Wagner, E.D.; Jazwierska, P.; Richardson, S.D.; Chen, P.H.; McKague, A.B. Halonitromethane drinking water disinfection byproducts: chemical characterization and mammalian cell cytotoxicity and genotoxicity. *Environ. Sci. Technol.* **2004**, *38*, 62-68.

6. Muellner, M.G.; Wagner, E.D.; McCalla, K.; Richardson, S.D.; Woo, Y.T.; Plewa, M.J. Haloacetonitriles vs. regulated haloacetic acids: Are nitrogen containing DBPs more toxic? *Environ. Sci. Technol.* **2007**, *41*, 645-651.
7. Nobukawa, T.; Sanukida, S. Effect of bromide ions on genotoxicity of halogenated by-products from chlorination of humic acid in water. *Water Res.* **2001**, *35*, 4293-4298.
8. Kargalioglu, Y.; McMillan, B. J.; Minear, R. A.; Plewa, M.J. Analysis of the Cytotoxicity and Mutagenicity of Drinking Water Disinfection By-Products in Salmonella typhimurium. *Teratog. Carcinog. Mutagen.* **2002**, 22, 113-128.
9. Echigo, S.; Itoh, S.; Natsui, T.; Araki, T.; Ando, R. Contribution of brominated organic disinfection by-products to the mutagenicity of drinking water. *Water Sci. Technol.* **2004**, *50*(5), 321-328.
10. Echigo, S.; Minear, R. A. Chemical quenching method for the kinetics of the reactions between hypobromous acid and phenolic compounds. *Natl. Meet. – Am. Chem. Soc., Div. Environ. Chem.* **2001**, *41*(2), 638-642.
11. Gallard, H.; Pellizzari, F.; Croué J. P.; Legube, B. Rate constants of reactions of bromine with phenols in aqueous solution. *Water Res.* **2003**, *37*, 2883-2892.
12. Westerhoff, P.; Chao, P.; Mash, H. Reactivity of natural organic matter with aqueous chlorine and bromine. *Water Res.*, **2004**, *38*, 1502- 1513.
13. Echigo, S.; Itoh, S.; Araki, T.; Ando, R. Toxicity of chlorinated water in the presence of bromide ion: contribution of brominated disinfection by-products to the toxicity of chlorinated water. *Environ. Eng. Res.* **2004**, *41*, 779-289 (in Japanese).
14. Echigo, S.; Zhang, X.; Plewa, M. J.; Minear, R. A. Differentiation of TOCl and TOBr in TOX measurement. In *Natural Organic Matter and Disinfection By-Products;* S. Barrett, S. Krasner and G. Amy Eds.; ACS Symposium Series 761; American Chemical Society: Washington, DC, 2000; pp 330-342.
15. Itoh, S.; Matsuoka, Y. Contributions of disinfection by-products to activity inducing chromosomal aberrations of drinking water. *Water Res.* **1996**, *30*, 1403-1410.
16. Tsujimura, Y.; Tabuchi, M.; Itoh, S. Estimation of the toxicity of unregulated haloacetic acids in drinking water, *Proceedings of the 36th Annual Conference of Japan Society on Water Environment*, Sapporo, Japan, March 17-19, 2004; p 439 (in Japanese).
17. Amy, G.; Siddiqui, M.; Zhai, W.; DeBroux, J.; Odem, W. *Survey of bromide in drinking water and impacts on DBP formation*; Report 90662; American Water Works Association Research Foundation: Denver, CO., 1994.

18. Kuwahara, M.; Miura, T.; Niwa, A.; Echigo, S., and Itoh, S. Bromide removal by hydrotalcite-like compounds. *Advances in Asian Environ. Eng.* **2006**, *5*(1), 47-52.
19. Johnson C.J.; Singer P.C. Impact of a magnetic ion exchange resin on ozone demand and bromate formation during drinking water treatment. *Water Res.* **2004**, *38*, 3738-3750.
20. Amy, G. L.; Siddiqui, M. S. *Strategies to control bromate and bromide*; Report 90751; American Water Works Association Research Foundation: Denver, CO., 1999.
21. Kimbrough D.E.; Suffet I.H. Electrochemical removal of bromide and reduction of THM formation potential in drinking water. *Water Res.* **2002**, *36*, 4902-4906.
22. Sanchez-Polo M.; Rivera-Utrilla J.; von Gunten, U. Bromide and iodide removal from waters under dynamic conditions by Ag-doped aerogels. *J. Colloid Interface Sci.* **2006**, *306*, 183-186.

Reactivity of Natural Organic Matter toward Disinfection By-Product Formation

Chapter 6

Disinfection By-Product Precursor Content of Natural Organic Matter Extracts

David A. Reckhow[1], Gladys Makdissy[1,2], and Paula S. Rees[1]

[1]Department of Civil and Environmental Engineering, University of Massachusetts, Amherst, MA 01003
[2]Current Address: Department of Civil and Environmental Engineering, University of Waterloo, Waterloo ON N2L 3G1

For more than 20 years, researchers have been using resin-based separation methods to isolate natural organic matter (NOM) for the purposes of assessing disinfection byproduct (DBP) precursors. Most have employed non-standard precursor methodologies, which makes comparison between studies very difficult. As new data of this type are acquired, it is important to be able to place them into context of the existing database. In this study, we critically reviewed and summarized this vast collection of literature data. Using some of the well-tested precursor models, we were able to normalize the data allowing cross comparison between studies using differing precursor testing regimes. These comparisons revealed some distinct differences between hydrophobic, hydrophilic and transphilic NOM. In addition, some of these distinctions are general, being evident across the various DBP classes (e.g,. trihalomethanes, di & trihaloacetic acids).

Introduction

Natural organic matter (NOM) reacts quickly with aqueous chlorine to form an extensive mixture of halogenated and non-halogenated byproducts. These reactions occur in US drinking water treatment plants where chlorine is still the disinfectant of choice. Toxicologists and epidemiologists have slowly amassed a convincing body of evidence supporting chlorination byproducts of NOM as a cause of cancer and other serious health problems in humans *(e.g., 1)*.

Fractionation of NOM based on hydrophobic properties is a powerful and widely-used approach for simplifying the study of NOM, its geographical variability and behavior in water treatment systems. Many researchers have assessed the tendency of their own NOM isolates to form DBPs upon chlorination. Most attempt to compare these values with the "typical values" reported in the literature. This is problematic for several reasons; (1) the large amount of data that exists in the literature, (2) the different conditions used for NOM extraction, and (3) the different conditions used for assessing precursor content. This current study was undertaken to summarize the NOM literature in a quantitative way that accounts for different experimental protocols.

Disinfection byproduct formation is explored through analysis of existing data on NOM extracts and whole waters. By far the majority of these data focus on THM formation, and for this reason, we begin our analysis on that particular class of DBP. The other DBPs that will be examined are the haloacetic acids. These are grouped into those having two halogens (the dihaloacetic acids, or DHAAs) and those having three halogens (the trihaloacetic acids, or THAAs). This is primarily in recognition of the differing precursor types that give rise to these groups *(e.g., 2)*. The monohaloacetic acids (MHAA) are not considered here, because of the generally poor quality of MHAA data. In a few cases, other precursors are considered (e.g., those that form dihaloacetonitriles, halonitromethanes), but the scarcity of these data makes the analysis less helpful.

The datasets on DBP formation from NOM fractions are almost all based on slightly different methodologies. For this reason, we have chosen to focus the first part of this analysis on NOM fractions only prepared and treated at the University of Massachusetts using the same general approach. The second part of this analysis considers all data pertaining to the NOM fractions based on partitioning with XAD resins. This particular fractionation protocol is judged sufficiently uniform as to yield useful results.

Experimental Methods

This work involved collection of new data in the UMass laboratories as well as a detailed analysis of existing literature data. For the UMass work, NOM isolation was more extensive, incorporating anion and cation exchange resins in addition to the more common hydrophobic resin. This resulted in a group of 8 fraction types *(3)*. NOM solutions[1] were chlorinated in the laboratory under a standard set of conditions (20 mg/L dose, pH 7.0, 72 hours, 20°C) as used with prior NOM and model compound studies. The high-level fixed-dose testing protocol was adopted to minimize the chances that the residual would dip down to levels below 1 mg/L, and it imposes a chlorine residual profile that is relatively constant from one sample to the next (i.e., residual is initially 20 mg/L and rarely drops below 10 mg/L). This was done to avoid variations between tests in the reactive chemical environment (i.e., chlorine residual). At the end of the 72 hour incubation period a sample was removed for measurement of residual chlorine, and the remainder was quenched with sulfite, partitioned into a set of identical aliquots and analyzed for haloacetic acids, trihalomethanes and other neutral extractable byproducts. Controls were prepared by chlorinating high-purity water as described above.

In an effort to render the diverse literature values for DBP precursor density directly comparable, we chose to adjust the data using a widely tested power function modeling approach. The power functions models are generally developed with terms that contain a single independent variable each, with few or no interaction terms (equation 1). The specific model coefficients used for this work are shown in Table 1.

$$DBP = a(Br + b)^{c}(pH + d)^{e}(temp + f)^{g}(Cl_2dose)^{h}(time)^{i} \quad (1)$$

Table I. Model Parameters used to Adjust DBP Yields[2]

DBP Group	*Bromide (μg/L)*		*pH*		*Temp. (C)*		*Cl_2 Dose (mg/L)*	*Time (hr)*
	Offset (b)	*Exp. (c)*	*Offset (d)*	*Exp. (e)*	*Offset (f)*	*Exp. (g)*	*Exp. (h)*	*Exp. (i)*
TTHM	1	0.036	-2.6	0.715	0	0.791	0.272	0.261
THAA	1	0.036	0	-1.495	0	0.307	0.378	0.188
DHAA	1	0.036	0	0.259	0	0.377	0.346	0.186

[1] Buffered with 10 mM phosphate. Most solutions were prepared at 5 mg/L DOC, although some of the less abundant fractions were as low as 3 mg/L DOC.

[2] SOURCE: Reproduced with permission from Reference 15. Copyright 2007 AWWARF.

The model parameters in Table Iwere arrived at in different ways. The literature was surveyed and models critically assessed. The TTHM model was supported by the largest number of studies and using the largest databases. In addition, all TTHM models encompassed the full set of 4 THM species. Several published models were identified whose prediction space covered the range of conditions in the NOM fraction database *(4-9)*. In addition, one unpublished model based on an aquatic fulvic acid database *(10)* was used. Since all were derived from a broad set of conditions, all of these models could be applied to the NOM fraction data. To avoid issues of bias, coefficients from all 7 models were averaged when they could be compared on an equivalent basis (i.e., when there were no "offsets' or when the offsets were identical). Because of the low bromide levels in most hydrophobic NOM fractions, it was necessary to have a model with a positive bromide offset (to avoid the problem of bromide levels near zero). Thus the bromide coefficients from Amy et al.'s 1987 model were adopted without attempting to blend in the other model predictions.

A different approach had to be taken for the haloacetic acids, because these models were less numerous, and generally not inclusive of all brominated forms. In general only 4 models were combined to get the final set of coefficients *(8,10-12)*. However, none of these models was based on a large dataset including the full suite of brominated forms. Instead, the bromide model for the THMs was incorporated without modification. Singer and colleagues *(13-14)* have shown that bromine incorporation into the HAAs follows a quantitative pattern quite similar to that of the THMs. This has been recently verified using data from the National Occurrence Survey, and is used as justification for adopting bromine incorporation factors (BIFs) from the THMs for the purpose of estimating HAA speciation.

In many of the graphical presentations that follow, DBP precursor levels are shown based on the standard conditions used at UMass (called the std-DBPFP). This was calculated with by means of the power function model (equation 2 shows the case for std-THMFP). To help put these numbers into context, a left offset scale is included that allows translation of these numbers into uniform formation condition values, sometimes used as standard SDS conditions. The particular conditions that these tests refer to are pH 7, 20C, 20 mg/L dose and 72 hr time for FP tests, and pH 8, 20C, 1.5 mg/L and 24 hr for the SDS test. In many of these figures there is also a right-had vertical scale with the "pre-exponential value". This is the "a" value in the power function model (equation 1).

$$Std-THMFP = THMFP\left(\frac{20}{Cl_2dose}\right)^{0.272}\left(\frac{72}{time}\right)^{0.261}\left(\frac{7-2.6}{pH-2.6}\right)^{0.715}\left(\frac{1}{Br+1}\right)^{0.036}\left(\frac{20}{temp}\right)^{0.791} \quad (2)$$

Results and Discussion

DBP PrecursorContent in an Internally Consistant set of NOM Fractions

In the following section a single set of internally consistent NOM fractions are compared. All were isolated at the University of Massachusetts using very similar protocols and tested for DBP formation using a single procedure. These NOM fractions were all derived from raw and treated waters collected at various locations in Massachusetts and Connecticut (Quabbin Reservoir, New Haven raw and treated waters, Boston Tap water, Forge Pond). All fractions were dissolved in buffered water at DOC concentrations between 3 and 5 mg/L. This DOC window was selected so that DBP formation was high enough for best analytical precision, yet not so high as to cause substantial depletion of the chlorine redsidual. The solutions were then treated under the standard chlorination conditions as described earlier.

Specific chlorine demand ranged from less than 0.5 mg/mg-C to as much as 7.8 mg/mg-C (data not shown). There is a weak positive correlation between specific chlorine demand and specific UV absorbance (SUVA). When looking at the THM formation, there appears to be a stronger positive correlation with SUVA (Figure 1). Although the slope of this line and its standard error (5.1 and 1.2, respectively) indicate that variations in SUVA can be used to explain a significant amount of the variability in the specific THMFP, there is still quite a bit of variability that remains unaccounted for (i.e., there is much scatter around the regression line). There are also clear clusters of NOM fractions, suggesting a common reactivity for the same fractions from different sources. In general, the neutrals and bases are quite unreactive. Most do not exceed 25 µg/mg-C, which places them at the very low end of the surface water data *(15)*. Most of the hydrophilic acids and weak hydrophobic acids are only slightly more reactive. In term of mean reactivity, it is the fulvic acids that most resemble the average whole surface waters. As expected, the humic acids are even more reactive, however, their abundance in surface waters is generally low.

Of all the DBP precursors, it is the trihaloacetic acids (THAAs) that exhibit the largest range in concentration (Figure 2b). They also show a strong correlation with SUVA. The mechanistic explanation for this is to be found in the literature *(e.g., 2)*. A high degree of conjugation facilitates electron transfer from site of THAA cleavage to the reactive sites undergoing oxidative attack. This leads to faster THAA formation, and therefore less time for hydrolysis (which yields THMs). Of course, the conjugated bonds also result in higher absorbance in the UV range (SUVA).

The dihaloacetic acid precursors (DHAA) also show a strong correlation with SUVA (Figure 2a). This is a bit more surprising based on observations with treated drinking waters, and not easily explained with mechanistic

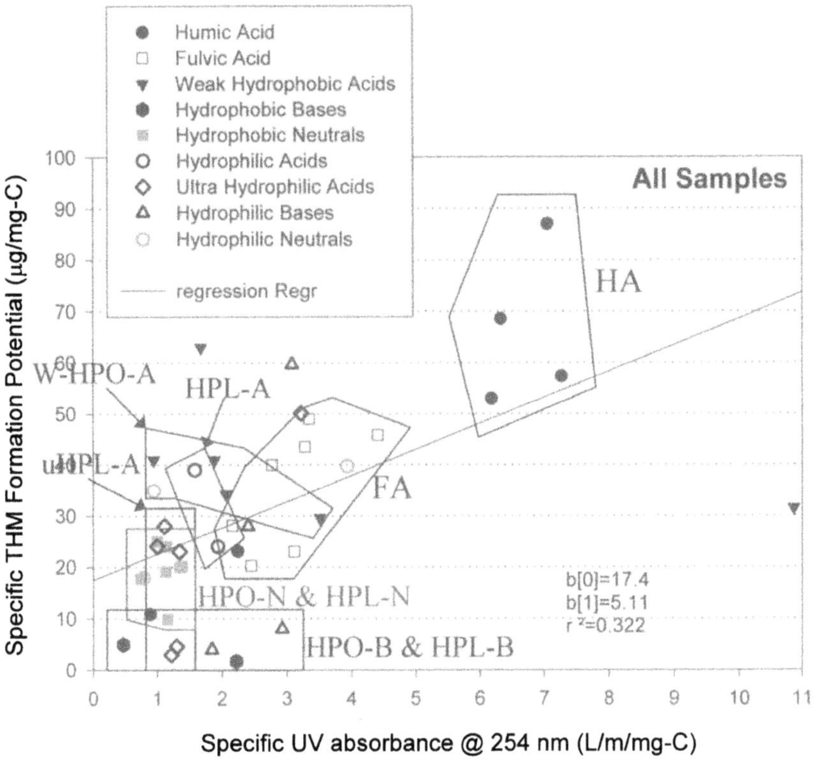

Figure 1. Relationship between Specific THM Precursor Content and SUVA for Isolated NOM Fractions from New England Waters. Including boxed NOM Domains. (Reproduced with permission from Reference 15. Copyright 2007 AWWARF.)

reasoning. Nevertheless, it's clear that for the DHAAs there is almost no difference between the yields for fulvic acid as compared to the hydrophilic and weak hydrophobic acids. This is in contrast to the THAAs, and seems to explain why coagulated waters (i.e., waters with the fulvic acid largely removed) produce higher ratios of DHAA/THAA than do the raw waters.

The ratio of TCAA to THM formation most directly addresses the dual pathway to these two trihalomethyl species as presented in *(2)*. Figure 3 shows that there is a strong positive correlation between SUVA and the TCAA/THM ratio. If a higly conjugated system encourages THAA (TCAA) formation as discussed above, higher UV absorbances should accompany a shift from THM to THAA.

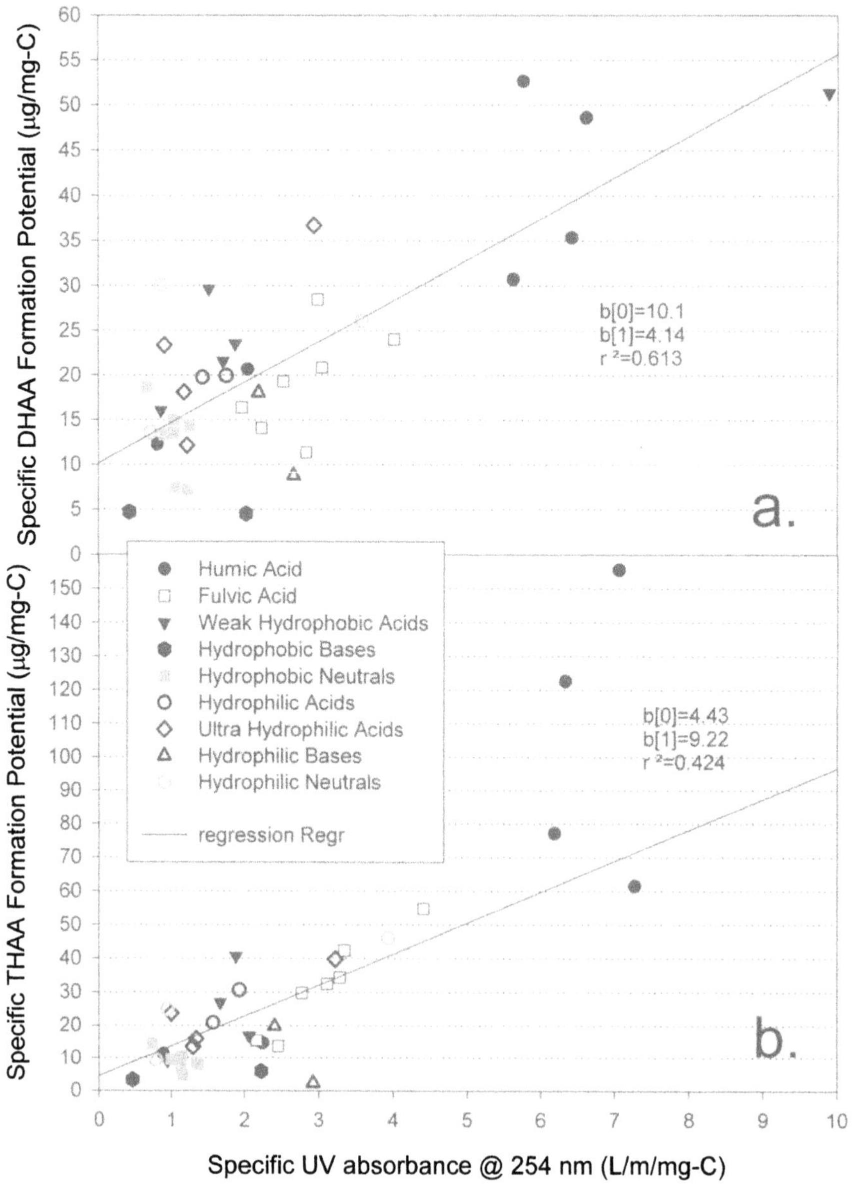

Figure 2. Relationship between Specific HAA Precursor Content and SUVA for Isolated NOM Fractions from New England Waters; a. DHAA, b. THAA (Reproduced with permission from Reference 15. Copyright 2007 AWWARF.)

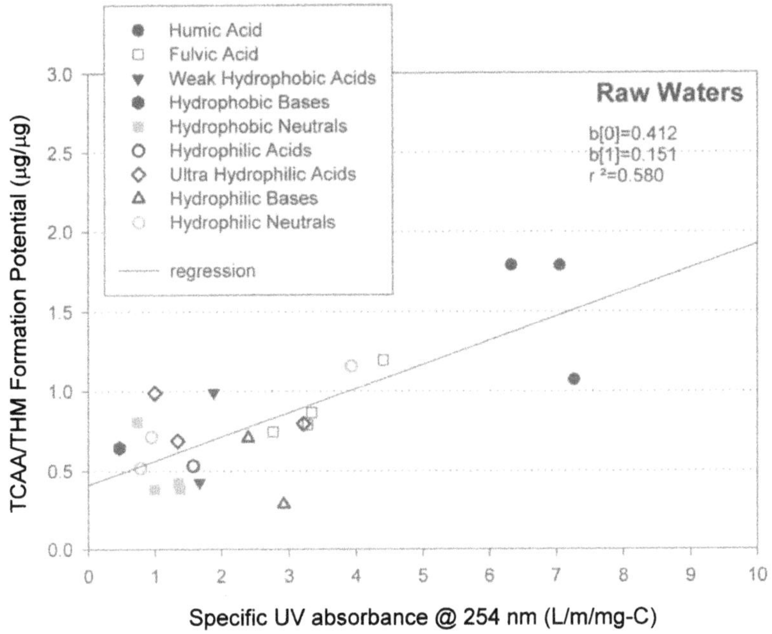

Figure 3. Relationship between TCAA/THM Precursor Ratio and SUVA for Isolated NOM Fractions from New England Waters (Reproduced with permission from Reference 15. Copyright 2007 AWWARF.)

Analysis of Published Data on Hydrophobic-based NOM Fractions

For the purpose of this data summary, we started with 32 well controlled, data-rich studies. These studies incorporated DBP precursor analyses on a total of 508 separate NOM samples. Of these, only 205 unique samples were accompanied by datasets of the proper quality, type and completeness to allow for their full inclusion in this summary. All studies covered in this analysis used XAD-8 resin to selectively remove the hydrophobic acids. Nearly all made this separation at pH 2 using a k' of about 50. Most of the comparisons were done with the major acid fractions (e.g., hydrophobic acids, hydrophilic acids and transphilic acids[3]). When hydrophobic acids were separated into humic acids and fulvic acids (or weak hydrophobic acids), only the fulvic acids were retained for this analysis. The others are generally of much lower abundance in natural waters, and therefore not representative of the bulk hydrophobic acids. In most cases, the hydrophilic fraction was left whole and not further fractionated into

---

[3] XAD-4 was used in all cases for the separation of transphilic acids from the XAD-8 effluent

acids, bases or neutrals. These were retained in this analysis, based on the objective of comparing hydrophobic organics with hydrophilic organics.

Comparisons were drawn between the major phobic and philic components of raw waters. The full database was sampled and only the major acid-bearing fractions were selected. These included hydrophobic NOM, hydrophobic acids, fulvic acids, hydrophilic acids, hydrophilic NOM, transphilic acids, transphilic NOM. These are believed to be dominated by the acidic compounds and are primarily distinguished by hydrophobic behavior. Excluded from this were fractions containing only bases, neutrals or humic acids. Also excluded are other fractions that are separated based on size or other properties (HMW, colloids, etc.).

In addition, fractions isolated from treated waters were excluded. Finally, whenever the same fraction was treated with chlorine under 2 or more different treatment conditions, only the one that was subject to those most closely resembling the standard FP conditions was retained for further analysis. For a complete list of the original literature sources for these data, the reader should consult *(15)*.

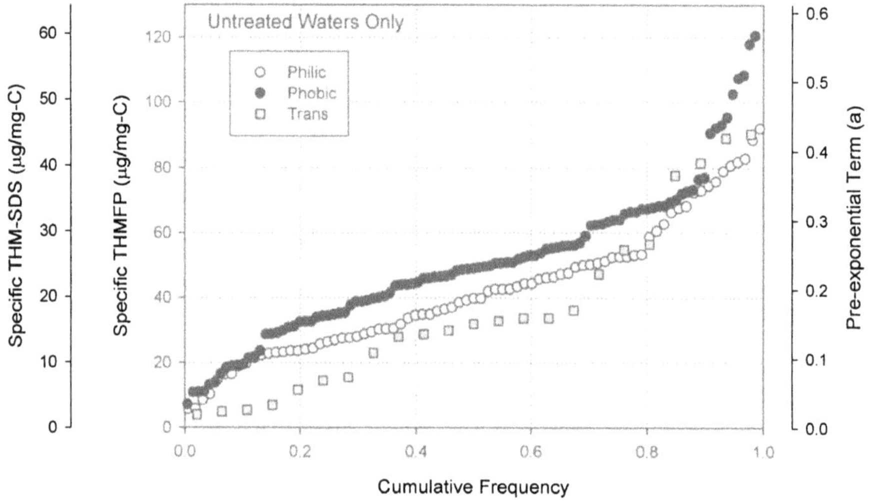

Figure 4. Cumulative Frequency Plot for THM Precursor Content in Major RW Fractions (Reproduced with permission from Reference 15. Copyright 2007 AWWARF.)

Figure 4 shows the summary of the full database for specific THM precursors. This type of presentation allows one to see the full range of values recorded and corresponding percentiles. The median values (50%ile) for hydrophobic and hydrophilic NOM are about 50 μg/mg-C and 40 μg/mg-C, respectively. Figure 4 also makes it clear that the hydrophobic fractions are shifted toward the higher THM precursor levels as compared to the hydrophilic ones. The transphilic fractions are shown as being low compared to the others, however, this might be attributed to a higher degree of uncertainty (i.e., fewer data) for this less-commonly isolated fraction. The fact that ambient bromide will tend to collect in hydrophilic fractions may artificially elevate the molar THM yield of these organics versus the hydrophobic acids. For this reason, the intrinsic differences in the THM precursor content of phobic and philic fractions may be somewhat larger than shown in Figure 4. Figure 4 and later figures also contain offset axes that facilitate direct readings of the data in µg/mg-C based on SDS conditions and translations of the data to the "a" value in equation 1 (see Experimental Methods section).

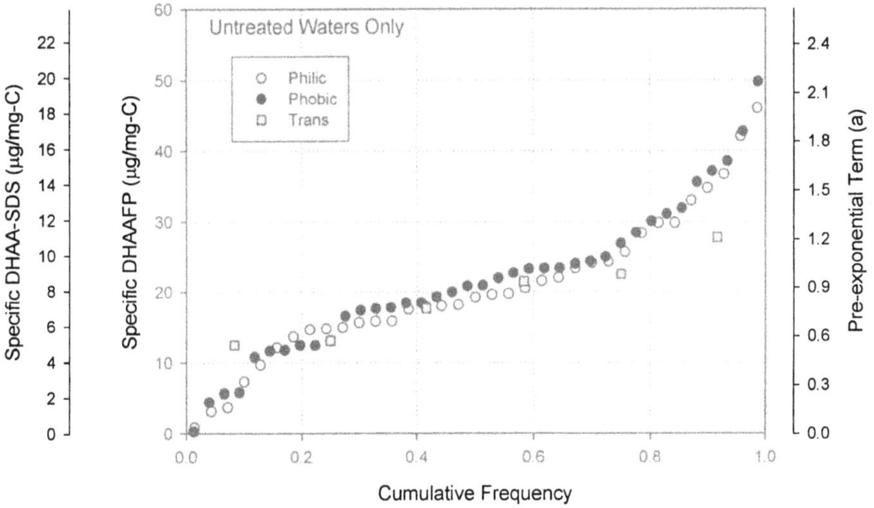

Figure 5. Cumulative Frequency Plot for DHAA Precursor Content in Major RW Fractions (Reproduced with permission from Reference 15. Copyright 2007 AWWARF.)

In contrast to the THMs, the dihaloacetic acid precursors show very little systematic difference between the three NOM fractions (Figure 5). All three show median values of about 20 µg/mg-C (FP conditions), with little separation at other percentiles. The trihaloacetic acids (Figure 6) exhibited at least as much displacement between the fractions as the THMs, if not more. Once again, it is the hydrophobic fraction that has the higher precursor density. The first part of this study indicates that humic acids are more productive than other fractions, and that they are especially rich in THAA precursors (Figures 1, 2a and 2b). Since humic acids are a part of the hydrophobic fraction, small amounts of these

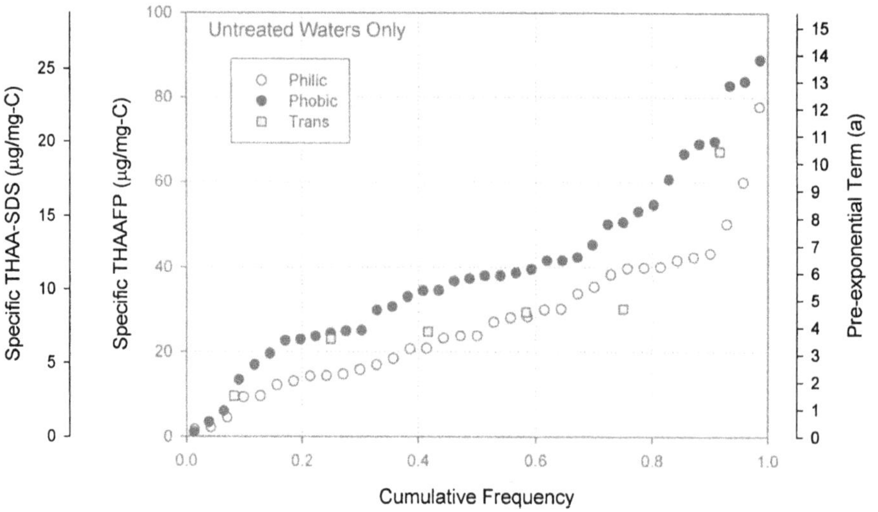

Figure 6. Cumulative Frequency Plot for THAA Precursor Content in Major RW Fractions (Reproduced with permission from Reference 15. Copyright 2007 AWWARF.)

acids may be at least partly responsible for the higher reactivity observed in Figures 4 and 6 for the hydrophobic fractions.

A more direct set of comparisons between complementary NOM isolates can be made by using the trihalomethanes as a benchmark (i.e., presenting the HAA values as ratios to the THMs). This analysis (not shown) indicates that even though the THMs are favored in hydrophobic fractions, the THAAs are favored even more. This is true even when comparing complementary fractions from the same water. However, the reverse is true for the DHAAs.

The literature database does show a substantial positive correlation between specific THMFP and SUVA for all NOM fractions (Figure 7). This correlation exists within each of the three major NOM fractions as well (i.e., philic, phobic and trans). While the tendency of hydrophobic fractions to have higher SUVA values is evident, differences in the relative precursor densities between these solubility-based fractions is not obvious. Figure 8 shows the specific THMFP ratios for each of the paired NOM fractions derived from the same raw water. The median value for these ratios is less than unity, although there are a substantial number of data with ratios greater than one. This indicates that hydrophilic fractions do tend to have slightly lower THM precursor densities than their corresponding hydrophobic fractions, but for any given water, the reverse might be true. The trend is similar for the transphilics, although less well defined, probably due to the smaller dataset.

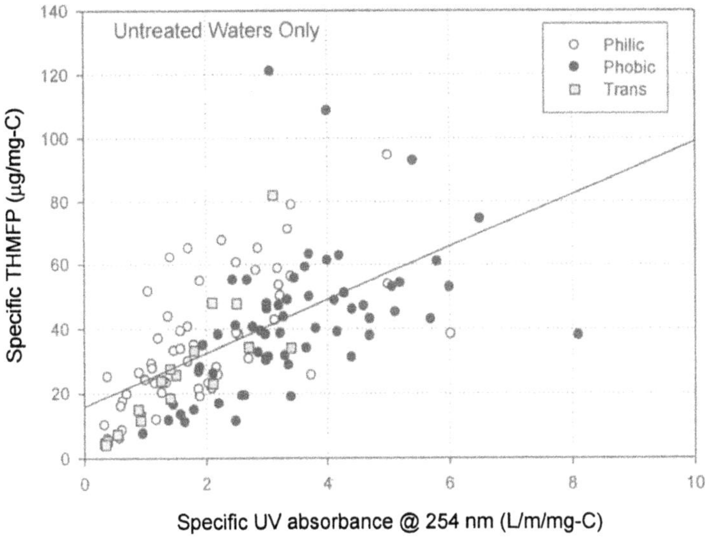

Figure 7. Specific THM Precursor Content of all Fractions plotted versus the SUVA: (Reproduced with permission from Reference 15. Copyright 2007 AWWARF.)

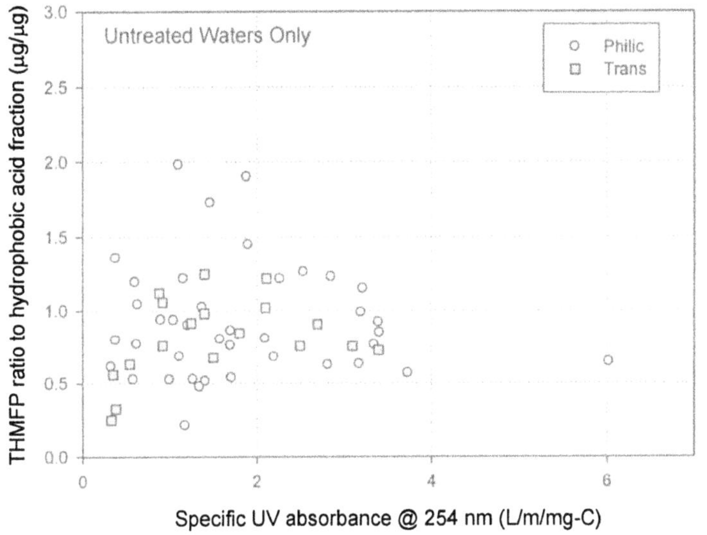

Figure 8. Relationship between Specific THM Precursor Content of Hydrophilic and Transphilic Fractions to the Specific Content in the Corresponding Hydrophobic Fraction plotted versus its SUVA: (Reproduced with permission from Reference 15. Copyright 2007 AWWARF.)

Figure 9 shows this ratio in the form of a frequency plot. It appears that about 70% of the data have higher THM yields in the hydrophobic fractions than in their corresponding hydrophilic fractions. This tendency also carries out to the fulvic and humic acids. Aquatic humic acids, which are excluded from this analysis due to their low natural abundance, are considered even more hydrophobic than then fulvic acids or bulk hydrophobic acids. When these two have been analyzed for precursors from the same source, it is almost always the humic acids that have the higher specific THM precursor content *(e.g., 16)*.

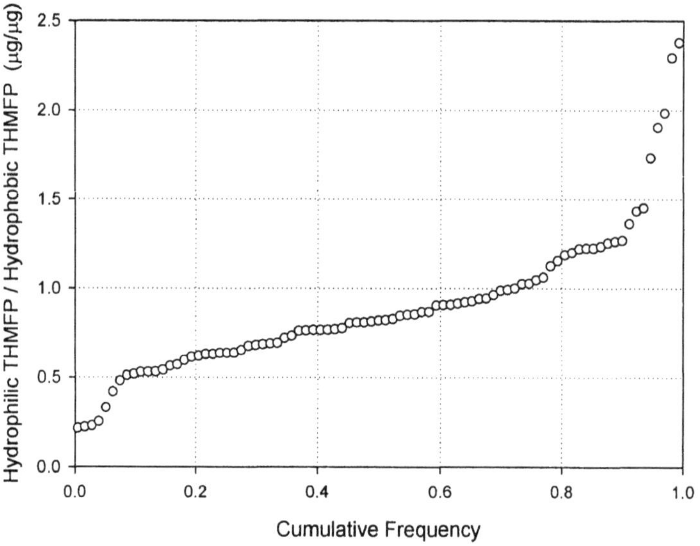

Figure 9. Cumulative Frequency Plot for the Ratio of the Specific THM Precursor Content for a Hydrophilic Fraction to its Corresponding Hydrophobic Faction (Reproduced with permission from Reference 15. Copyright 2007 AWWARF.)

Conclusions

The formation of disinfection byproducts is heavily dependent on the characteristics of the precursor natural organic matter (NOM). Although, specific UV absorbance has been used successfully as a measure of NOM reactivity with oxidants/disinfectants, it falls short of capturing the full complexity of these reactions. This study documents the range of variability in DBP yields from a broad spectrum of NOM fractions and sourcess. Specifically, the follow conclusions can be drawn:

1. THM precursor levels in NOM fractions show a high level of uniformity. The median THM precursor level in hydrophobic NOM (based on the UMass FP protocol) is about 50 μg/mg-C, and the bulk of the population falls within 20-80 μg/mg-C (10th–90th percentiles). This translates to about 25 μg/mg-C under SDS conditions with a range of about 10-40 μg/mg-C. The precursor level in hydrophilic NOM is on average slightly lower at 40 μg/mg-C.
2. Hydrophobic acids tend to produce only slightly more THM on a per-carbon basis than hydrophilic acids from the same source.
3. Hydrophilic acids produce a higher level of dihaloacetic acids as compared to the other DBPs. As a result, they are almost identical to the hydrophobic acids in their per-carbon yield of DHAAs (median ~20 μg/mg-C).
4. Hydrophobic NOM produces substantially more trihaloacetic acids than hydrophilic NOM. The median yield for the two groups is 38 μg/mg-C and 25 μg/mg-C, respectively.
5. The greater THM and HAA yields from hydrophobic NOM may be accentuated in waters with a high humic/fulvic ratio.
6. The specific formation potential for all 3 major classes of byproducts (THMs, DHAAs, THAAs) exhibits a positive correlation with SUVA. In addition, as SUVA goes up there is a tendency for the trihalo-species to shift from the THMs to the THAAs.

Acknowledgments

The authors would like to thank the American Waterworks Association Research Foundation (project #2868) for support of this research. We would also like to acknowledge the assistance of Drs. Boijayanta Bezbarua, John McClellan, and Benedicte Mousset.

References

1. Xie YF. *Disinfection Byproducts in Drinking Water: Formation, Analysis and Control*. CRC Press, **2003**.
2. Reckhow DA, Singer PC. In: Jolley RL, Bull RJ, David WP, Katz S, Roberts MHJr, Jacobs VA, editors. *Water Chlorination: Environmental Impact and Health Effects* Lewis Publishers, Chelsea, MI, **1985**, pp. 1229-1257.
3. Leenheer, J.A. and Noyes, T. I. *A Filtration and Column-Adsorption System for Onsite concentration and Fractionation of Organic Substances from*

Large Volumes of Water. **1984**. Washington, D.C., U.S. Government Printing Office. U.S. Geological Survey Water Supply Paper.
4. Amy GL, Minear RA, Cooper WJ. *Water Research* **1987**; 21: 649-659.
5. Owen, D.M., Amy, G.L., and Chowdhury, Z.K. *Characterization of Natural Organic Matter and Its Relationship to Treatability* AWWA Research Foundation, Denver, CO. **1993**.
6. Rodriguez MJ, Serodes JB. *Water Research* **2001**; 35: 1572-1586.
7. Rathbun RE. *Science of the Total Environment* **1996**; 191: 235-244.
8. Amy, G. L., Siddiqui, M., Ozekan, K., Zhu, H., and Wang, C. *Empirically Based Models for Predicting Chlorination and Ozonation By-products: Trihalomethanes, Haloacetic Acids, Chloral Hydrate, and Bromate*. **1998**. Washington, DC, US Government Printing Office. US EPA Report, Office of Water.
9. Malcolm Pirnie, Inc, report to US EPA, **2001**
10. Reckhow, D.A., Hua, G., Kim, J., Hatcher, P.G., Caccamise, S.A. L., and Sachdeva, R.. *Characterization of TOX Produced During Disinfection Processes*. **2007**. Denver, CO, American Water Works Association Research Foundation
11. Reckhow, D.A., McClellan, J.N., Mousset, B., and Bezbarua, B.K. American Chemical Society, Preprint of Extended Abstracts. 39(1), pp. 217-218. **1999**. Washington, ACS.
12. Malafronte JA. *Disinfection By-product Models for the Stamford, CT, Water System*. MS Thesis, University of Massachusetts, Amherst, MA, **2003**.
13. Cowman, G.A. and Singer, P. C. *Environmental Science and Technology* 30[1], 16-24. **1996**.
14. Roberts MG, Singer PC, Obolensky A. *Journal American Water Works Association* **2002**; 94: 103-114.
15. Reckhow, D.A., Rees, P.L., Nusslein, K., Makdissy, G., Devine, G., Connelly, T., Boutin, A., and Bryan, D. *Watershed Sources and Long-Term Variability of BDOM and NOM as Precursors*. **2007**. Denver, CO, American Water Works Association Research Foundation.
16. Reckhow, D.A., Singer, P.C., and Malcolm, R.L. *Environmental Science and Technology* 24[22], 1655-1664. **1990**.

Chapter 7

Parameters Affecting Haloacetic Acid and Trihalomethane Concentrations in Treated UK Drinking Waters

Cynthia M.M. Bougeard[1], Imran H.S. Janmohamed[2], Emma H. Goslan[1], Bruce Jefferson[1], Jonathan S. Watson[2], Geraint H. Morgan[2], and Simon A. Parsons[1]

[1]Centre for Water Science, Cranfield University, Cranfield, Bedfordshire MK43 0AL, UK
[2]Planetary & Space Sciences Research Institute, The Open University, Milton Keynes, Buckinghamshire MK7 6AA, UK

Parameters (pH, bromide, and temperature) affecting the formation of trihalomethanes (THMs) and haloacetic acids (HAAs) were investigated by chlorinating two geographically different waters in the UK: lowland and upland water. It was expected that THM levels would increase as pH increased whereas HAA levels would decrease with an increase in pH. The lowland water reacted as expected to changes in pH but the upland water did not. The greatest impact of pH was observed in the formation of THMs in the lowland water. The difference in natural organic matter (NOM) structure may account for such differences. Concentration of bromide had a greater impact in the upland water. Reducing the temperature from 20°C to 7°C resulted in a mean decrease of DBP concentration by 50%.

Natural organic matter (NOM) is described as an intricate mixture of organic compounds that occurs universally in ground and surface waters. Whilst NOM itself is not problematic, it can be converted to disinfection by-products (DBPs) when disinfectants are used during water treatment (*1*). In the UK, regulated DBPs include trihalomethanes (THMs - chloroform, bromoform, dibromochloromethane and bromodichloromethane). It is required by law that total THMs do not exceed 100 µg/L (2). Frequency of sampling will depend on the population size. For example if the population is below 100 inhabitants one sample will be taken per year, whereas if the population is 5000 to 100000 inhabitants eight samples will be taken per year. Recently it has been observed that THMs may not be the major representative of the chlorinated DBPs (*3*). For instance, in the US, levels of haloacetic acids (HAAs) have been reported at similar or higher levels than THMs in finished drinking waters (*3*).

In the US, four THMs and five of the HAAs are regulated at values of 80 µg/L and 60 µg/L respectively (*4*) and it is possible that HAAs may be regulated in Europe (*5*). Currently HAAs are not routinely measured in the UK and very little is known about their formation in UK waters. To our knowledge, only one UK study has been published that reported HAA levels up to 244 µg/L (*6*).

The formation of HAAs is influenced by a number of factors including disinfectant type, its concentration, contact time, water temperature and pH. The type and concentration of natural organic matter (NOM) and the concentration of bromide will directly impact on the level of HAAs formed (*7, 8*).

In chlorination, DBP formation increases with increasing contact time and chlorine dose applied (*9*). It is known that the formation of THMs is enhanced at high pH (*10*). However, the effect of pH on the formation of HAAs is equivocal. Overall, HAA formation increases with decreasing pH (*11*). Specifically, it is well established that dihalogenated acetic acid (DXAA) concentration remains constant, while trihalogenated acetic acid (TXAA) concentration decreases with increasing pH (*12*). High NOM concentrations have generally been associated with high DBP concentrations (*13, 14, 15*). In the UK, chlorination tends to occur after the water has been treated by a coagulant and filtered (*16, 17*), which is different from US treatment practices where pre-chlorination is widely used (*12*). After conventional treatment processes, NOM is mainly hydrophilic in character and low in concentration (*18*). However, hydrophilic NOM has been reported to contribute substantially to the formation of DBPs especially for waters with a low humic (hydrophobic) content (*7*). The presence of bromide in water will influence the speciation of the DBPs and the amount of bromide present will also affect the concentration of the DBPs. Waters with levels of bromide as low as 100 µg/L have been reported to form brominated HAAs and THMs (*8*). It should also be remembered that precursor removal by coagulation will increase the ratio of bromide to DOC and may result in increased formation of brominated DBP species (*12*).

HAAs are highly water-soluble DBPs that exist as ions at ambient pH. For analysis by gas chromatography (GC), they must first be converted to their protonated forms before extraction from water with organic solvent and then derived to form more volatile methyl esters (*19*). As HAAs exist as ions, it is possible to analyse them directly using ion chromatography (IC) (*20*) or capillary electrophoresis (CE) (*21*) but at higher levels than with GC coupled with electron capture detector (ECD). In this study, GC with mass spectrometry (GC-MS) and GC-ECD were investigated as published analysis techniques.

This study reports the concentration of THMs and HAAs when two distinctly different waters are chlorinated under controlled conditions and investigates the impacts of pH, temperature and bromide concentration on their formation. The waters are from different geographical regions in the UK and were collected after treatment but before disinfection. The aims of this work were to examine the sensitivity of DBP formation to differences in water character and establish whether treated UK waters follow the trends identified for untreated US waters regarding THM and HAA formation.

Material and Methods

Water Samples and Characterisation

All experiments were undertaken with samples collected from two water utilities: Anglian Water from East Anglia (lowland water) and Yorkshire Water (upland water). The water treatment works (WTWs) were selected because of their different organic matter content as well as their different geographical locations. The lowland water reservoir is situated in the East Anglian region of England in the South East. The reservoir is on a plateau with nearly all of its water being extracted from a local river. The upland water reservoir is situated in northern England. It is fed from a range of reservoirs set in peat-rich moorlands. The lowland water was ozonated, coagulated, sand filtered and was collected after contact with biologically active granular activated carbon. The upland water was coagulated and was collected after sand filtration. Both samples did not contact with any disinfectant. A large volume of each water was collected (≥100 L) and stored at 5°C until used. Periodic measurements of pH, non purgeable organic carbon (NPOC), ultraviolet absorbance at a wavelength of 254 nm (UV_{254}) and bromide concentration were carried out and results were consistent.

NPOC was measured using a Shimadzu TOC-5000A analyser (Shimadzu, Milton Keynes, UK). Samples were acidified and purged to convert the inorganic carbon to CO_2. UV_{254} was measured using a Jenway 6505 UV/VIS spectrophotometer (Patterson Scientific Ltd., Luton, UK). Analysis of bromide was carried out with an ion chromatography (IC) system, (Dionex DX500 series, Dionex, UK).

Fractionation

To determine the hydrophilic/hydrophobic NOM ratio, 50 litres of the treated waters were fractionated by XAD and cation exchange resin adsorption techniques into their hydrophobic neutral (HPO-N), hydrophobic acid (HPO-A), transphilic dissolved organic matter (TPI-DOM), hydrophilic base (HPI-B) and hydrophilic acid + neutral (HPI-A+N) fractions. The method used was adapted from Leenheer et al. (*22*). The column capacity factor (k') was 60 for both samples. The recovery of NOM was 90% and 113% for the lowland and upland water respectively.

The resins used were Amberlite XAD-7HP resin and Amberlite XAD-4 resin (Rohm & Haas, Germany). Amberlite XAD-7HP is an acrylic ester polymer and is equivalent to XAD-8; Amberlite XAD-4 is a styrene divinylbenzene polymer. Amberlite 200 strongly acidic cation exchanger has a sulfonated polystyrene/DVB matrix (Sigma-Aldrich, UK). The XAD resins were precleaned by sequentially Soxhlet extracting for 48 hours each with methanol, acetonitrile and methanol again to remove impurities. Before use the resins were packed into columns and rinsed with deionised water (DI) until the column effluent DOC was < 2 mg/L (*23*).

Formation Potential Experiments

The treated waters were chlorinated at pH 6, 7 and 8 and room temperature to determine their disinfection by-product formation potential (DBP-FP). Phosphate buffer was used to adjust the pH. In addition one set of samples for each water was chlorinated at pH 7 with addition of bromide (200 µg/L) and another set at pH 7 and a temperature of 7°C. Hypochlorite solution was prepared using the Standard Method 4500-Cl B (*24*). The chlorine dose required was determined by preliminary chlorine demand experiments to have the free chlorine residual higher than 1 mg/L as Cl_2 after seven days of contact time. A 100 mL bottle was partly filled with the water sample, the buffer and the chlorine solution were added, and the bottle was filled up and capped headspace free with a PTFE-lined cap. Samples were incubated for 168 hours at 20°C in the dark with the exception of the samples incubated at 7°C. At the end of the incubation period the chlorine residual was measured (*24*) and 100-150 µL of sulphur-reducing agent (sodium sulphite, 100 g/L) was added to the samples to destroy the chlorine residual whilst not degrading the five HAAs measured (*12*). Samples were prepared in duplicate independently.

For the measurement of THMs, 5 mL of water sample was transferred into a 10 mL vial allowing 5 mL of headspace. Following this, the samples were analysed by headspace GC-MS. Samples were prepared in duplicate and analysed in triplicate. HAA samples were first converted to their protonated forms before processing the extraction with organic solvent and deriving to form methyl esters. The method used for the derivatisation was adapted from USEPA Method 552.3.

Analytical Methods

THMs were analysed using a Varian Saturn 2200 (ion-trap) gas chromatograph-mass spectrometer (GC-MS). The samples were heated and agitated by CTC CombiPal to 60°C for 30 minutes. 500 μL of headspace was removed by heated syringe and injected with a 10:1 split, separation was performed by a BPX5 column (SGE; 30 m × 0.25 mm id × 0.25 μm film thickness) with a helium carrier gas at a column flow rate of 1.1 ml/min. The injector temperature was 250°C; the initial oven temperature was 45°C for 2 minutes followed by a 10°C per minute temperature ramp to 90°C. The MS was operated in the electron ionisation (EI) mode. The ion-trap temperature was set at 230°C and the electron energy was 70 eV. Mass spectra were collected in full scan mode (33-300 amu). The ions of 83, 129 and 173 m/z were selected as quantification ions. Quantification of THMs was achieved by comparing the chromatograms of the samples with the calibration curves from standards.

HAA standards were run with a GC Perkin Elmer AutoSystem XL coupled with a TurboMass Gold MS using the method reported in a study of Xie (*25*). HAAs were also measured on a gas chromatograph with a micro electron capture detector (Agilent 6890 GC-μECD). A volume of 1 μL was injected with the injector at 200°C with a 5:1 split, separation was performed by a BPX5 column (SGE; 30 m × 0.25 mm id × 0.25 μm thickness) with a helium carrier gas at a column flow rate of 1.1 ml/min. The initial oven temperature was 35°C followed by a 5°C per minute temperature ramp to 220°C and held for 1 minute. The detector temperature was 230°C and the rate of data collection 20 Hz.

HAA samples were run in parallel by comprehensive two dimensional GC-MS utilising a Leco Pegasus VI GC×GC-time of flight mass spectrometer (GC×GC-TOFMS). GC×GC separation was performed using an Agilent 6890 GC with a Leco GC×GC modulator fitted coupled to a Pegasus IV time-of-flight mass spectrometer (LECO Corporation). The GC injector was operated in splitless mode with a column flow rate of 1.0 ml/min and held at 200°C. GC×GC separation utilised a non-polar column and a polar column a BPX5 (SGE; 30 m × 0.25 mm × 0.25 μm) and a BPX50 (SGE; 1.8 m × 0.1 mm × 0.1 μm) respectively. The GC oven temperature was held for 1 minute at 35°C and ramped to 220°C at a rate of 5°C/min and then held for 1 minute, the second column was ramped at 30°C above the first column. Modulation time was 4 seconds. Mass spectra were acquired in electron ionisation mode from 33 to 400 amu with an acquisition rate of 133 spectra per second.

Results and Discussion

Comparison of Analytical Devices for HAA Measurements

Published methods for HAA analysis include the use of GC-MS (*25*) and GC-ECD (*13*). Here we have evaluated both methods. Firstly, GC-MS (Perkin Elmer Turbomass) was used to analyse six out of nine HAA standards (monochloroacetic acid (MCAA), monobromoacetic acid (MBAA), dichloroacetic acid (DCAA), trichloroacetic acid (TCAA), bromochloroacetic acid (BCAA) and dibromoacetic acid (DBAA)) that had been derived to form their methyl esters. However it was difficult to quantify HAA methyl esters using this method (*25*). The peaks were not well resolved nor was the S/N ratio sufficient. The GC-MS was run in the selective ion monitoring (m/z 59) mode but this did not significantly improve the sensitivity. It was not possible to determine limits of detection for this method.

In order to confirm the findings, samples were run in parallel using another GC-MS (Agilent 5973). The results were comparable.

To investigate the difficulties further, samples were run using a Leco Pegasus 4D GC×GC-TOFMS. This machine uses two GC columns to separate analytes based on volatility as well as polarity. The derived HAA methyl ester peaks could be well observed as they had been separated from the interfering material. The interfering material had a greater intensity than some of the derived methyl esters and also eluted at retention times that overlapped with the derived HAA methyl esters (Figure 1).

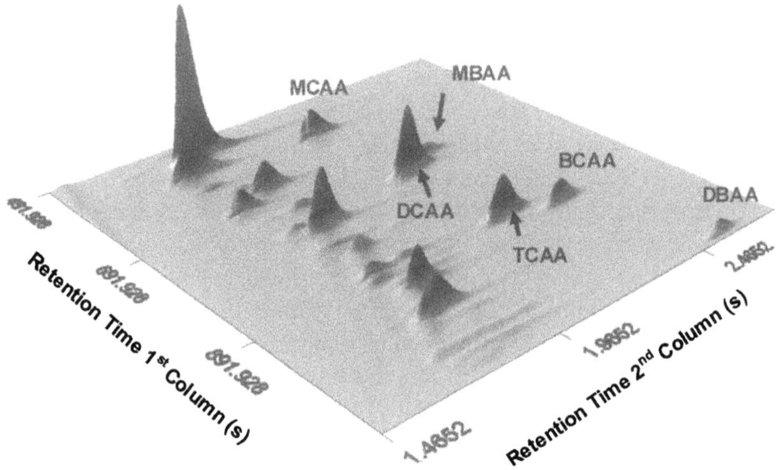

Figure 1. A partially reconstructed mass chromatogram (m/z 59) of a derivatised HAA_6 standard (Leco Pegasus 4D GC×GC-TOFMS)

The interfering peaks are thought to be incurred from the derivatisation procedure but, to date, have not been identified. Due to the difficulties encountered using the GC-MS method in this study all the HAA analyses have been undertaken using GC-ECD. MCAA was not quantified because of results inconsistency and analytical difficulties but the limit of detection for five HAAs was 1.1 µg/L and 1.0 µg/L for the four THMs.

Water Characterisation

Characteristics of the waters are summarised in Table I. The concentration of organic matter was greater in the lowland (4.7 mg/L) than in the upland water (2.1 mg/L). NOM fractionation indicated that organic matter in the upland water had a higher hydrophilic content than the lowland water which had significant transphilic content (Table I). Hwang et al. (*26*) reported that the transphilic fraction of intermediate polarity is generally more hydrophobic than hydrophilic but this statement was highly dependent on the water source.

Table I. Water Characterisation

Parameters	*Upland water*	*Lowland water*
pH	6.7	8.0
NPOC[a] (mg/L)	2.1	4.7
UV_{254} (m^{-1})	4.8	5.9
$SUVA_{254}$[b] (m^{-1} L mg^{-1} C)	2.3	1.3
Alkalinity (mg/L of $CaCO_3$)	6	188
Bromide content (μg/L)	34	206
THM-FP[c] (μg/L)	72	89
HAA-FP[c] (μg/L)	104	84
Hydrophobic – Neutral (%)	4	2
Hydrophobic – Acid (%)	19	23
Transphilic (%)	8	31
Hydrophilic – Base (%)	2	4
Hydrophilic – Acid + Neutral (%)	67	40

[a] Non purgeable organic carbon

[b] Specific ultraviolet absorbance

[c] pH 7 and at 20°C

The reactivity with respect to DBP formation potential can be characterised with $SUVA_{254}$ (*27*). A high $SUVA_{254}$ value is an indicator of a high reactivity toward DBP production (*12*). The lowland water had a lower $SUVA_{254}$ value than the upland water (1.3 and 2.3 respectively), but both waters had overall relatively low $SUVA_{254}$ values as compared to other fresh waters reported in the literature (*27*).

As expected, the two waters differed not only in their alkalinity but also in their bromide concentration. The bromide concentration of the lowland water (206 μg/L) was six times higher than that of the upland water (34 μg/L). It is therefore likely that the lowland water will produce more brominated species.

HAA_5 (MBAA, DCAA, TCAA, BCAA, DBAA) and THM_4 concentrations were similar in the lowland water after 168 hours contact time at pH 7, whereas the upland water had the potential to form more HAA_5 than THM_4. It should be considered that the quantification of the three remaining HAAs (not measured here) might contribute to higher concentration of the total HAAs in the lowland water considering the high level of bromide. Malliarou et al. (*6*) found that some regions of the UK produced an average total level of THMs higher than the HAAs, while the contrary was found in other regions. This highlights the differences observed in different geographical locations in the UK.

Parameters Affecting the Formation of THMs and HAAs

Impact of pH

It is well known that the formation of DBPs is strongly dependent on the chlorination pH (*11*, *28*, *29*). Here we have shown (Figure 2) that increasing pH from 6 to 8 decreased HAA_5 in the lowland water by 15%. DCAA and BCAA were found to be more affected by pH than TCAA. DCAA and BCAA concentrations at pH 8 were significantly lower than at pH 6 and 7. Liang and Singer (*13*) reported that increasing pH from 6 to 8 had a very little effect on the formation of the monohaloacetic acid (XAA) and dihaloacetic acid (X_2AA) species, but significantly decreased the formation of the trihaloacetic acid (X_3AA) species. In the literature, DCAA formation was reported to be highest at pH 7 (*11*) which again is true of the results found here.

For the upland water (Figure 3) HAA_5 formation was 14% greater at pH 6 than at 8 and the lowest concentration was found at pH 7. In the upland water, only TCAA, DCAA and BCAA were detected due to the low bromide content in this water. The DCAA concentration was higher at pH 6 and similar at pH 7 and 8. TCAA increased by 28% with increasing pH, which is contrary to the literature (*13*, *30*) and the results found for the lowland water.

Trussell and Umphres (*31*) reported that the formation of THMs consists of alternate hydrolysis and halogenation steps. All these reactions are favoured under alkaline condition, thus more THMs are formed at higher pH, which is illustrated by the results found here (Figure 4). The impact of pH is limited in the upland water compared to the lowland water which could be explained by the difference of organic matter responsible for the THM formation and its likelihood to undergo hydrolysis and halogenation reactions.

Impact of bromide

The effect of bromide concentration on HAA and THM formation and speciation was investigated by spiking the lowland and the upland water with 200 µg/L of bromide. In the lowland water, the addition of bromide had a slight impact (10% decrease) on the total concentration of HAAs measured. Less

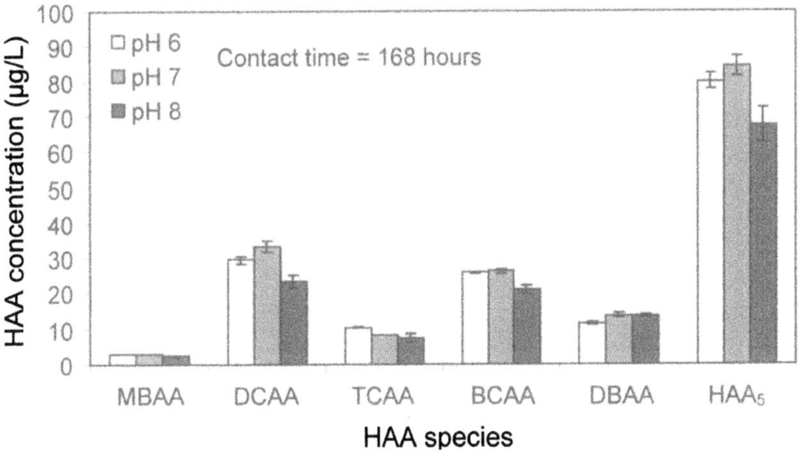

Figure 2. Comparison of pH effect on the formation of measured HAAs in the lowland water

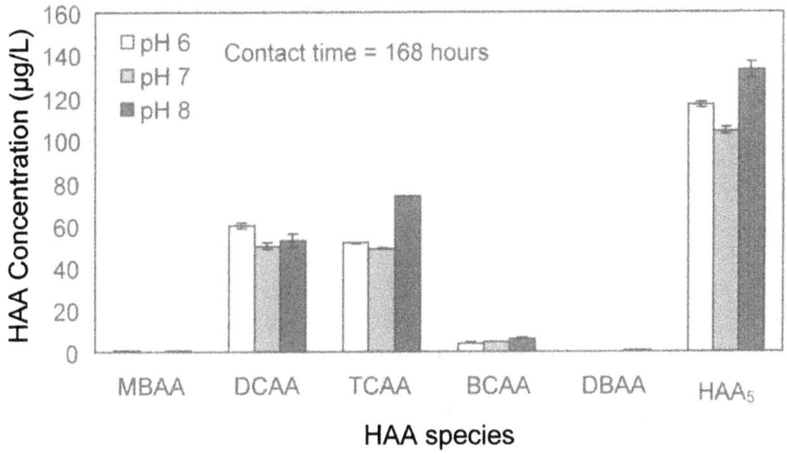

Figure 3. Comparison of pH effect on the formation of measured HAAs in the upland water

DCAA and TCAA were formed, whereas more brominated HAA species were produced as expected. The addition of bromide had a greater impact in the upland water than in the lowland water. With the upland water the concentration of HAA_5 decreased (Figure 5) with a switch from DCAA and TCAA to brominated species MBAA, BCAA and DBAA.

It was reported by Hua et al. (*32*) that the total concentration of the five regulated HAAs in the US (MCAA, MBAA, DCAA, TCAA and DBAA) decreased as bromide concentration increased because of the number of brominated species measured. This applies here but the exception is that BCAA is included in the total HAAs and not MCAA. However the same study reported that addition of bromide increased the total HAA_9 (five HAAs regulated in the US, plus BCAA, bromodichloroacetic acid (BDCAA), chlorodibromoacetic acid (DBCAA) and tribromoacetic acid (TBAA)) yield between 0 and 35%.

Bromine was reported by Cowman and Singer (*8*) to be more reactive than chlorine in substitution and addition reactions that form HAAs, thus the inclusion of bromine shifts the speciation of the HAA towards the brominated species.

The formation of THM is also affected by the addition of bromide. Hua et al. (*32*) reported that increasing initial bromide levels resulted in a substantially increased THM molar concentration between 14% and 74%. Here the total THM weight concentration increased by 60% in the lowland water and by 54% in the upland water. In the upland water, only the brominated species increased, whereas all the brominated species and chloroform were augmented in the lowland water. The difference observed in bromide incorporation is likely to be due to the number of HAA species measured.

Impact of temperature

Reducing the incubation temperature from 20°C to 7°C, resulted in a reduction of both HAAs and THMs. The concentration of HAAs and THMs dropped by 59% and 43% respectively in the lowland water (Figure 6) and by 43% and 53% respectively in the upland water.

El-Dib and Ali (*33*) reported that the effect of temperature (rise between 0 and 30°C) on the THM yield was rather limited compared with data reported by other investigators (*34*) and concluded that the differences were due to the nature of organic precursors liable to be found in the water.

Dojlido et al. (*35*) reported that the concentrations of HAAs were seasonally dependant. During the winter season (1°C) they found levels of ~ 0.63 μg/mg C whereas in the summer (23°C), concentrations reached ~ 7.4 μg/mg C. In the UK, the effect of season on HAA formation has not been determined but the results shown here (Figure 6) indicate there may be a seasonal effect. A UK study by Malliarou et al. (*6*) concluded that THM concentrations were not correlated with temperature but the correlation between HAA levels and temperature was significant. However, no actual temperature values were reported.

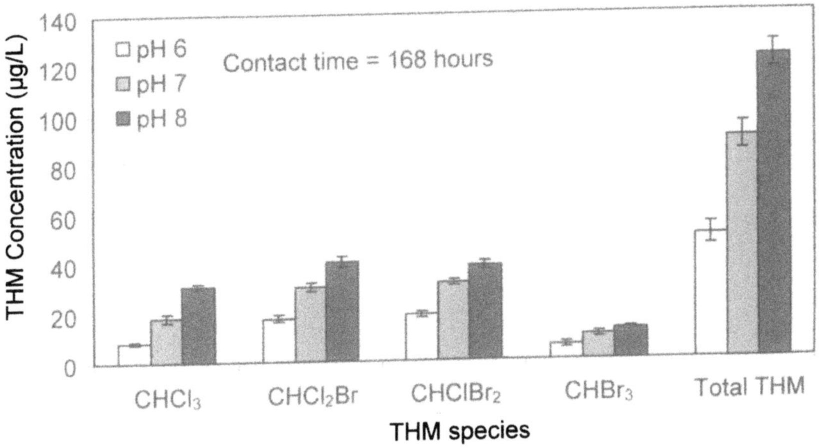

Figure 4. Comparison of pH effect on the formation of THM_4 in the lowland water

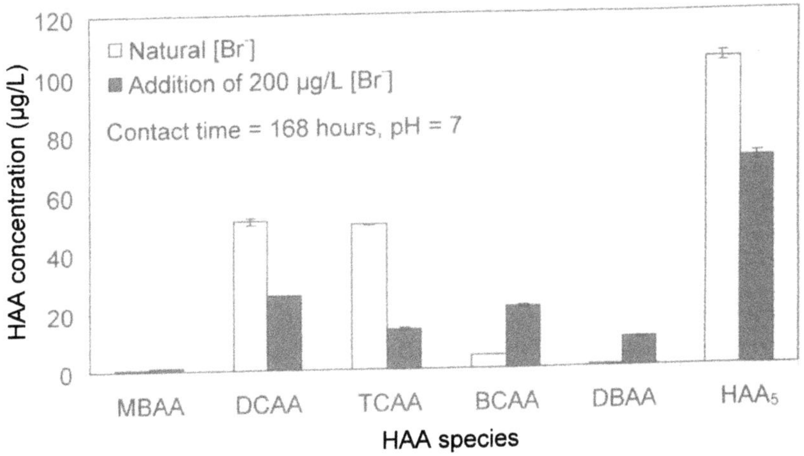

Figure 5. Impact of bromide on the formation of measured HAAs in the upland water

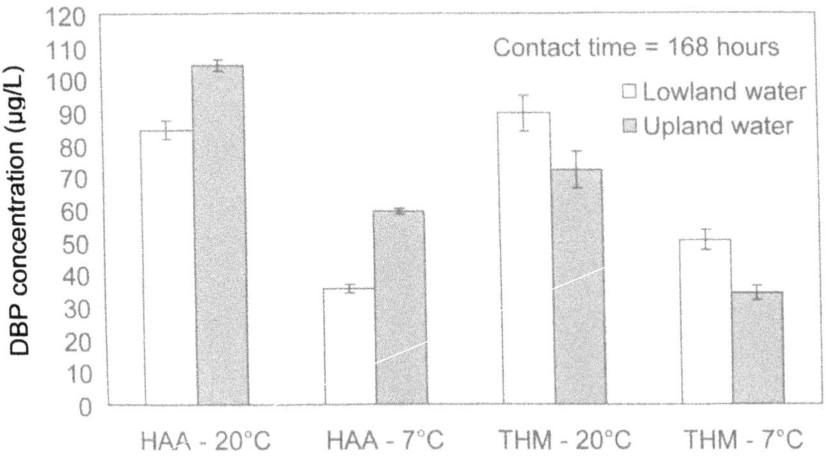

Figure 6. Temperature effect on the DBP formation in lowland and upland water

Conclusions

Water was collected from two different geographical locations. The upland water NOM was primarily hydrophilic, whereas the lowland water had higher transphilic content.

It was not possible to use GC-MS for analysis of HAAs at low µg/L levels. Thus GC-ECD was regarded as the most suitable technology. Levels of HAA_5 were almost identical to the level of THMs in the lowland water, whereas the concentration of HAA_5 was higher than the concentration of THMs in the upland water.

The greatest pH impact was observed in the formation of THMs in the lowland water. Although THM formation was significantly affected in both waters, HAA_5 formation did not exhibit a strong pH effect. When changing the pH, the lowland water behaved as predicted in the reported literature with regard to HAA formation. However the behaviour of the upland water did not follow the same pattern. The differences in the precursor characteristics may account for these observations.

Addition of bromide to the water leads to a higher percentage of brominated HAAs and THMs and a total increase in concentration if all THMs and HAAs are measured. The impact on HAAs will vary depending on the number of brominated species measured.

A reduction in temperature resulted in a major decrease in DBP formation.

The majority of the data reported here followed the literature trends reported in the US. Future work at Cranfield will focus on DBP formation by each of the isolated fractions. This will give an insight into the specific organics responsible for the DBP formation in UK drinking water. Following this, treatment could be adapted to remove HAA precursors should regulation occur in the UK. Future research in the UK should build on the information obtained here by investigating different water sources.

Acknowledgements

This work was funded by Anglian Water, Northumbrian Water Limited, Severn Trent Water, United Utilities and Yorkshire Water Services.

References

1. Krasner, S. W.; McGuire, M. J.; Jacangelo, J. G.; Patania, N. L.; Reagan, K. M.; Marco Aieta, E. *J. Am. Water Works Assoc.* **1989**, *81*, 41-53.
2. Drinking Water Inspectorate **2005**, available at: www.dwi.gov.uk/regs/pdf/GuidanceMay05.pdf.
3. Weinberg, H. S.; Krasner, S. W.; Richardson, S. D.; Thruston, A. D. J., The Occurrence of Disinfection By-Products (DBPs) of Health Concern in Drinking Water: Results of a Nationwide DBP Occurrence Study, Report EPA/600/R-02/068, 2002.
4. Lin, C. F.; Huang, Y. J.; Hao, O. J. *Wat. Res.* **1999**, *33*, 1252-1264.
5. WEKNOW **2003**, available at: www.weknow-waternetwork.com.
6. Malliarou, E.; Collins, C.; Graham, N.; Nieuwenhuijsen, M. J. *Wat. Res.* **2005**, *39*, 2722-2730.
7. Hua, G.; Reckhow, D. A. *Environ. Sci. Technol.* **2007**, *41*, 3309-3315.
8. Cowman, G. A.; Singer, P. C. *Environ. Sci. Technol.* **1996**, *30*, 16-24.
9. Fleischacker, S. J.; Randtke, S. J. *J. Am. Water Works Assoc.* **1983**, *75*, 132-138.
10. Carlson, M.; Hardy, D. *J. Am. Water Works Assoc.* **1998**, *90*, 95-106.
11. Krasner, S. W. Chemistry of Disinfection By-Product Formation. In *Formation and Control of Disinfection By-Products in Drinking Water*; AWWA, CO, 1999.
12 Singer, P. C.; Weinberg, H. S.; Brophy, K.; Liang, L.; Roberts, M.; Grisstede, I.; Krasner, S. W.; Baribeau, H.; Arora, H.; Najm, I. Relative Dominance of HAAs and THMs in Treated Drinking Water; Report 90844, AWWA, CO, 2002.
13. Liang, L.; Singer, P. C. *Environ. Sci. Technol.* **2003**, *37*, 2920-2928.
14. Fearing, D. A.; Goslan, E. H.; Banks, J.; Wilson, D.; Hillis, P.; Campbell, A. T.; Parsons, S. A. *J. Environ. Eng.* **2004**, *130*, 975-982.
15. Sharp, E. L.; Parsons, S. A.; Jefferson, B. *Environ. Pollution* **2006**, *140*, 436-443.
16. Parsons, S. A.; Jefferson B. *Introduction to Potable Water Treatment Processes*; Blackwell Publishing Ltd Ed., Oxford, UK, 2006.

17. Sharp, E. L.; Parson, S. A.; Jefferson, B. *Wat. Sci. Technol.* **2006**, *53*, 67-76.
18. Goslan, E. H.; Fearing, D. A.; Banks, J.; Wilson, D.; Hills, P.; Campbell, A. T.; Parsons, S. A. *J. Wat. Supply: Research and Technology - AQUA* **2002**, *51*, 475-482.
19. Randtke, S. J. Disinfection By-Product Precursor Removal by Coagulation and Precipitative Softening Formation and Control of Disinfection By-Products in Drinking Water; AWWA, CO, 1999.
20. Liu, Y.; Mou, S. *J. Chromato. A* **2003**, *997*, 225-235.
21. Hozalski, R. M.; Zhang, L.; Arnold, W. A. *Environ. Sci. Technol.* **2001**, 35, 2258-2263.
22. Leenheer, J. A.; Noyes, T. I.; Rostad, C. E.; Davisson, M. L., *Biogeochem.* **2004**, *69*, 125-141.
23. Malcolm, R. L.; MacCarthy, P. *Environ. Int.* **1992**, *18*, 597-607.
24. Greenberg, A. E.; Clesceri, L. S.; Eaton, A. D., Ed.; *Standard Method for the Examination of Water and Wastewater,* American Public Health Association; American Water Works Association and Water Environment Federation 18th ed., DC, 1992.
25. Xie, Y. *Wat. Res.* **2001**, *35*, 1599-1602.
26. Hwang, C. J.; Amy, G. L.; Bruchet, A.; Croué, J.; Krasner, S. W.; Leenheer, J. A. Polar NOM: characterization, DBPs, treatment; AwwaRF and AWWA, Denver, CO, 2001.
27. Edzwald, J. K.; Tobiason, J. E. *Wat. Sci. Technol.* **1999**, *40*, 63-70.
28. Singer, P. C. Formation and Control of Disinfection By-Products in Drinking Water, AWWA, CO, 1999.
29. Xie, Y., Ed. *Disinfection by-products in drinking water: Formation, analysis and control,* Lewis Publishers, MI, 2003.
30. Zhuo, C.; Chengyong, Y.; Junhe, L.; Huixian, Z.; Jinqi, Z. *Chemosphere* **2001**, *45*, 379-385.
31. Trussell, R. R.; Umphres, M.D. *J. Am. Water Works Assoc.* **1978**, *70*, 604-612.
32. Hua, G.; Reckhow, D. A.; Kim, J. *Environ. Sci. Technol.* **2006**, *40*, 3050-3056.
33. El-Dib, M. A.; Ali, R. K. *Wat. Res.* **1995**, *29*, 375-378.
34. Urano, K.; Takemasa, T.*Wat. Research* **1986**, *20*, 1555-1560.
35. Dojlido, J.; Zbiec, E.; Swietlik, R. *Wat. Res.* **1999**, *33*, 3111-3118.

Chapter 8

Relationship between Brominated THMs, HAAs, and Total Organic Bromine during Drinking Water Chlorination

Guanghui Hua[1], and David A. Reckhow[2]

[1]Jones Edmunds and Associates, Inc., 730 NE Waldo Road, Gainesville, FL 32641
[2]Department of Civil and Environmental Engineering, University of Massachusetts, Amherst, MA 01003

Naturally occurring bromide present in drinking water can be quickly oxidized by chlorine to bromine, which can react with natural organic matter to form brominated disinfection byproducts (DBPs). This study investigated the relationship between known specific brominated DBPs and total organic bromine (TOBr) formed during chlorination of NOM isolates and natural waters in the presence of various levels of bromide. Unknown TOBr (UTOBr) is determined as the difference between TOBr and bromine incorporated into measured specific DBPs. The unknown TOBr fraction, as represented by the ratio of UTOBr to TOBr increased with increasing initial bromide concentrations during chlorination. The majority of organic bromine was incorporated into known specific DBPs during chlorination of low bromide containing waters. Hydrophilic and low molecular weight (MW) NOM was more reactive with bromine as measured by the formation of trihalomethanes and haloacetic acids than corresponding hydrophobic and high MW NOM. However, hydrophobic and high MW NOM formed more TOBr than hydrophilic and low MW NOM. Water utilities should work to remove both hydrophobic and hydrophilic NOM in the water sources to reduce the formation of chlorinated and brominated DBPs.

Introduction

Bromide ions are nearly ubiquitous in natural waters. During drinking water chlorination, bromide can be quickly oxidized to hypobromous acid (HOBr) by chlorine (*1*). Both hypochlorous acid (HOCl) and HOBr react readily with natural organic matter (NOM) to form a series of chlorinated and brominated disinfection byproducts (DBPs) (*2, 3*). Trihalomethanes (THMs) and haloacetic acids (HAAs) are the two most abundant known specific DBPs identified in chlorinated waters. It has been shown that the formation of THMs and HAAs shifts to more brominated species in the presence of increasing initial bromide concentrations (*2-4*). In addition to THMs and HAAs, many other halogenated DBPs have also been identified in chlorinated drinking waters, such as haloketones, haloacetonitriles, chloropicrin, cyanogen halide, and chloral hydrate (*5*). These specific DBPs are usually present at much lower concentrations than the THMs and HAAs. The known specific DBPs together account for roughly 50% of the total organic halogen (TOX) formed during drinking water chlorination (*6, 7*).

Bromide concentrations can affect the formation and distribution of TOX during chlorination. It has been shown that the fraction of TOX that is attributed to THMs and HAAs increased substantially with increasing bromide concentrations (*4, 8*). Conventional TOX analysis is a non-specific measurement of the total amount of halogenated organic DBPs. A method that combines adsorption, combustion and ion chromatography detection has been developed to measure total organic chlorine (TOCl) and total organic bromine (TOBr) in drinking waters (*9*). Toxicological studies have shown that brominated DBPs may be more toxic than chlorinated analogues (*10*). Therefore, it is important to understand the impact of bromide on the formation and distribution of TOCl and TOBr during chlorination. This work was designed to investigate the relationship between brominated THMs, HAAs, and TOBr during chlorination at various bromide levels. The relationship between chlorinated THMs, HAAs, and TOCl was also examined.

Experimental Methods

Three sets of experiments were conducted in this research. First, two natural waters were collected and chlorinated after being spiked with various levels of bromide. Secondly, NOM in two natural waters was isolated using XAD-8 resin and ultrafiltration membranes. Chlorination was conducted on each NOM fraction solutions after being spiked with various levels of bromide. Finally, six natural waters were collected and chlorinated at ambient bromide concentrations.

Chlorination of Bromide Spiked Natural Waters

Raw waters were collected from the drinking water treatment plant intakes at the city of Winnipeg, Manitoba, and Tulsa, OK. Table I presents the characteristics of these two waters. Specific ultraviolet absorbance (SUVA) was calculated from UV absorbance at 254 nm (UV_{254}) divided by the dissolved organic carbon (DOC) concentration.

Table I. Characteristics of Winnipeg and Tulsa Waters

Sample	*DOC (mg/L)*	*UV_{254} (cm^{-1})*	*SUVA (L/mg/m)*	*Br^- (μg/L)*
Winnipeg	8.5	0.131	1.6	9
Tulsa	5.1	0.160	3.1	63

These two waters were spiked with bromide ions at levels of 0, 2, 10, and 30 μM (0, 160, 800, and 2400 μg/L, respectively). Chlorination was conducted in 1 L chlorine demand-free, glass bottles on each sample buffered with 1 mM phosphate at pH 7. The chlorine doses were determined by preliminary chlorine demand tests on raw waters. The target free chlorine residual was 0.5 mg/L after 48 h at 20 °C. The requisite doses were 6.2 and 5.0 mg/L for Winnipeg water and Tulsa water, respectively. A stock solution of sodium hypochlorite (Fisher Scientific, Fairlawn, NJ) was standardized by DPD ferrous titrimetric method according standard method 4500-Cl F (*11*). After being dosed with chlorine, samples were stored headspace-free at 20 °C in the dark for 48 h.

Chlorination of NOM Fractions

Raw waters were collected from the drinking water treatment plant intakes at the city of Springfield, MA, and Tampa, FL. A portion of raw water was acidified to pH 2 using concentrated sulfuric acid and then passed through XAD-8 resin (Rohm and Haas, Philadelphia, PA). The column distribution coefficient was kept at 50 (*12*). Effluent from the XAD-8 resin was referred to as the hydrophilic fraction. The fraction referred to as the hydrophobic NOM was retained by XAD-8 resin and eluted with 0.1 N NaOH in the reverse direction. The pH of the two fractions was adjusted to 7 using sulfuric acid or sodium hydroxide immediately after extraction. The DOC concentration of the hydrophobic fraction was adjusted to the same level as the hydrophilic fraction using deionized water.

Another portion of raw water was fractionated using Millipore YM3 ultrafiltration (UF) membranes with a molecular weight (MW) cut-off 3 kDa (Amicon, Beverly, MA). UF was performed with stirred 200 mL Amicon cells. The nitrogen pressure was maintained at 50 psi. For an initial sample volume of 200 mL, filtration was stopped when the volume of retentate decreased to 50 mL. Permeate was referred to as the MW <3 kDa fraction. Deionized water was added to the cell to bring the volume back to 200 mL and the filtration was continued until the volume decreased to 50 mL again. This flushing process was repeated twice further to remove compounds with MW lower than the membrane cut-off. The retentate was referred to as the MW> 3 kDa NOM. The DOC of the MW>3 kDa fraction was adjusted to the same level as the MW<3 kDa fraction. Each fraction was chlorinated at pH 7 after being dosed with a high concentration of bromide (Springfield and Tampa fractions: 150 and 800 μg/L, respectively). Table II presents the characteristics of the NOM fractions prior to chlorination. Chlorine doses were 3.5 and 10.0 mg/L for Springfield and Tampa NOM fractions, respectively. Chlorination was conducted in 300 mL glass-stoppered bottles. After being dosed with chlorine, samples were stored headspace-free at 20 °C in the dark for 48 h.

Table II. Characteristics of NOM Fractions Prior to Chlorination

Sample	*Fraction*	*DOC (mg/L)*	*UV_{254} (cm^{-1})*	*SUVA (L/mg/m)*	*Br^- (μg/L)*
Springfield	Hydrophobic (64%)	1.2	0.049	4.2	150
	Hydrophilic (36%)	1.2	0.030	2.5	150
	MW>3 kDa (82%)	0.6	0.026	4.2	150
	MW<3 kDa (18%)	0.6	0.016	2.5	150
Tampa	Hydrophobic (67%)	4.3	0.199	4.7	800
	Hydrophilic (33%)	4.3	0.109	2.5	800
	MW>3 kDa (67%)	4.2	0.189	4.5	800
	MW<3 kDa (33%)	4.2	0.129	3.0	800

Chlorination of Natural Waters

Six geologically distinct natural waters were collected from the drinking water plant intakes at the following locations: Dallas, TX; Elgin, IL; Newport News, VA; Waco, TX; Cambridge, MA and Repentigny, Quebec. The chemical characteristics of these waters are summarized in Table III. Chlorination was conducted in 300 mL chlorine demand-free, glass-stoppered bottles. All samples were buffered with 1 mM phosphate at 7. Chlorine doses were determined by chlorine demand tests on each sample. The target free chlorine residual was 0.5 mg/L after 48 h at 20 °C. After being dosed with chlorine, samples were stored headspace-free at 20 °C in the dark for 48 h.

Table III. Characteristics of Natural Waters

Sample	*DOC (mg/L)*	*UV_{254} (cm^{-1})*	*SUVA (L/mg/m)*	*Br^- (µg/L)*
Dallas, TX	4.5	0.074	1.7	89
Elgin, IL	8.1	0.193	2.4	38
Newport News, VA	4.3	0.120	2.8	32
Waco, TX	4.0	0.114	2.9	45
Cambridge, MA	4.2	0.141	3.4	95
Repentigny, Quebec	7.1	0.313	4.4	46

Analytical Methods

All samples were analyzed for THMs, HAAs, dihaloacetonitriles (DHANs), haloketones (HKs), chloropicrin (CP), TOX, TOCl, and TOBr. Four THMs ($CHCl_3$, $CHBrCl_2$, $CHBr_2Cl$, and $CHBr_3$), three DHANs (dichloro-, bromochloro- and dibromoacetonitriles (DCAN, BCAN, and DBAN)), two HKs (dichloro- and trichloropropanone), and chloropicrin were quantified by liquid/liquid extraction with pentane followed by gas chromatography with electron capture detection (GC/ECD) according to USEPA Method 551.1. Nine HAAs (monochloro-, monobromo-, dichloro-, bromochloro-, dibromo-, bromodichloro-, dibromochloro-, trichloro- and tribromoacetic acid (MCAA, MBAA, DCAA, BCAA, DBAA, BDCAA, DBCAA, TCAA, and TBAA, respectively)) were analyzed by liquid/liquid extraction with methyl *tertiary*-butyl ether (MTBE) followed by derivatization with acidic methanol and analysis by GC/ECD according to USEPA Method 552.2. TOX was determined by an adsorption-pyrolysis-titrimetric method using a Euroglas 1200 TOX Analyzer (Euroglas, Delft, Netherlands) based on Standard Method 5320 B (*11*). TOCl and TOBr were analyzed by an adsorption, combustion, and off-line ion chromatographic method (*9*). The DOC concentrations were determined with a Shimadzu TOC-5000 Analyzer (Shimadzu Corp., Kyoto, Japan) according to Standard Method 5310 B (*11*). UV_{254} was measured using a Hewlett Packard 8425A Diode Array Spectrophotometer based on Standard Method 5910 B (*11*). Bromide ion concentrations were analyzed with an AS14A anion separation column and an AG14A guard column using a DX-500 ion chromatography system (Dionex, Sunnyvale, CA).

Results and Discussion

Chlorination of Bromide Spiked Natural Waters

Figure 1 shows the effect of bromide on the concentrations of chlorine and bromine incorporated into THMs and HAAs during chlorination of Tulsa water. Concentrations of chlorine and bromine were normalized to their equivalent value as μg Cl/L, removing the bias imposed by differing atomic weights of chloride and bromide. Increasing initial bromide concentration shifted the THM and HAA distribution from chlorinated species to mixed species and then to brominated species. Brominated species became dominant for THMs and HAAs at the level of 30 μmol/L bromide addition. The mole ratios of bromine to chlorine incorporated into THMs (THM-Br/THM-Cl) and HAAs (HAA-Br/HAA-Cl) were 1.7 and 1.3, respectively, for a bromide level of 10 μmol/L. The corresponding initial Cl_2 to Br mole ratio was 6.5. This suggests that bromine incorporation into THMs and HAAs is more efficient than chlorine incorporation. The ratio of THM-Br to THM-Cl increased to 11 at the bromide addition of 30 μmol/L. The corresponding HAA-Br to HAA-Cl ratio was only 2.5. It appears that bromine incorporation into THMs is more efficient than into HAAs. Similar results were also observed for Winnipeg water.

The distribution among mono, di, and trihalogenated HAAs at varying levels of bromide is shown in Table IV. The trihalogenated HAAs (THAAs) constituted the largest fraction of the total HAAs (49-64%), the mole fraction of the dihalogenated HAAs (DHAAs) was between 31 and 47% of the total HAAs, and the monohalogenated HAAs (MHAAs) were below 5% of the total HAAs for both waters. The mole fractions of DHAAs and THAAs were not

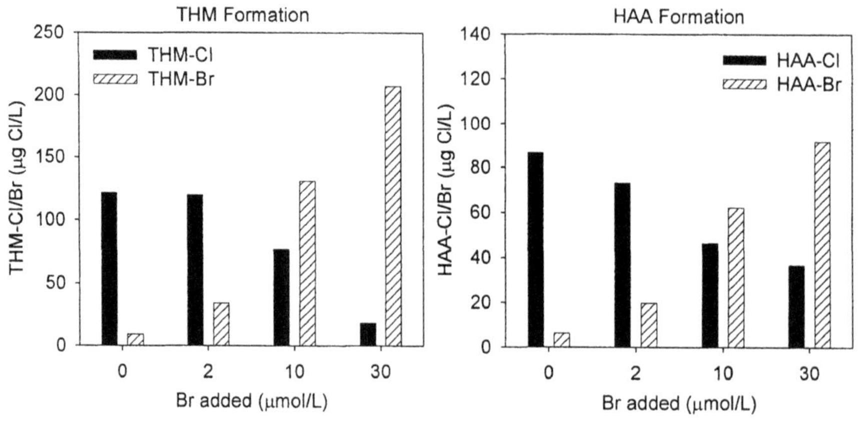

Figure 1. Effect of bromide on the formation of THMs and HAAs (Tulsa water)

Table IV. Effect of Bromide on the Mole Fraction of HAAs

Sample	*HAAs*	*Br⁻ (0 μM)*	*Br⁻ (2 μM)*	*Br⁻ (10 μM)*	*Br⁻ (30 μM)*
Winnipeg Water	MHAAs	5%	4%	4%	5%
	DHAAs	45%	47%	42%	31%
	THAAs	50%	49%	54%	64%
Tulsa Water	MHAAs	4%	4%	3%	3%
	DHAAs	43%	47%	43%	35%
	THAAs	53%	49%	54%	62%

significantly affected when the added bromide was increased from 0 to 10 μM. However, there was a clear tendency toward higher fractions of THAAs but lower fractions of DHAAs when the bromide addition was further increased to 30 μM. This result is different from the work of Cowman and Singer (*3*), where the distribution of HAAs into mono, di, and trihalogenated HAAs was independent of bromide concentration (0-25 μM) during chlorination of humic substances. Differences in organic matter and chlorination conditions may be responsible. A shorter reaction time (24 h) was used in Cowman and Singer's study. It has been shown that dihalogenated HAAs form faster than trihalogenated HAAs during chlorination (*13*).

Figure 2 shows the effect of bromide on the formation of DHANs. The distribution and yield of DHANs were greatly influenced by the initial bromide levels. Increasing bromide concentrations shifted the distribution of DHANs from DCAN to BCAN and then to DBAN. The molar yields of the total DHANs increased with elevated bromide concentrations up to 101% for Winnipeg water

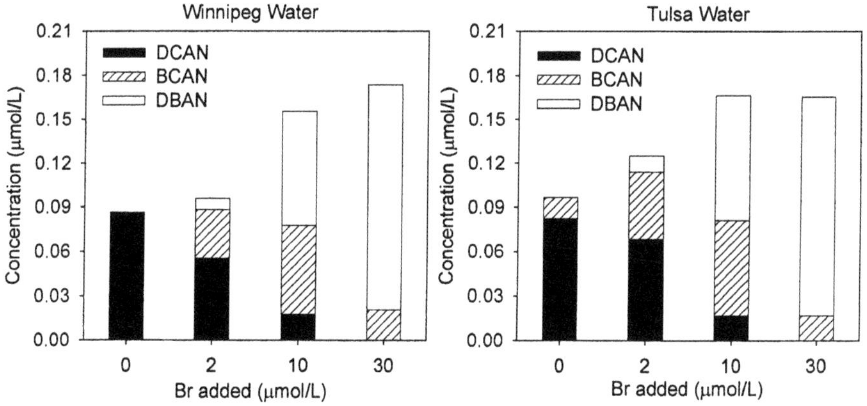

Figure 2. Effect of bromide on the formation of DHANs

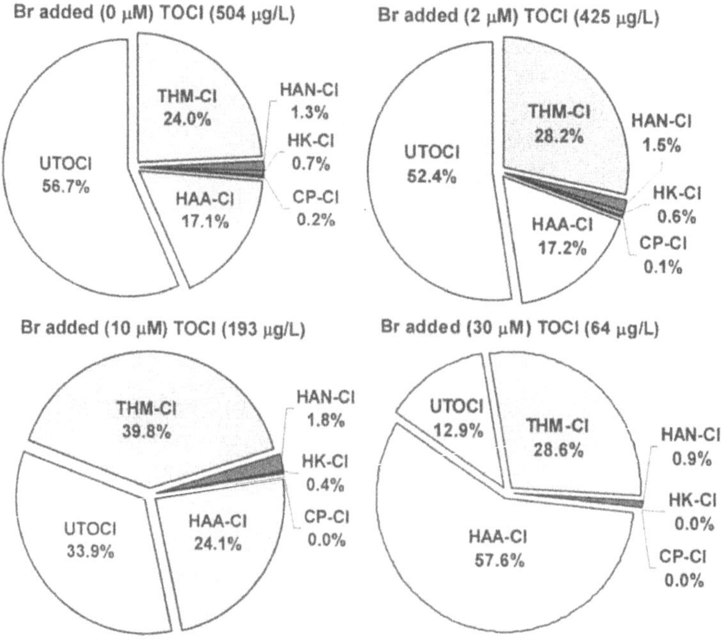

Figure 3. Effect of bromide on the formation of TOCl (Tulsa water)

and 73% for Tulsa water. The magnitude of the increases of DHANs was higher than that of THMs (58-74%) and HAAs (18-35%) at the highest bromide addition (30 μmol/L) for the two waters.

Figure 3 shows the impact of bromide on the formation and distribution of TOCl during chlorination of Tulsa water. Unknown TOCl (UTOCl) was calculated as the difference between TOCl and chlorine incorporated into measured specific DBPs. TOCl concentrations decreased from 504 μg/L to 64 μg/L when increasing bromide addition from 0 to 30 μmol/L. Approximately 57% of the TOCl was attributed to UTOCl during chlorination at ambient bromide concentration (63 μg/L) Chlorinated THMs and HAAs together accounted for 41% of the TOCl. The percentage of UTOCl decreased with increasing bromide concentrations. UTOCl accounted for only 13% of the TOCl at a level of 30 μmol/L bromide addition, whereas 86% of the TOCl was attributed to THMs and HAAs. The majority of TOCl was present as known specific DBPs during chlorination at high levels of bromide.

The reverse results were observed for the TOBr formation of Tulsa water (Figure 4). Unknown TOBr (UTOBr) was determined as the difference between TOBr and bromine incorporated into measured specific DBPs. TOBr concentration increased from 40 μg/L (as bromide) to 1064 μg/L (as bromide) when bromide was increased from 0 to 30 μmol/L. THMs, HAAs and DHANs together accounted for 87% of the TOBr while UTOBr accounted for only 13% of the TOBr during chlorination without bromide addition. The majority of TOBr was present as known specific DBPs during chlorination at low levels of bromide. However, the percentage of UTOBr increased and the percentage of THMs and HAAs decreased with increasing bromide concentrations. The percentage UTOBr increased to 34% at a level of 30 μmol/L of bromide addition. Similar trends were also observed for the chlorination of Winnipeg water.

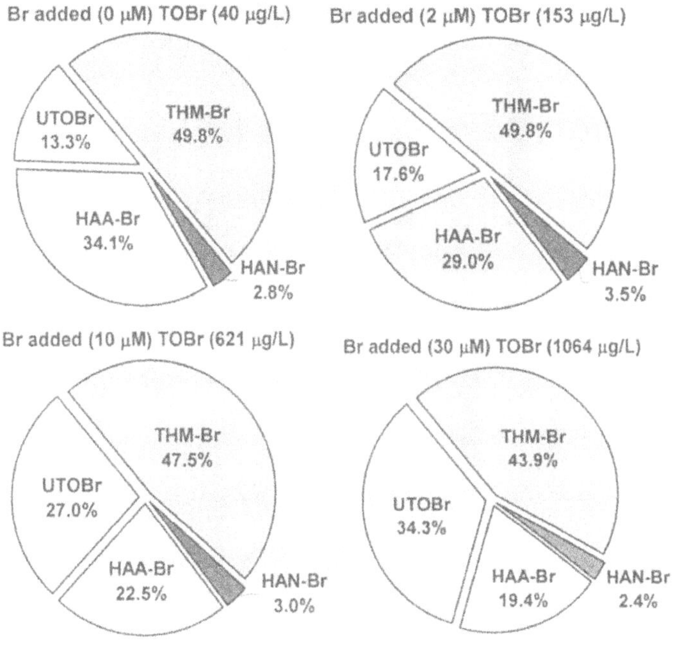

Figure 4. Effect of bromide on the formation of TOBr (Tulsa water)

Chlorination of NOM Fractions

Figure 5 shows the formation of THMs and HAAs from chlorination of Tampa NOM fractions at a bromide level of 800 μg/L. Hydrophobic and high MW NOM showed a different THM and HAA formation pattern as compared to hydrophilic and low MW NOM. Hydrophobic and MW>3 kDa fractions formed more fully chlorinated species (chloroform, DCAA, and TCAA) than hydrophilic and MW<3 kDa fractions. However, hydrophilic and MW<3 kDa fractions produced more fully brominated species (bromoform, DBAA, and TBAA). For example, chloroform formation by hydrophobic NOM was 67% higher than that

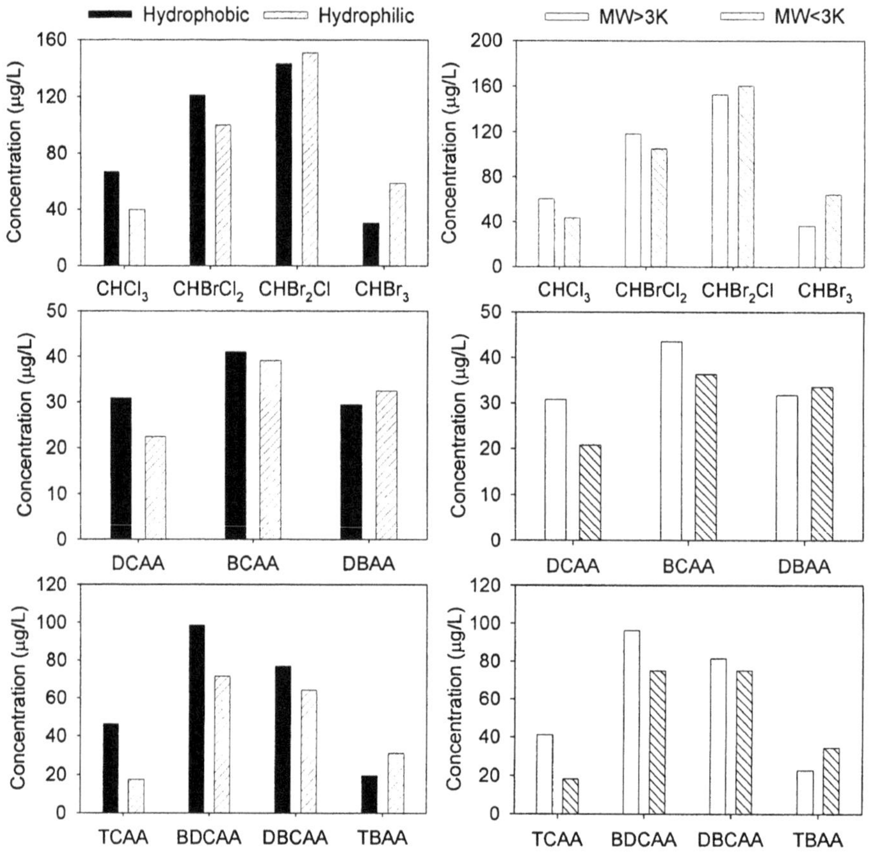

Figure 5. THM and HAA formation from Tampa NOM fractions (Br=800 μg/L)

formed by hydrophilic NOM. Bromoform formation by hydrophilic NOM was 93% more than that formed by hydrophobic NOM. Similar results were also observed for Springfield NOM fractions. Many researchers have observed that hydrophobic NOM, which is typically high in humic content and aromatic carbon, is more reactive with chlorine and forms more THMs and HAAs than the corresponding hydrophilic NOM (*14, 15*). However, the results of this study suggest that bromine is more reactive with hydrophilic NOM than with hydrophobic NOM with regards to the formation brominated THMs and HAAs.

Figure 6 shows the distribution of TOBr formed by Tampa NOM factions. Although hydrophilic and low MW NOM formed more brominated THMs and HAAs than hydrophobic and high MW NOM, less TOBr was produced by hydrophilic and low MW NOM. The percentage of UTOBr ranged from 27% to39% for the four fractions. Hydrophobic and MW>3 kDa fractions produced

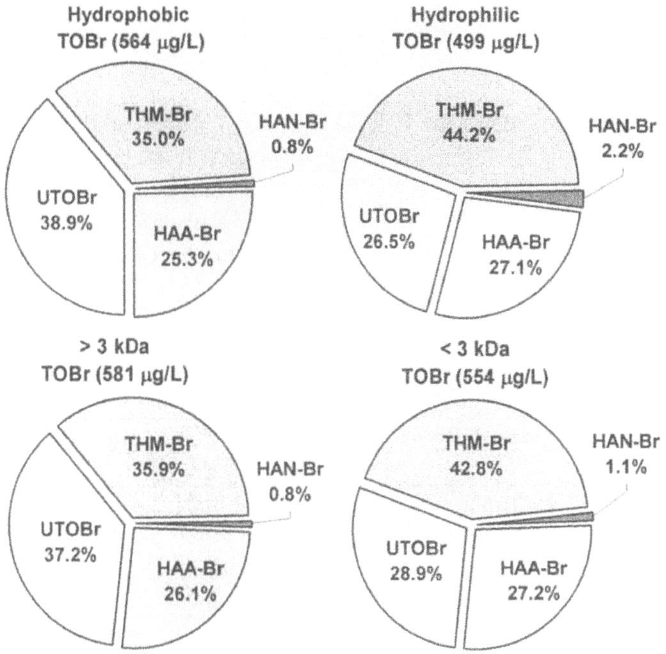

Figure 6. TOBr formation from Tampa NOM fractions

higher percentages of UTOBr than hydrophilic and MW< 3 kDa fractions. These results suggest that hydrophobic and high MW NOM is the primary precursor to the formation of unknown brominated DBPs during chlorination.

Bromine substitution factor (BSF) has been used to compare the degree of bromination of DBPs. It is defined as the ratio of the molar concentration of bromine incorporated into one class of DBPs to the total molar concentration of chlorine and bromine in that class (BSF = DBP-Br/(DBP-Cl + DBP-Br)). The BSF of each DBP ranges from 0 to 1. Equations (1), (2), (3), and (4) present the calculations of the BSFs for THMs, DHAAs, THAAs, and TOX, respectively.

$$\mathrm{BSF(THMs)} = \frac{[\mathrm{CHCl_2Br}] + 2[\mathrm{CHBr_2Cl}] + 3[\mathrm{CHBr_3}]}{3 \times ([\mathrm{CHCl_3}] + [\mathrm{CHBrCl_2}] + [\mathrm{CHBr_2Cl}] + [\mathrm{CHBr_3}])} \quad (1)$$

$$\mathrm{BSF(DHAAs)} = \frac{[\mathrm{BCAA}] + 2[\mathrm{DBAA}]}{2 \times ([\mathrm{DCAA}] + [\mathrm{BCAA}] + [\mathrm{DBAA}])} \quad (2)$$

$$\mathrm{BSF(THAAs)} = \frac{[\mathrm{BDCAA}] + 2[\mathrm{DBCAA}] + 3[\mathrm{TBAA}]}{3 \times ([\mathrm{TCAA}] + [\mathrm{BDCAA}] + [\mathrm{DBCAA}] + [\mathrm{TBAA}])} \quad (3)$$

$$BSF(TOX) = \frac{[TOBr]}{[TOCl]+[TOBr]} \tag{4}$$

Table V shows the BSFs for THMs, DHAAs, THAAs, and TOX of chlorinated NOM fractions of Springfield and Tampa waters. It is clear that hydrophilic and low MW NOM fractions produced higher BSFs than their corresponding hydrophobic and high MW NOM fractions. DBPs formed by hydrophilic and low MW NOM showed a relatively higher degree of bromination than those formed by hydrophobic and high MW NOM.

Table V. BSFs of Chlorinated NOM Fractions

Sample	*Fraction*	*BSF (THM)*	*BSF (DHAA)*	*BSF (THAA)*	*BSF (TOX)*
Springfield	Hydrophobic	0.30	0.31	0.23	0.20
	Hydrophilic	0.35	0.39	0.34	0.23
	MW>3 kDa	0.46	0.37	0.35	0.28
	MW<3 kDa	0.53	0.47	0.46	0.33
Tampa	Hydrophobic	0.39	0.41	0.38	0.31
	Hydrophilic	0.48	0.47	0.48	0.39
	MW>3 kDa	0.41	0.41	0.40	0.33
	MW<3 kDa	0.49	0.47	0.48	0.40

Chlorination of Natural Waters

Figure 7 shows the TOCl and TOBr concentrations from chlorination of the collected natural waters. All waters showed much higher TOCl concentrations than TOBr concentrations due to the relatively low bromide levels in these waters. A comparison of UTOX/TOX, UTOCl/TOCl, and UTOBr/TOBr ratios of these chlorinated waters is shown in Figure 8. The UTOX/TOX ratios ranged from 55% to 62%, with an average of 57%. The UTOCl/TOCl ratios of these chlorinated waters were very close to UTOX/TOX ratios. The UTOBr/TOBr ratios ranged from 12% to 26%, with an average of 19% for the chlorination of collected waters. The percentages of UTOBr were much lower than those of UTOCl. Most bromine was incorporated into known specific DBPs during chlorination of low bromide containing waters.

Figure 9 shows UTOCl/TOCl and UTOBr/TOBr ratios as a function of the mole ratio of bromide to chlorine consumption for the data collected in this study. It is clear that UTOCl/TOCl and UTOBr/TOBr exhibited different trends with increasing bromide to chlorine ratios. The percentage of UTOCl decreased as the bromide to chlorine consumption ratio increased. The UTOCl/TOCl ratio

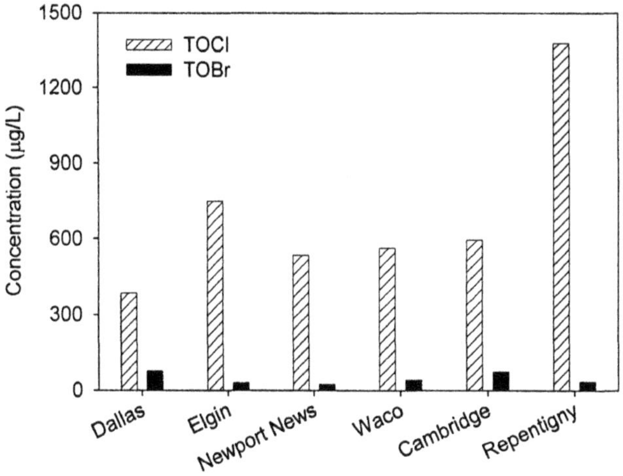

Figure 7. TOCl and TOBr formation from natural waters

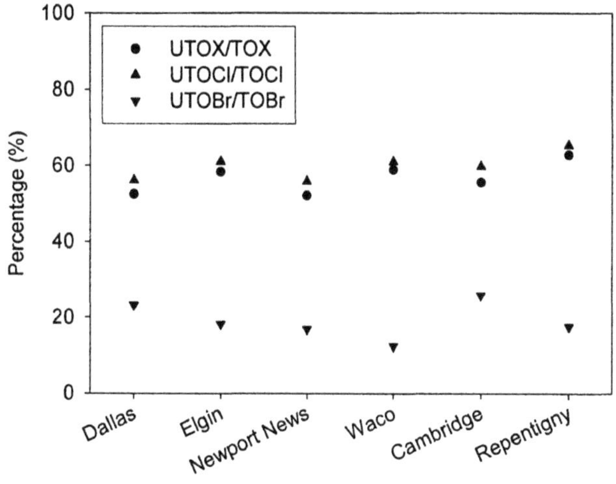

Figure 8. UTOX/TOX ratios from chlorination of natural waters

was close to 60% when the bromide to chlorine consumption was lower than 50 μM/mM. The UTOCl/TOCl ratio decreased to less than 20% when the bromide to chlorine consumption was higher than 400 μM/mM. The reverse trend was observed for the UTOBr/TOBr ratio. The UTOBr/TOBr ratio was close to 20% when the bromide to chlorine consumption was lower than 50 μM/mM. The UTOBr/TOBr ratio increased to 40% at very high bromide to chlorine consumption ratios (> 300 μM/mM). Because limited data were obtained from this study, more work needs to be done to confirm these observations.

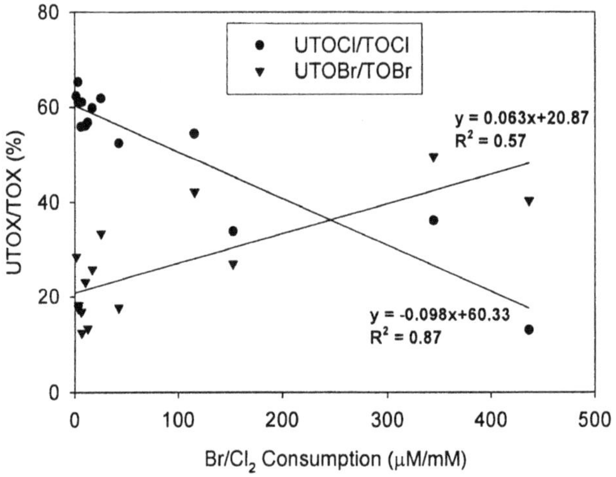

Figure 9. Correlation between Br/Cl_2 and UTOCl/TOCl, UTOBr/TOBr ratios

Implications on Drinking Water Treatment

The results of this study suggest that hydrophilic and low MW NOM is highly reactive with bromine as measured by THMs and HAAs. Hydrophilic NOM is generally far less amenable to removal by coagulation. Water utilities with high natural bromide levels should endeavor to improve the removal of both hydrophilic and hydrophobic NOM to minimize the formation of DBPs.

The percentages of UTOCl and UTOBr formed during chlorination are related to the bromide to chlorine consumption ratios. Most TOBr was in the form of THMs and HAAs when the bromide level was low. The UTOBr to TOBr ratio increased as the bromide to chlorine consumption ratio increased. The reverse was true for the UTOCl to TOCl ratio. These results can be used to help estimate the UTOCl and UTOBr levels in chlorinated drinking waters.

Conclusions

The ratio of UTOBr to TOBr increased, and the ratio of UTOCl to TOCl decreased with increasing bromide to chlorine ratios. Hydrophilic and low MW NOM formed more brominated THMs and HAAs than hydrophobic and high MW NOM when chlorinated in the presence of high levels of bromide. However, more TOBr was formed by hydrophobic and high MW NOM. Most organic bromine was incorporated into THMs and HAAs during chlorination of low bromide containing waters. Most organic chlorine was incorporated into THMs and HAAs during chlorination of high bromide containing waters.

Acknowledgments

This research was funded by AwwaRF through Project No. 2755. The authors thank the assistance from participating water utilities.

References

1. Kumar, K.; Margerum, D. W. *Inorg. Chem.* **1987**, *26*, 2706-2711.
2. Symons, J. M.; Krasner, S. W.; Simms, L. A.; Sclimenti, M. *J. Am. Water Works Assoc.* **1993**, *85(1)*, 51-62.
3. Cowman, G. A.; Singer, P. C. *Environ. Sci. Tech.* **1996**, *30*, 16-24.
4. Hua, G.; Reckhow, D. A.; Kim, J. *Environ. Sci. Tech.* **2006**, *40*, 3050-3056.
5. Krasner, S.W.; McGuire, M. J.; Jacangelo, J. G.; Patania, N. L.; Reagen, K. M.; Aieta, E. M. *J. Am. Water Works Assoc.* **1989**, *81*(8), 41-53.
6. Singer, P. C.; Obolensky, A.; Greiner, A. *J. Am. Water Works Assoc.* **1995**, *87*(10), 83-92.
7. Hua, G.; Reckhow, D. A. *Water Res.* **2007**, *41*, 1667-1678.
8. Pourmoghaddas, H.; Stevens, A. A. *Water Res.* **1995**, *29*, 2059-2062
9. Hua, G.; Reckhow, D. A. *Anal. Bioanal. Chem.* **2006**, *384*, 495-504.
10. Plewa, M. J.; Wagner, E. D.; Richardson, S. D.; Thruston, A. D.; Woo, Y.; McKague, A. B. *Environ. Sci. Technol.* **2004**, *38*, 4713-4722.
11. APHA, AWWA, WEF *Standard Methods for the Examination of Water and Wastewater*, 20th ed.; Washington, DC, **1998**.
12. Aiken, G. R.; McKnight, D. M.; Thorn, K. A.; Thurman, E. M. *Org. Geochem.* **1992**, *18*, 567-573.
13. Hua, G.; Reckhow, D. A. In *Proceedings of the AWWA Water Quality Technology Conference*, Quebec City, Quebec, **2005**.
14. Liang, L.; Singer, P. C. *Environ. Sci. Technol.* **2003**, *37*, 2920-298.
15. Hua, G.; Reckhow, D. A. *Environ. Sci. Technol.* **2007**, *41*, 3309-3315.

Chapter 9

HAA Formation and Speciation during Chloramination

Tanju Karanfil[1], Ying Hong[2], Hocheol Song[1]

[1]Department of Environmental Engineering and Science, Clemson University, Clemson, SC 29634
[2]Black & Veatch, 201 South Orange Avenue, Suite 500, Orlando, FL, 32801

The main objective of this study was to systematically examine the roles of dissolved organic matter characteristics, pH and bromide in the formation and speciation of dihalogenated acetic acids, the major class of haloacetic acids (HAA) formed during chloramination. This chapter also summarizes briefly the potential reaction routes of HAA formation, HAA formation kinetics, and the effect of quenching agent use during chloramination.

Introduction

The use of chloramine, in the form of monochloramine (NH_2Cl), has gained increasing popularity in the United States drinking water industry during the last two decades, primarily due to the increasingly stringent regulations imposed by the United States Environmental Protection Agency (USEPA) under the Disinfectants/Disinfection Byproducts Rule (D/DBPR). Stage 2 of the D/DBPR, published in January 2006, requires water utilities to comply with the maximum contaminant levels (MCLs) of 80 μg/L total trihalomethanes (THMs: chloroform, bromodichloromethane, dibromochloro-methane, and bromoform) and 60 μg/L five haloacetic acids (HAA5: the sum of mono-, di-, and trichloroacetic acids, and mono- and dibromoacetic acids) at each individual monitoring location in a distribution system (i.e., locational running annual averages) *(1)*. The most recent survey of 363 drinking water utilities across all 50 states showed that 68 and 29% used chlorine and chloramines as a secondary disinfectant, respectively, and 3% did not use any secondary disinfectant *(2)*. Of those 257 utilities using chlorine or no secondary disinfectant, seven planned to convert to chloramines, and another seventy seven were considering switching to chloramines with the implementation of the Stage 2 D/DBPR.

Chloramination (using preformed NH_2Cl) almost completely suppresses THM formation and reduces HAA concentrations to 3-30% of the amounts formed during chlorination *(3-6)*. Dihalogenated haloacetic acids (DXAAs) are the primary regulated DBP species formed during chloramination, whereas the formation of monohalogenated (MXAAs) and trihalogenated haloacetic acids (TXAAs) is significantly suppressed *(5,7,8)*. Although chloramination does not completely eliminate HAA formation, it may provide a viable alternative to some water utilities complying with the Stage 2 D/DBPR. This and control of THMs are the main reasons for the increased use of chloramination by the drinking water treatment industry in the US in recent years.

The main objective of this chapter is to examine the formation and speciation of HAA during chloramination of two natural waters with different dissolved organic matter (DOM) characteristics (i.e., one with a high specific ultraviolet absorbance (SUVA) and the other with a low SUVA value) under similar experimental conditions. In addition, the potential reaction routes of HAA formation, HAA formation kinetics, and the effect of quenching agent use during chloramination are also briefly summarized.

Materials and Methods

Reverse Osmosis (RO) Isolation of DOM

Two DOM solutions were concentrated from the influents of drinking water treatment plants in South Carolina, Greenville (GV) and Myrtle Beach (MB), using a RO system, described elsewhere *(9)*. The selected water quality characteristics of the source waters are provided in Table 1. The RO isolation enabled performing the chloramination experiments of the two different source waters at the same organic carbon concentration. Good DOM recoveries (88% for GV and 99% for MB) were obtained based on dissolved organic carbon (DOC) during the isolation. The reactivity tests confirmed that there was no impact of RO isolation on the DOM reactivity during chlorination and chloramination *(10,11)*.

Table 1. Selected physicochemical characteristics of the source waters

Parameter	*Unit*	*Greenville*	*Myrtle Beach*
DOC	(mg/L)	0.8	10.7
UV_{254}	(cm^{-1})	0.015	0.522
$SUVA_{254}$	(L/mg-m)	1.9	4.9
Bromide	(μg/L)	<MRL	68
pH	(-)	6.9	7.0
Alkalinity (as $CaCO_3$)	(mg/L)	8	32
Ionic Strength	(as mM NaCl)	0.49	3.94

All values are for samples filtered using a pre-washed 0.45-μm filter (either Osmonics PES or Gelman Supor) *(12)*. MRL: Minimum Reporting Level.

Chloramination Experiments

The RO isolate of each water was diluted to prepare a DOM solution of 7.5 mg/L DOC using distilled and deionized water (DDW) (Super-Q^{TM} plus, Millipore). Solution pH was adjusted by addition of 4 mM carbonate buffer and 1 M HCl/NaOH solutions. Three pHs (6, 7.5 and 9) and three Br/DOC ratios (low (~50 μg/mg), intermediate (~130 μg/mg) and high (~350 μg/mg)) were selected for chloramination experiments. Bromide was spiked into DOM solutions to achieve the target bromide levels. The selected pH values and Br/DOC ratios cover typical drinking water treatment conditions.

All samples were chloraminated using preformed NH_2Cl following the uniform formation potential condition (UFC) protocol *(13)* with some modifications: (1) Experiments were conducted at three different pHs, not only at pH 8 as prescribed in the UFC protocol, to investigate the pH effect. (2) The total combined chlorine residual after 24-hr was maintained around 2±0.4 mg/L as Cl_2. (3) Carbonate buffer was used instead of borate buffer. These modifications of the UFC protocol cover the typical conditions for chloramination in water treatment practice in the US. Fresh NH_2Cl stock solution was prepared by mixing sodium hypochlorite (5-6% free chlorine) and ammonium sulfate solutions in DDW at a Cl_2/N mass ratio of 3.5:1 and pH 9. Preliminary experiments were conducted for each DOM to determine the ratio of NH_2Cl/DOC that would provide a total combined residual of about 2 mg/L as Cl_2 after 24-hr of contact time.

Kinetics data were collected during the UFC experiments for reactions times of 0.5, 1, 2, 4, 8, 24, 48 and 72 hr. Usually, three reaction times (i.e., 1, 4 and 24 hr) were chosen for duplicate samples. The variation of duplicate samples was about 1 or 2 μg/L, and the averages of duplicate samples were used in the data presentation. At the end of each target contact time, samples were collected for pH, free and combined chlorine measurements. Samples were then immediately quenched with a stoichiometric amount of sodium sulfite (Na_2SO_3) and stored in a refrigerator (4-6°C). All the samples were analyzed together for THM and all nine HAA species (HAA9) after the last sample (i.e., 72 hr) of the kinetics experiment was quenched.

Analytical Methods

A number of analytical methods were used to measure several parameters. Free and combined chorine concentrations were measured according to Standard Method 4500-CI F (DPD Ferrous Titrimetric Method). Dissolved (DOC) and total organic carbon (TOC) analyses were performed according to Standard Method 5310B using a TOC analyzer. (TOC-V CHS & TNM-1, Shimadzu Corp., Japan). UV_{254} absorbance was measured using a Beckman DU640 Spectrophotometer. Bromide concentration was measured according to USEPA Method 300 employing an ion chromatography system (Dionex DX-600, Dionex Corporation). USEPA Method 551.1 (liquid-liquid extraction and gas chromatography) and Standard Method 6251B with some modification as described by Singer et al. *(14)* were employed for THM and HAA measurements, respectively. Detailed information about the analytical method can be found elsewhere *(11)*.

Results and Discussion

Formation and Speciation of HAA

DXAA was the main HAA species in both DOM solutions, comprising at least 80% of HAA (as mass or molar concentrations) formation after 72-hr for all bromide concentrations at pH 6 (Figure 1) and pH 7.5 (data not shown). At pH 6, the increase in Br/DOC ratio from ~50 to ~130 μg/mg increased brominated DXAA formation, while decreasing monochloroacetic acid (MCAA) formation. The further increase in Br/DOC increased monobromoacetic acid (MBAA) and brominated TXAA formation, resulting in a decrease in the DXAA percentage. At pH 7.5, no MBAA and brominated TXAA were formed, and the DXAA percentage increased with bromide concentration. At pH 9, dichloroacetic acid (DCAA) was the only HAA specie formed at all bromide concentrations (data not shown).

Cowman and Singer *(4)* reported that DXAA, MXAA and TXAA constituted about 65, 30 and 5% of HAA formed, respectively, as a result of chloramination of a DOM extract ($SUVA_{254}$= 4.2 L/mg.m) at pH 8. Diehl et al.

(8) reported that at pH 6, 8 and 10, DXAA constituted typically more than 80% of the HAA6 (i.e., HAA5 + bromochloroacetic acid). MCAA formation was observed at pH 6 and 8 in some waters, while formation of trichloroacetic acid (TCAA) and MBAA was negligible at all three pH conditions. Speitel et al. *(15)* reported that 72-hr DXAA formation represented at least 70% of HAA formation at pH 7 and 9 for different natural waters. MXAA formation was also observed but TXAA was, in general, negligible.

Overall, these results obtained with preformed NH_2Cl indicate that at least 80% of HAA to be formed from chloramines(s) in natural waters will be in the DXAA form, and the contribution of DXAA may reach or exceed over 90% in some cases. Small amount of MXAA formation may also be observed, while formation of TXAA is usually negligible.

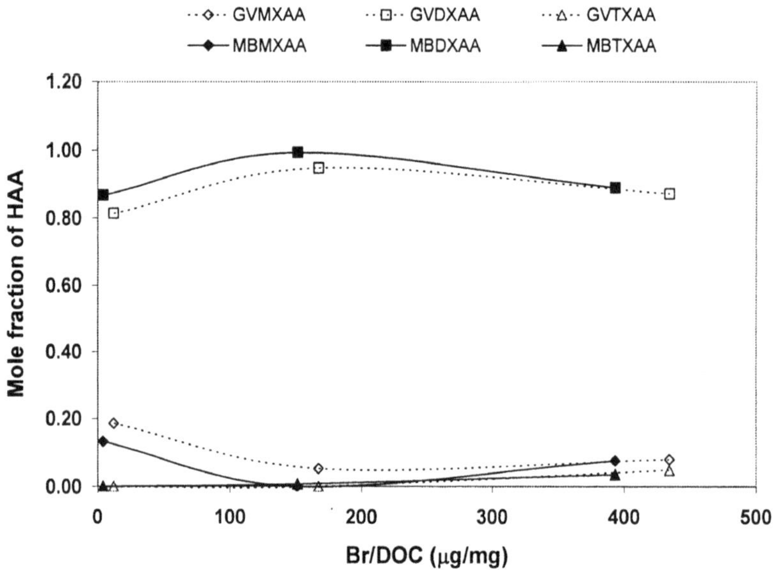

Figure 1. Distribution of MXAA, DXAA and TXAA at 72-hr during chloramination of GV and MB DOM solutions at pH 6, adapted from (11).

The Effect of pH

The results of NH_2Cl decay during the kinetics experiments are provided in Figure 2. The NH_2Cl decay rate increased with decreasing pH. At pH 6, NH_2Cl decay increased significantly with increasing bromide concentration, while no major impact of bromide concentration on the NH_2Cl decay was observed at pH 7.5 and 9. Formation of $NHCl_2$ was only observed at pH 6.

Table 2. Comparison of HAA formation patterns as a function of pH during chlorination and chloramination

HAA	*Chlorination*			*Chloramination**		
	pH 6	*pH 7.5*	*pH 9*	*pH 6*	*pH 7.5*	*pH 9*
MXAA	Essentially the same			Decreases with increasing pH		
DXAA	Essentially the same			Decreases with increasing pH		
TXAA	Decreases with increasing pH			Usually negligible (may occur at low pH)		

* using preformed NH_2Cl

SOURCE: Reproduced with permission from Reference 11. Copyright 2007 Awwa Research Foundation.

The formation of DXAA was sensitive to pH. The effect of pH on DXAA formation was consistent with that observed for NH_2Cl decay. DXAA formation decreased with increasing pH at all three bromide concentrations for both DOM solutions (Figure 3). Figure 4 shows DXAA formation after the 72-hr chloramination at three pHs. The lower pH favors HAA formation dramatically; however, the pH effect was not proportional to the change in pH. There was a larger decrease in DXAA yield when pH was increased from 6 to 7.5 than from 7.5 to 9. When pH was increased from 6 to 7.5, DXAA formation decreased by 18 and 22 μg/L for GV and MB DOM, respectively, while it only dropped 7 and 9 μg/L when pH was increased another 1.5 unit from 7.5 to 9. In recent research, Duirk *(16)* and Speitel et al. *(15)* reported similar dependency of DXAA formation on pH. In this study, it was also found that HAA speciation (i.e., the formation of DXAA, MXAA and TXAA) depends on pH. However, this trend was different from those observed during chlorination (Table 2), suggesting that chloramination had its own specific reaction mechanism(s) of HAA formation.

Overall, these results indicate that water utilities intending to decrease HAA formation during chloramination should maintain pH above 7.5.

The Effect of Bromide

To assess the extent of bromine substitution in HAA, the bromine incorporation factor (BIF) was calculated *(17)*:

$$BIF=HAABr6\ (\mu mol/L)/HAA9\ (\mu mol/L)$$

where HAABr6 is the sum of the molar concentrations of bromine incorporated in the six brominated HAA species. HAA9 represents the sum of molar concentrations of all nine HAA species.

During chloramination, the presence of bromide shifted HAA speciation to more brominated species (Figure 5). This observation was similar to that

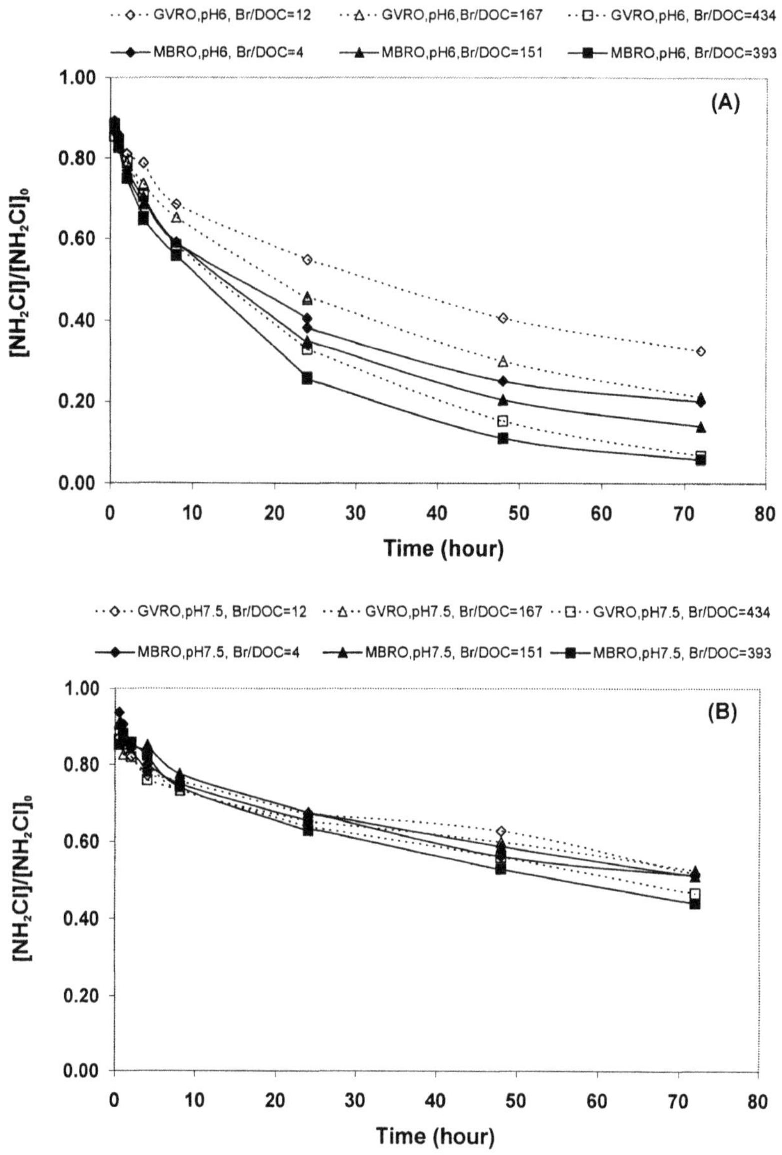

Figure 2. NH_2Cl consumption by GV and MB DOM solutions at (A) pH 6, (B) pH 7.5, and (C) pH 9, adapted from (11).

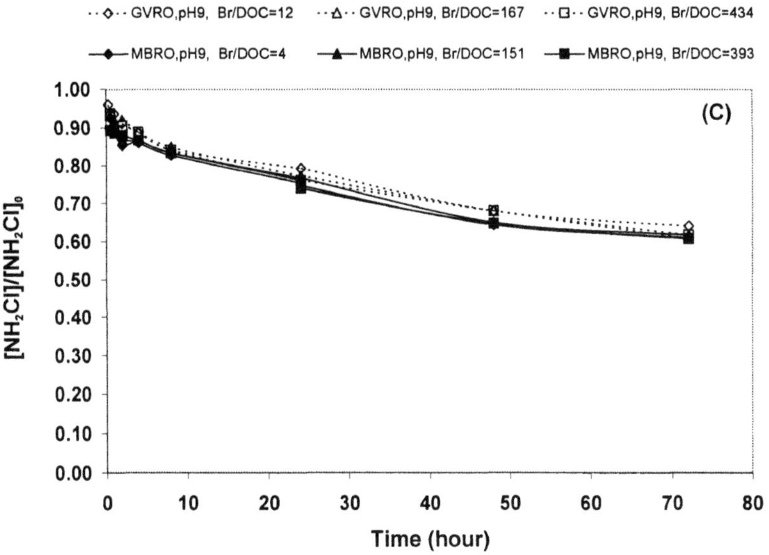

Figure 2. Continued.

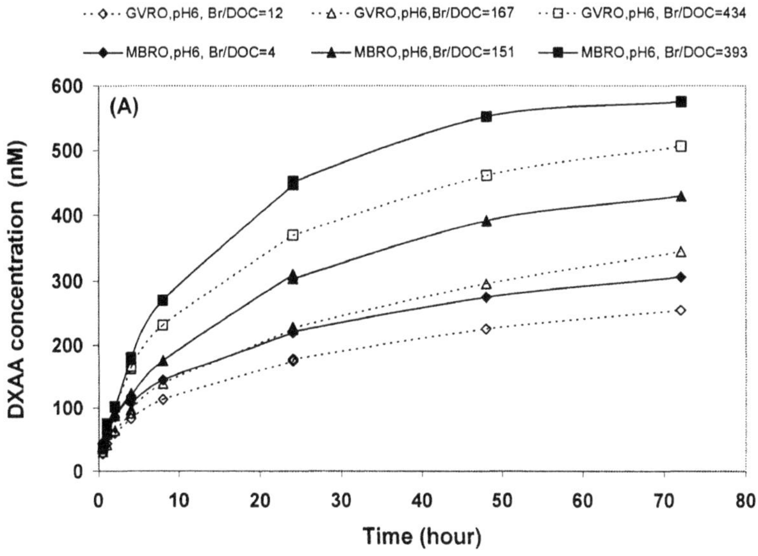

Figure 3. DXAA formation of GV and MB DOM solutions at (A) pH 6, (B) pH 7.5, and (C) pH 9, adapted from (11). Continued on next page.

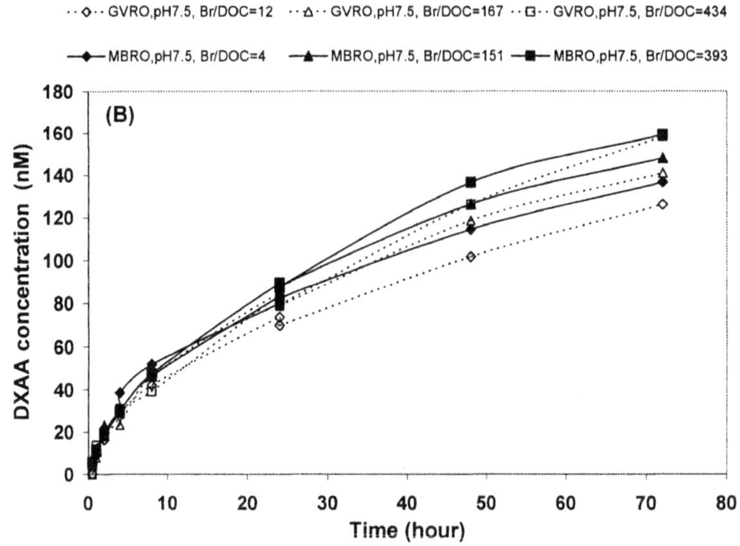

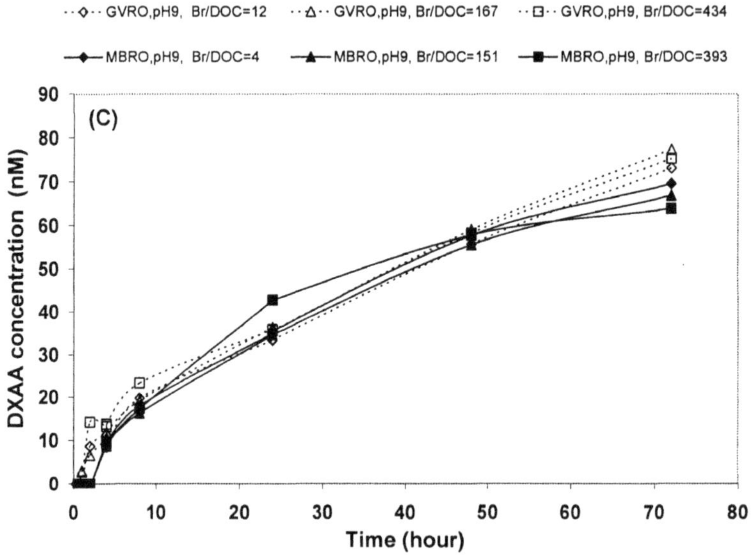

Figure 3. Continued.

observed for chlorination *(18)*, but to a much lesser extent. The effect of bromide depended on pH. Bromine incorporation was higher and occurred more rapidly at pH 6 than pH 7.5. The BIF increased significantly with increasing bromide concentration for both DOM. No brominated specie was formed at pH 9 (data not shown).

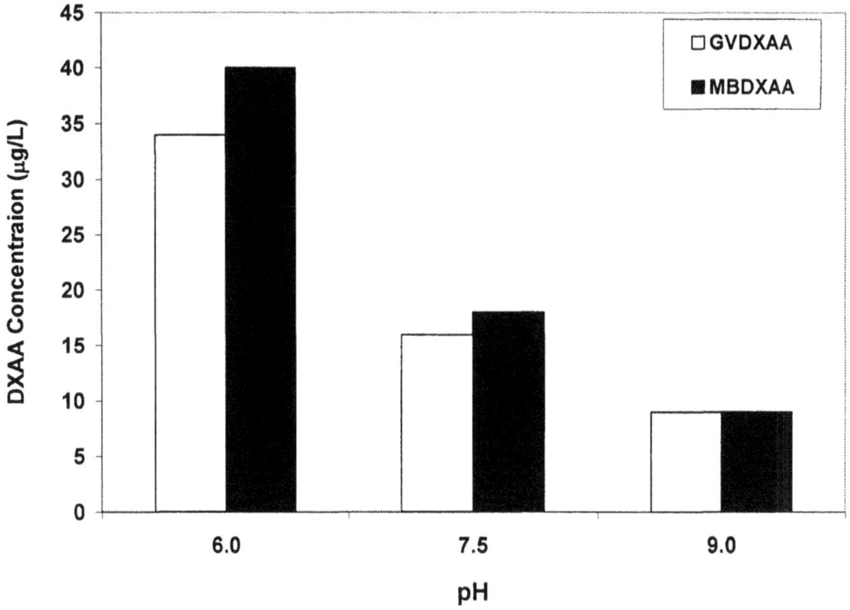

Figure 4. pH effect on 72-hr DXAA formation for GV and MB DOM solutions at ambient Br/DOC (<12 μg/mg) level, adapted from (11).

At pH 6, DXAA formation and NH_2Cl decomposition increased with increasing bromide concentration (Figures 2 and 3). At the same time, the DXAA speciation shifted toward brominated species. Figure 6 (A) shows the change in distribution of DXAA species (DCAA, bromochloroacetic acid (BCAA) and dibromoacetic acid (DBAA)) with bromide concentration after 72-hr reaction time. DCAA formation decreased significantly with increasing bromide concentration to the intermediate level (Br/DOC=150 μg/mg), while DBAA and BCAA increased dramatically. However, the further increase in bromide concentration did not increase the contribution of BCAA to DXAA even though overall DXAA formation increased with bromide concentration. This was because high bromide conditions favored DBAA formation over BCAA. At the highest bromide concentration after 72-hr, the mole percentage of DBAA was about 50%, while BCAA and DCAA accounted for about 30 and 20%, respectively. The distribution of DXAA species was almost independent of DOM source at pH 6 in spite of the significantly different $SUVA_{254}$ value of the two DOMs. The presence of bromide resulted in the formation of trace amounts of brominated TXAA only at the highest Br/DOC ratios for all waters. At the ambient bromide levels, MCAA was the only MXAA species formed, and its

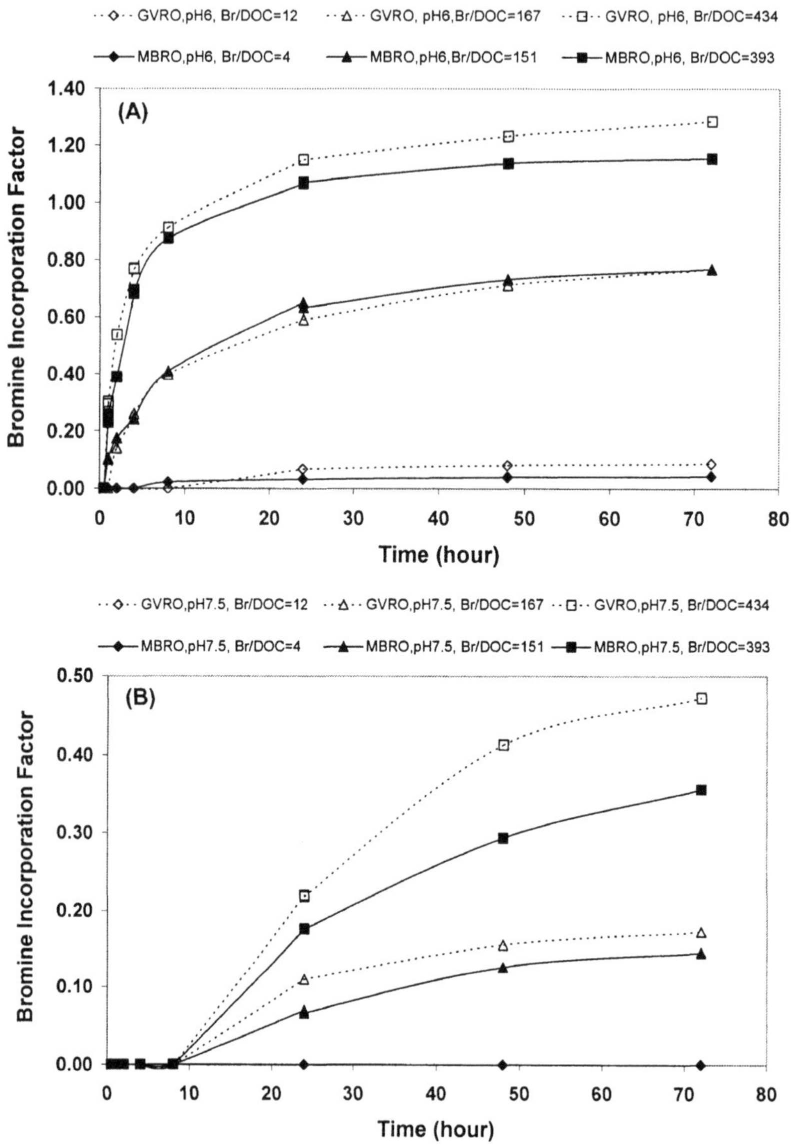

Figure 5. Bromine incorporation of GV and MB DOM solutions at (A) pH 6 and (B) pH 7.5, adapted from (11).

formation decreased with increasing bromide concentration. At the highest Br/DOC ratio, MBAA was the only MXAA formed.

At pH 7.5, the impact of bromide on DXAA formation and NH_2Cl decay was much less than that at pH 6 (Figures 2 and 3). The major HAA specie was DCAA even at the highest bromide concentration, accounting for 60-70% of total DXAA. Formation of DBAA was not observed until the highest bromide concentration and its contribution was no more than 10% even after 72-hr reaction time, indicating that DBAA formation was more sensitive to pH than BCAA and DCAA when compared with the results at pH 6 (Figure 6). These results show that one additional advantage of having operational pH of 7.5 or above for chloramination is to control the formation of brominated HAA species, which are shown to be more toxic than chlorinated species *(19)*.

The Effect of DOM Characteristics

The characteristics of DOM are expected to play a role in the formation and speciation of DBPs. At pH 6, the formation of DXAA in MB water was about 10-30% higher (on a mass and molar basis) than in GV water at three bromide levels (Figure 3). Since the DOC concentrations of the experiments were almost identical, these results indicate that a higher degree of DXAA reactivity was exhibited by the high-$SUVA_{254}$ DOM than the low-$SUVA_{254}$ DOM. However, the difference between the DXAA yields was rather small in comparison to the difference in the $SUVA_{254}$ values of MB and GV DOM. The NH_2Cl decay patterns at pH 6 were consistent with the DXAA formation trends; more NH_2Cl consumption was measured in MB than GV water (Figure 2). This was in agreement with the previous reports that NH_2Cl decomposition increased with the $SUVA_{254}$ of the DOM *(20)*. In terms of bromine effect, relatively similar bromine incorporation factors and trends were obtained for both DOMs at pH 6 in spite of significantly different SUVA values (Figure 5). At the highest Br/DOC ratio, the GV DOM had a slightly higher bromine incorporation factor than the MB DOM. At pH 7.5 and 9, the impact of DOM aromaticity (i.e., $SUVA_{254}$) on NH_2Cl decay was negligible at each bromide condition (Figure 2). The DXAA formation patterns were consistent with their NH_2Cl decay patterns (Figure 3). Bromine incorporation at pH 7.5 was significantly smaller compared to pH 6 and it was somewhat higher for GV than MB water (Figure 5). No brominated HAA species were formed at pH 9.

Speitel et al. *(15)* studied DXAA formation for two surface waters, Lake Austin water ($SUVA_{254}$=2.49 L/mg-m and DOC=3.74 mg/L), and Metedeconk River water ($SUVA_{254}$=3.48 L/mg-m and DOC=3.05 mg/L) at pH 7 and 9. The normalized DXAA yields with respect to DOC were very similar for both waters at each pH in spite of one unit SUVA difference between the waters. In contrast, Duirk *(16)* reported almost three times higher DXAA yield from Upper

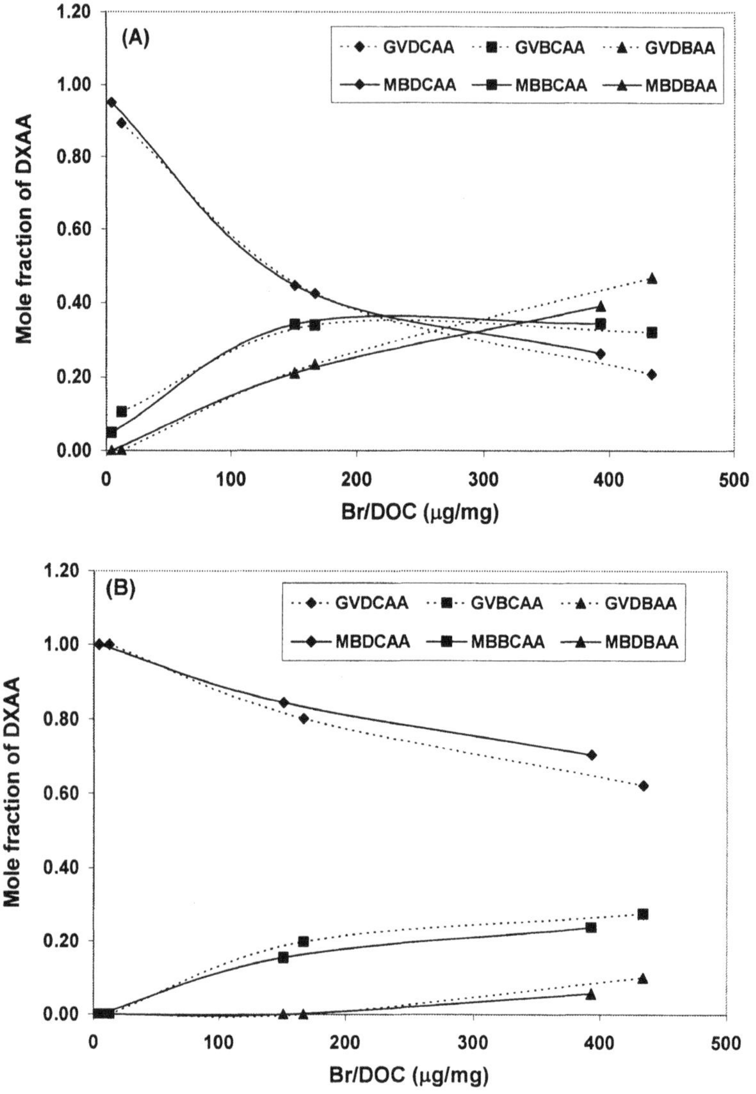

Figure 6. Distribution of individual DXAA species at 72-hr in GV and MB DOM solutions at (A) pH 6 and (B) pH 7.5, adapted from (11).

Peninsula water ($SUVA_{254}$=3.05 L/mg-m) than from Iowa River water ($SUVA_{254}$=2.78 L/mg-m) at pH 8.3, although both had similar $SUVA_{254}$ values. Both waters had the same DOC concentration and NH_2Cl dose. This suggests the existence of highly reactive sites which were not reflected by the $SUVA_{254}$ of the Upper Peninsula water. This also implies that non-humic DOM may react with chloramines to form DXAAs.

Overall, these results show that the impact of DOM and bromide were significantly less pronounced with increasing pH, indicating that pH is a more important factor controlling the extent of DXAA formation and speciation than bromide and $SUVA_{254}$ of DOM under typical drinking water conditions.

HAA Formation Routes and Kinetics, and the Effect of Quenching Agent

The HAA formation routes during chloramination are still open to debate in the literature and more research is needed to further obtain mechanistic information. Due to space limitation, a brief summary about these studies will be presented here. In the absence of bromide, there are three oxidants, NH_2Cl and its two decomposition products, free chlorine (HOCl) and dichloramine ($NHCl_2$), that can be present in water during chloramination potentially reacting with DOM to form HAA. The recent research of the authors have shown that the direct reaction between NH_2Cl and DOM plays the major role and is responsible for about 80% of HAA formation *(10,11)*. The remaining HAA formation was attributed to HOCl that results from NH_2Cl decomposition. No HAA formation was observed from preformed $NHCl_2$. In contrast, Valentine and co-workers recently published work indicating that DXAA formation was mostly attributed to a direct reaction of HOCl in quasi-equilibrium with NH_2Cl, with DOM *(21,22)*. The rate of HAA formation correlated with the rate of DOM oxidation by this mechanism. The presence of bromide significantly complicates the solution chemistry during chloramination *(11,15)*. Bromamines, chloramines, and bromochloramines would co-exist in a chloraminated water with bromide. Among several species, previous research suggests that bromochloramine (NHBrCl) will be the dominant specie during water treatment *(23, 24)*. Overall, understanding the HAA formation and speciation pathways during chloramination and the roles of the bromo/chloroamine species in HAA formation warrant further research.

This research also provided some insight to the different HAA formation kinetics patterns reported for chloramination in the literature *(10,11)*. Speitel and co-workers have reported that approximately 73% of the 72-hr DXAA formed after only 0.5 hr and almost 82% after 1 hr at pH 8 *(15,25,26)*. Hua & Reckhow *(27)* reported similar rapid kinetics (85% of 72-hr yield formed within 30 min) during chloramination, which was faster than chlorination for a surface water. These rapid formation rates were unexpected because chlorine is a much stronger oxidant than chloramines, and about 80 to 90% of HAA formation is not completed during the first 24-hr of chlorination. Singer et al *(7)* observed relatively fast formation of DCAA at pH 8 during the first hour of reaction (i.e., approximately 50% of the 72-hr DCAA yield). In contrast to these trends, gradually increasing formation of DCAA (~ 10% of 72-hr HAA yield within 30 min.) was observed during the experiments conducted in this study. It was found that the use of ammonium chloride (NH_4Cl), the quenching agent recommended by USEPA Method 552.3 for DBP samples, appears to be the main reason for the false high HAA determination at early reaction periods *(10,11),* and significantly different kinetic patterns reported for HAA formation during chloramination in the literature. Since NH_4Cl is not a quenching agent for NH_2Cl, it minimizes NH_2Cl decomposition without eliminating it from sample vials. The residual NH_2Cl can continue to react with the DOM and bromide until the sample is extracted. Therefore, the use of NH_4Cl may create an artifact on the HAA determination in chloraminated samples and during kinetics experiments, where all the samples for different reaction times are typically collected and processed together at the end of an experiment. Five different quenching approaches, (1) stoichiometric amount of Na_2SO_3, (2) 100 mg/L NH_4Cl, (3) 100 mg/L sodium meta-arsenite ($NaAsO_2$), (4) increasing pH to 10 using borate acid buffer solution, and (5) increasing pH to 10 followed by 100 mg/L ascorbic acid addition, were also tested for chloraminated samples. The advantages and disadvantages of these approaches are discussed elsewhere *(11)*. Among all approaches tested, the stoichiometric amount of Na_2SO_3 and $NaAsO_2$ were found to be the best quenching agent, showing no artifact on the determination of HAA even after 14 days. Stoichiometric amount of Na_2SO_3 addition can be used in controlled laboratory experiments, however it is not feasible to use for distribution samples because such samples have varying residual concentrations and excess amount of Na_2SO_3 addition results in decomposition of some brominated HAA species. Na_2SO_3 also does not have biocide capability and requires the addition of a separate biocide. On the other hand, $NaAsO_2$ serves as a biocide to protect samples from bacterial growth and it does not have any impact on the stability of HAA in DDW *(28)*. However, due to its toxic nature, its use in practical applications can be problematic. More research is needed to further evaluate the use of quenching/preservation agents for laboratories studies and water distribution system DBP samples.

Acknowledgments

This study was supported by American Water Works Association Research Foundation (AwwaRF) (Project 2993). The AwwaRF is the joint owner of the technical information upon which this publication is based. The comments and views detailed herein may not necessarily reflect the views of the AwwaRF, its officers, directors, affiliates or agents.

References

1. USEPA. National Primary Drinking Water Regulations: Stage 2 Disinfectants and Disinfection Byproducts Rule; Final Rule, Federal Register. Part II, 40 CFR Part 9, 141 and 142, 71: 2: 388, January 4, 2006.
2. Seidel, C.; McGuire, M.J.; Summers, R.S.; Via. S. *J. AWWA* **2005**, *97*(10), 87.
3. Long, B.W.; Longsdon, G.S.; Neden, D.G. *Proc. of Water Quality Technology Conference, Part I.*, 1992, Denver, CO.
4. Cowman, G.A.; Singer, P.C. *Environ. Sci. Technol.* **1996**, *30*(1), 16.
5. Symons, J.M.; Speitel, G.E.; Hwang, C.J.; Krasner, S.W.; Barrett, S.E.; Diehl, S.C.; Xia. R. *Factors Affecting Disinfection By-Product Formation during Chloramination.* AwwaRF, Denver, CO, 1998.
6. Zhang X.; Echigo, S.; Minear R.A.; Plewa. M.J. In *Natural Organic Matter and Disinfection By-Products: Characterization and Control in Drinking Water*; Barrett, S.E.; Krasner, S.W.; Amy G.L., Eds. ACS Symposium Series No 761, Washington, DC., 2000; p. 299.
7. Singer, P.C.; Harrington, G.; Cowman, G.; Smith, M.; Schechter, D.; Harrington, L. *Impacts of Ozonation on the Formation of Chlorination and Chloramination By-Products.* AwwaRF, Denver, CO, 1999.
8. Diehl, A.C.; Speitel, G.E.; Symons, J.M.; Krasner, S.W.; Hwang, C.J.; Barrett, S.E. *J. AWWA* **2000**, *92*(6), 76.
9. Kitis M.; Kilduff, J.E.; Karanfil. T. *Wat. Res.* **2001**, *35*(9), 2225.
10. Hong, Y.; Liu, S.; Song, H.; Karanfil, T. *J. AWWA* **2007**, *99*(8) 57.
11. Karanfil, T.; Hong, Y.; Song, H.; Orr, O. *Exploring the Pathways of HAA Formation during Chloramination.* AwwaRF, Denver, CO, 2007.
12. Karanfil, T.; Erdogan, I.; Schlautman, M.A. *J. AWWA* **2003**, *95*(3) 86.
13. Summers, R.S.; Hooper, S.M.; Shukairy, H.M.; Solarik, G.; Owen. D. *J. AWWA* **1996**, *88*(6), 80.
14. Singer, P.C.; Weinberg, H.S.; Brophy, K.; Liang, L.; Roberts, M.; Grisstede, I.; Krasner, S.W.; Baribeau, H., Arora, H.; Najm, I. *Relative Dominance of Haloacetic Acids and Trihalomethanes in Treated Drinking Water.* AwwaRF, Denver, CO, 2002.
15. Speitel, G.E.; Pope, P.G.; Collins, M.R.; Martin-Doole, M. *Disinfection By-Product Formation and Control During Chloramination.* AwwaRF, Denver, CO, 2004.
16. Duirk, S.E. Ph.D. Dissertation., The University of Iowa. Iowa City, IA, 2003.
17. Symons, J.M.; Krasner, S.W.; Sclimenti, M.J.; Simms, L.A.; Sorenson, H.W.; Speitel, G.E., Jr.; Diehl, A.C. In *Disinfection By Products in Water Treatment: The Chemistry of their Formation and Control.* Minear, R.A.; Amy, G.L., Eds. CRC Press Inc. 1996; p 91.
18. Stevens, A.A.; Leown A.M; Miltner, R.J. *J. AWWA* **1989**, *81*(8), 54.

19. Plewa, M.J.; Wagner, E.D.; Muellner, M.G.; Hsu, K-M.; Richardson, S.D. In *Occurrence, Formation, Health Effects and Control of Disinfection By-Products in Drinking Water*. Karanfil, T.; Krasner, S. W.; Westerhoff, P.; Xie, Y., Eds. American Chemical Society: Washington, DC, 2007 (in press).
20. Valentine, R.L.; Ozekin, K.; Vikesland, P.J. *Chloramine Decomposition in Distribution System and Model Waters*. AwwaRF, Denver, CO, 1998.
21. Duirk, S. E.; Gombert, B.; Croue, J-P.; Valentine, R. L. *Wat. Res.* **2005**, *39*(14), 3418.
22. Duirk, S. E.; Valentine, R. L. *Wat. Res.* **2006**, *40*(14), 2667.
23. Trofe, T.W.; Inman, G.W.Jr., Johnson, J.D. *Environ. Sci. Technol.* **1980**, *14*(5) 544.
24. Valentine, R.L. *Environ. Sci. Technol.* **1986**, *20*(2), 166.
25. Pope, P.G.; Speitel, G.E.; Martin-Doole, M.; Collins, M.R. In *Proc. of the Water Quality and Technology Conference*, 2002, Seattle, WA.
26. Pope, P.G.; Speitel, G.E.; Martin-Doole, M.; Collins, M.R. In *Proc. of the AWWA Annual Conference*, 2003, Anaheim, CA.
27. Hua, G.; Reckhow, D.A. In *Proc. of AWWA Water Quality Technology Conference*, 2005, Quebec, Canada.
28. Brophy, K.S.; Weinberg, H.S.; Singer, P.C. In *Natural Organic Matter and Disinfection By-Products: Characterization and Control in Drinking Water*. Barrett, S.E.; Krasner, S.W.; Amy G.L., Eds. ACS Symposium Series No 761, Washington, DC., 2000; p 343.

Chapter 10

Pre-Chlorination Induced DOC and DBPs Formation from *Microcystis aeruginosa* in Treatment Processes

Chao-An Chiu[1] and Gen-Shuh Wang[1,2,*]

[1]**Institute of Environ. Health, National Taiwan University, Taipei, Taiwan**
[2]**Department of Public Health, National Taiwan University, Taipei, Taiwan**

The effect of pre-chlorination on the trihalomethanes (THMs) and haloacetic acids (HAAs) formation from *Microcystis aeruginosa* was investigated. *M. aeruginosa* was cultivated under both batch and chemostat modes and harvested at different growth phases, and the formation of disinfection byproducts (DBPs) from the algal suspensions and extracellular organic matter (EOM) in the water treatment processes with and without pre-chlorination were measured. The results showed that pretreatment with 4 mg/L of chlorine increased chloroform formation potential by 62~113 μg/L and 12~23 μg/L from *M. aeruginosa* cultivated in batch culture and in chemostat, respectively. After conventional treatment processes, the pre-chlorination resulted in 10~50% decrease in overall DBPs precursor removal. When 0.5 mg/L of bromide was spiked into the algal suspension, the DBPs formation shifted from chlorinated to brominated species. Furthermore, the results of THM formation potential (THMFP) tests showed that the algae cultivated at the lower temperature water released less intracellular organic matter and less amounts of THMs precursors after pre-chlorination than that cultivated at the higher temperature water.

Introduction

Chlorine is widely used as the primary disinfectant in water treatment process for protection of public health. In addition, chlorine may also be used as a pre-oxidant for the oxidation of many reduced inorganic species and organic pollutants in raw water. When eutrophicated water is used as the raw water, algae cells and their excreted metabolic products may be present in the water and contribute to the formation of DBPs (*1*). In general, factors affecting DBP formation include algae species, algal growth phase, and reaction condition such as pH, chlorine dose, and contact time (*2*). It has been reported that algae cells and biomass play the major role in THM production while the extracellular products (ECP) produces only a small fraction of DBPs (*3*). The ECP of algae may not be effectively removed by the traditional water treatment processes while algae cells can be removed in the coagulation and filtration units. Compared to the humic substances in water, the importance and significance of algae as a DBP precursor has been reported (*4*).

Furthermore, related studies about the association between algal activity, THMFP and total organic halides (TOX) have also been investigated. It showed that per 100 ASU/mL of algal cells produced about 25μg/L of THMFP when chlorine was used to control algal growths (*5*). It has been shown that, after ozonation, the low apparent molecular weight fractions of EOM from algae was slightly increased; which comes with a corresponding decrease in the higher apparent molecular weight fractions (*6*).

In this study, the effects of prechlorination on DBPs formation from *M. aeruginosa* were assessed. The specific objectives were (i) to determine the effects of incubation modes on algal growth; (ii) to determine DBPs formation from algal suspensions and EOM; (iii) to determine the effects of prechlorination on DBPs formation from algal suspensions after water treatment processes.

Materials and Methods

Algal Culturing

Axenic culture of *M. aeruginosa* (cyanobecterium, strain 4044) obtained from Academic Sinica in Taiwan was cultured and investigated in batch mode and chemostat. In batch culture, *M. aeruginosa* was grown in a 22-L custom-made cultivation tank containing 15 L of the synthetic sterilized algal growth media (Table 1) that was modified from M-11(*7*). Cultures received 1500~2000 Lux of light on a 14-h light/10-h dark cycle and the water temperatures were maintained at 24±1°C. Samples were taken for analysis of chlorophyll a and nonpurgeable dissolved organic carbon (NPDOC) every five days to monitor the growth of algae and the ECP in water; cell appearance was examined with microscope before and after pre-chlorination. Cultures were harvested in the lag phase, early exponential phase, and late exponential phase (at the 10th, 20th, and 45th day of cultivation, respectively). For chemostat mode, *M. aeruginosa* was cultured in a continuously mixed, 15-L custom-made cultivation tank. The chemostat reactor was supplied with a constant inflow of synthetic sterilized algal growth media. The chemostat was cultivated at a dilution rate of 0.3/day at 24±1 °C, supplied with filtered air, and provided with 1500~2000 Lux of illumination on a 14-h light/10-h dark cycle. The growth phase of algae attained steady-state condition around the 30th day of the operation and was monitored by chlorophyll a and NPDOC every five days.

Analysis

Analysis of chlorophyll a follows the standard method promulgated by National Institute of Environmental Analysis, EPA, Taiwan (*8*). The NPDOC concentration was measured with a Total Organic Carbon analyzer (Shimadzu TOC 5000A). The water samples were filtered through a 0.45 μm syringe filter to remove particles and then acidified with one drop of 2 N HCl solution prior to the NPDOC analysis. The DBPFP measurement follows the procedures described in section 5710B of the Standard Methods (*9*). Pre-tested chlorine doses (20-100 mg/L) were chosen to ensure that a substantial residual (3-5 mg/L or more) was present at the end of 7 days contact time. The four THMs and the nine HAAs were quantified by a GC/MS (Agilent 6890GC/5973MSD). The detection limits are 0.1 μg/L for the four THMs, 1 μg/L for the mono- and di-HAAs and TCAA, and 3 μg/L for the brominated tri-HAAs, respectively.

Simulation of Water Treatment Processes

For pre-chlorination treatment, 4 mg/L chlorine (prepared with NaOCl) was added into 1500 mL of the incubated algal suspensions (denoted as the raw water), stirred evenly and kept still for thirty minutes. For coagulation, reagent grade aluminum sulfate (Nacalai Tesque, Kyoto, Japan) was used as coagulant in jar-test experiments. Prior to addition of the coagulants, the pH of the samples was adjusted to 5 with H_2SO_4 and/or NaOH. Following the addition of the 20 mg/L of alum, the jars were rapidly mixed at 100 rpm for 1 minute, flocculation occurred while stirring at 20 rpm for 30 minutes and quiescent settling for thirty minutes. The supernatant was withdrawn and filtered with 1 μm membrane filter (denoted as the filtered water) for further analysis. In order to differentiate the EOM contribution to DBP precursors, the algal suspensions were filtered with 0.45 μm filters to remove the algal cells.

Results and Discussion

NPDOC and Chlorophyll a Concentration

Figure 1 shows the NPDOC and chlorophyll a concentration measured at different growth phases of *M. aeruginosa*. In batch culture, the growth condition of algae corresponded to the lag phase (days 1-10), exponential phase (days 20-45), and stationary phase (days 45-65) (*10*). As shown in Figure 1, the chlorophyll a concentration can be used to represent the trends of algal growth at different phases, and the NPDOC concentration represents the organic compounds metabolized from the algal cells. The NPDOC increased moderately from 1.1 mg/L to 4.7 mg/L during the cultured period in batch mode. In the chemostat mode, the NPDOC and chlorophyll a concentration also show the similar growth trends as observed in the batch mode. However, the chlorophyll a concentration obtained in the steady-state of chemostat mode (day 30-40) was between 90 and 120 μg/L, which was much lower than those measured in the batch culture (~1400 μg/L) due to the continuous dilution in chemostat mode. The NPDOC was about 1.0 mg/L to 2.1 mg/L in the steady-state in chemostat mode, which was also much lower than those in batch mode (~4.5 mg/L). The results in Figure 1 also showed that higher water temperature gives higher NPDOC and chlorophyll a concentrations (higher algal growth) when steady-state condition is reached in chemostat mode.

Table 1. Composition of culture media for *Microcystis aeruginosa*

Chemicals	*Concentration (mg/L)*
K_2HPO_4	20
$NaNO_3$	100
$MgSO_4 \cdot 7H_2O$	75
$CaCl_2 \cdot 2H_2O$	20
Na_2CO_3	20
$FeSO_4 \cdot 7H_2O$	2
EDTA(2Na)	2.678

Effect of Pre-chlorination on Algal Suspensions and EOM

Table 2 shows the effect of pre-chlorination on the NPDOC concentration of the bulk algal suspensions. For *M. aeruginosa* growing in batch culture and collected at the 10th, 20th, and 45th day, the treatment processes resulted in a NPDOC removal from 1.96, 2.31, and 4.10 mg/L to 0.83, 1.57, and 3.44 mg/L, respectively, when pre-chlorination was not applied. On the other hand, the pre-

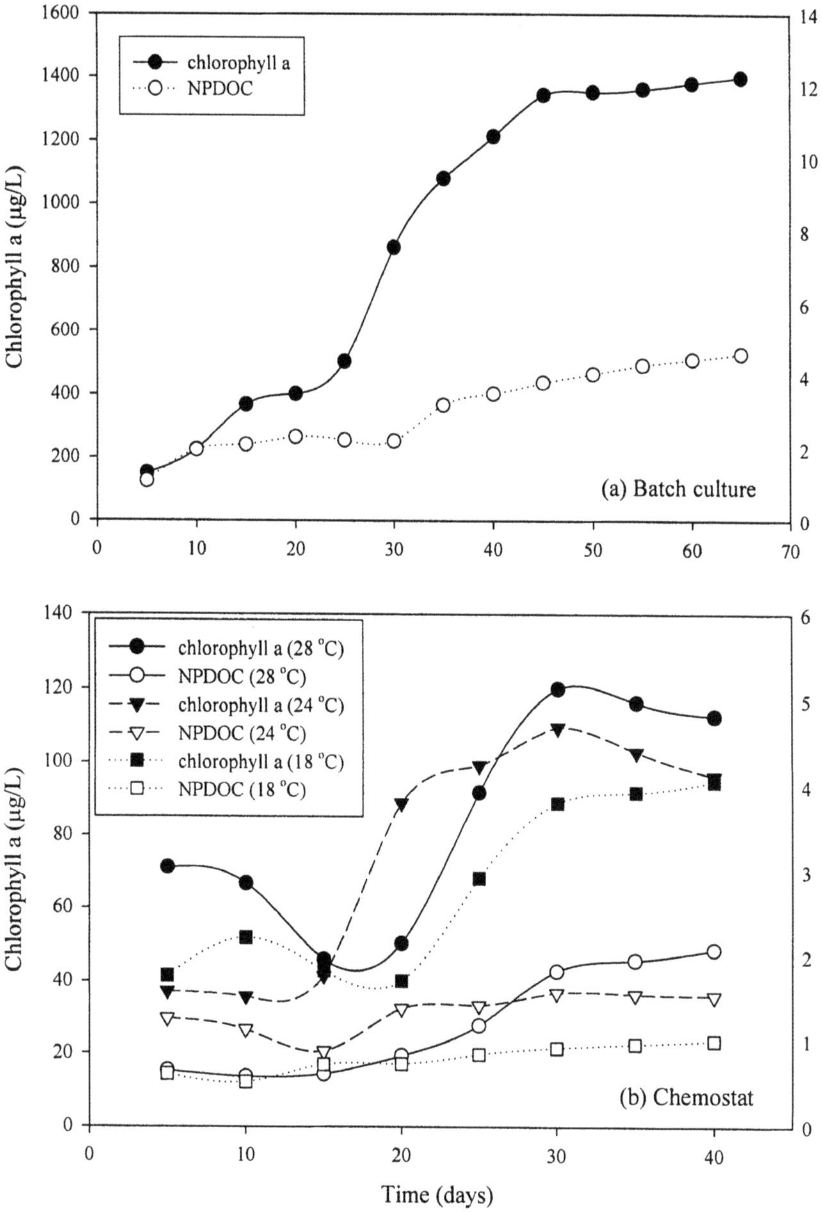

Figure 1. Growth of M. aeruginosa in batch culture and chemostat.

chlorination increased the NPDOC of raw water to 3.53, 4.82 and 7.64 mg/L, respectively, at different growth phases; and the NPDOC was 2.95, 3.51, and 5.48 mg/L, respectively, after treatments. After pre-chlorination, the NPDOC of raw water almost doubled for *M. aeruginosa* growing in batch mode, and similar results were observed for chemostat mode. From the data in Table 2, it appeared that the increase of NPDOC in raw water resulted from the liberation of the intracellar organic matter from algal cells. The lysing of algal cells due to chlorine oxidation will contribute to the NPDOC and hence the DBP precursors in the chlorinated raw water, and it is more apparent for algae in batch culture.

Table 2. NPDOC of algal suspensions with and without pre-chlorination

Cultivation time	Without pre-chlorination		With pre-chlorination	
	Raw	Filtered	Raw	Filtered
Batch culture				
10th day	1.96	0.83	3.53	2.95
20th day	2.31	1.57	4.82	3.51
45th day	4.10	3.44	7.64	5.48
Chemostat (24±1 °C)				
10th day	1.13	0.92	1.82	1.45
20th day	1.38	1.25	2.18	1.63
30th day	1.57	1.36	2.27	1.75

NOTE: The unit of NPDOC is mg/L.
The filtered water has gone through coagulation/flocculation and sedimentation units.

THMs Formation from Algal Suspensions

THMFP from *M. aeruginosa* suspension in batch culture with and without pre-chlorination is shown in Figure 2. Without pre-chlorination, it is observed that 498, 1397 and 2623 µg/L of chloroform formation were obtained from water samples taken on the 10th, 20th and 45th day of cultravation in batch mode. A significant increase of THM precursors accompanying the algal growth was observed. After the coagulation and sedimentation processes, a 70~90% reduction of chloroform precursors was obtained. However, the filtration process only exhibited a limited ability to remove additional THM precursors. When 4 mg/L of pre-chlorine was applied in the raw water, the chloroform formation was about the same as obtained without pre-chlorination. However, a slight increase of chloroform yield was noticed, especially in the filtered water.

Figure 3 shows the chloroform formation from the algal suspensions in chemostat mode that was collected on the 10th, 20th and 30th day of cultivation. Without pre-chlorination, the three raw water samples produced 80, 123, and 115 µg/L of chloroform, respectively. Unlike that in the batch culture, the algae

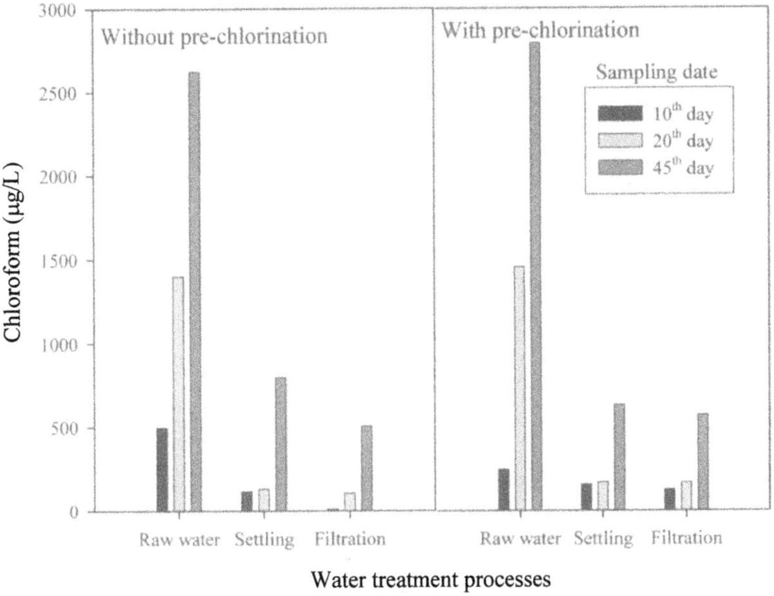

Figure 2. Effect of pre-chlorination on chloroform formation for algal suspensions in different growth phase (in batch culture).

cultivated in chemostat did not produce more chloroform precursors with the increasing growth time. The coagulation and sedimentation treatments reduces 53~77% of the chloroform precursors for samples taken from the chemostate. For filtered water, however, an additional 10% of THM precursors removal was achieved. On the other hand, when the samples were pretreated with 4 mg/L of chlorine, the chloroform formation was about the same as what was obtained without pre-chlorination in the raw and settled water. But no additional THM precursor removal was achieved after filtration when pre-chlorination was applied.

The results showed that the chloroform precursors increase with the growth of the algae cells in batch culture, and the coagulation and sedimentation treatments can effectively remove more than 50% of chloroform precursors. In the chemostat, however, a minor increase of chloroform precursors in the filtered water was observed after pre-chlorination. This result may be attributed to the cell lysing that raised the THM precursors which were not easily removed.

THMs Formation from EOM

Figure 4 shows the chloroform formation from EOM (ECP plus released intracellular products after pre-chlorination) in batch culture treated with the

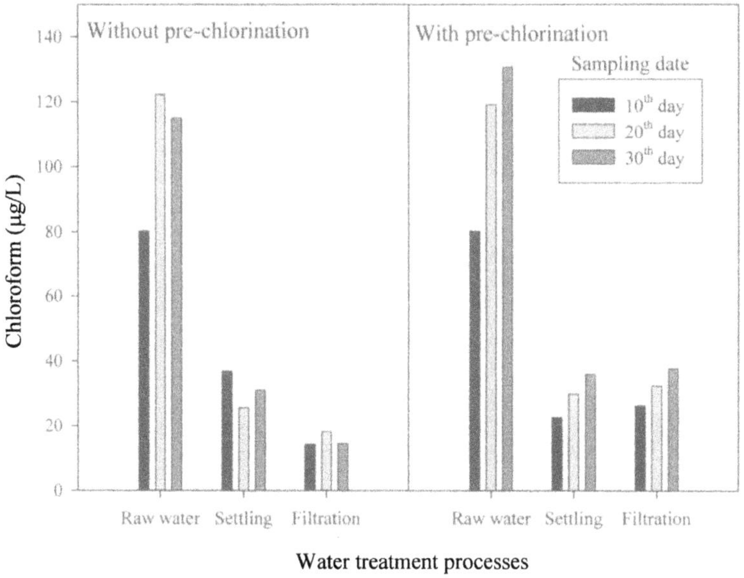

Figure 3. Effect of pre-chlorination on chloroform formation for algal suspensions in different growth phase (in chemostat).

same processes as described in the previous section for algal suspensions. The chloroform formation in raw water were lower than that obtained from algal suspensions; however, the chloroform formation in filtered water were about the same as those obtained from algal suspensions (see Figure 2). For comparison, Figure 5 shows the chloroform formation from EOM in chemostat mode. The effects of pre-chlorination on chloroform formation are much more apparent in Figure 5. Without pre-chlorination, the EOM (ECP only) in chemostat mode produced a nearly constant quantity of chloroform throughout the treatment train. With 4 mg/L of pre-chlorine addition in the raw water, however, the chloroform formation was tripled for the raw water samples. The sedimentation unit removed about 50% of the chloroform precursors, but no further removal resulted from filtration.

Results in Figures 2 to 5 indicate that the primary source of chloroform precursors in the filtered water of the algal suspensions comes from the EOM (ECP only) of the algal cells, and the ECP can not be effectively removed by the conventional water treatment process. When raw water is treated with pre-chlorination, the algal cells was oxidized and released the intracellular organic matter into the water because of cell lysing. The cell lysing contributes to the DBP precursors and increase the THM formation potential in the treated water.
Table 3 summarizes the percentage removal of THM precursors for algal suspension and EOM samples after treatment processes (filtered water THMFP

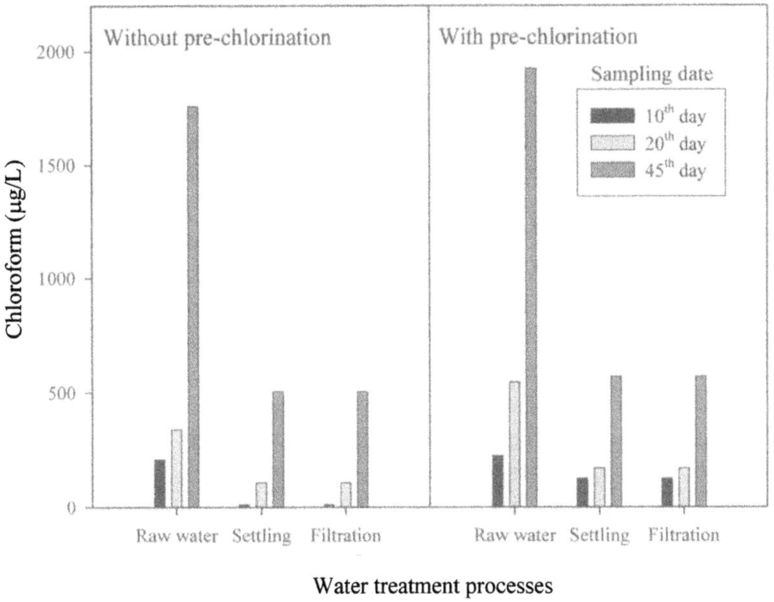

Figure 4. Effect of pre-chlorination of chloroform formation for EOM in different growth phase (in batch culture).

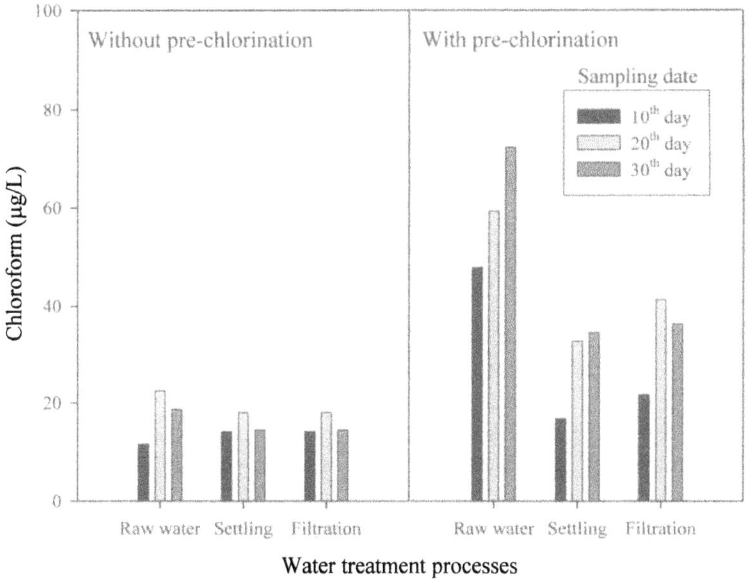

Figure 5. Effect of pre-chlorination on chloroform formation for EOM in different growth phase (in chemostat).

vs. raw water THMFP). For EOM in chemostat mode, the treatment processes remove 22.7% of the THM precursors when pre-chlorination was not applied; and the THM precursors percentage removal increased to 45~48% when samples were pre-chlorinated. It has been shown that the intracellular matters are comprised of high molecular weight organic compounds like humic acid while ECP were composed of low molecular weight organic matter (*11*). The high molecular weight intracellular organic matters produced from cell lysing after pre-chlorination could be partially removed by coagulation and settling units when it is transferred into water. After pre-chlorination, however, it should be noted that the DBPFP is still higher than those without pre-chlorination (Figure 5). For batch mode, the EOM didn't show this phenomenon since the 4 mg/L pre-chlorine dosage is not high enough to oxidize the algal cells in raw water. When the bulk algal suspensions were used for treatments, essentially all of the samples without pre-chlorination gave better percentage removal of THM precursors than that of the samples with pre-chlorination.

Table 3. THMFP percentage removal for algal suspensions after water treatment processes with and without pre-chlorination

Cultivation time	EOM		Algal suspensions	
	No pre-Cl_2	With pre-Cl_2	No pre-Cl_2	With pre-Cl_2
Batch culture				
10th day	94.3	43.7	97.6	49.3
20th day	68.6	69.3	92.4	88.5
45th day	71.2	70.3	80.7	79.5
Chemostat (24±1 °C)				
10th day	0.0*	45.0	82.3	67.2
20th day	19.5	45.3	85.2	72.8
30th day	22.7	47.7	87.4	71.1

*No apparent THMFP removal was observed.

Effects of Pre-chlorination on THM and HAA Species Distribution

Table 4 shows the DBP species distribution for algal suspension when 0.5 mg/L of bromide was spiked into the solutions prior to the DBPFP tests. The algal suspensions cultivated in chemostat produced 17, 26, and 22 μg/L chloroform, respectively, in each growth stage before pre-chlorination. After pre-chlorination, the chloroform production increased to 35, 40, and 62 μg/L, respectively. The percentage of chloroform of the total THMs also increased from 9~13% to 18~22%. HAA analysis also showed similar trends in chemostat samples. The yield of chlorinated HAA (MCAA, DCAA, and TCAA) did not significantly increase after pre-chlorination, but the proportion of the chlorinated-HAA to total HAAs did increase. The results was not apparent when batch cultured samples were used for experiments; the 4 mg/L pre-chlorination dosage was not enough for the oxidation of the high algal cells and ECP concentrations and the results of DBPFP tests are similar to those without pre-chlorination.

It has been reported that the aliphatic precursors favor the formation of brominated DBPs (*12*). Since the extracellular and intracellular degradation products of *M. aeruginosa* belong to aliphatic precursors (*13*), the presence of bromide could increase the DBPs formation and shift the DBP species to brominated DBPs. Figure 6 shows the effects of bromide on THMs and HAAs species distribution from *M. aeruginosa* with and without pre-chlorination. After addition of 0.5 mg/L bromide, the DBP formation shifts to brominated DBP, and this phenomenon is more apparent at THMs analysis.

Table 4. Speciation of THMs and HAAs from *M. aeruginosa* with and without pre-chlorination

Cultivation time	Chloroform		Chlorinated haloacetic acids	
	No pre-Cl_2	With pre-Cl_2	No pre-Cl_2	With pre-Cl_2
Batch culture				
10th day	146 (47%)	92 (33%)	102 (31%)	108 (49%)
20th day	1195 (77%)	1525 (82%)	—	—
45th day	2805 (81%)	2852 (80%)	—	—
Chemostat (24±1°C)				
10th day	17 (10%)	35 (21%)	56 (39%)	23 (72%)
20th day	26 (13%)	41 (18%)	105 (45%)	110 (50%)
30th day	22 (9%)	62 (22%)	43 (26%)	66 (45%)

NOTE: The units are μg/L. Chlorinated haloacetic acids denoted the summation of monochloro-, dichloro- and trichloro-acetic acids. The values in parenthesis stand for the percentage of chlorinated DBPs in total THMs and total HAAs.

Effect of Pre-chlorination for Algae Cultivated at Different Temperature

In order to examine the effect of incubation temperature on the characteristics of algal cells and EOM, colonies of *M. aeruginosa* were cultivated in chemostat mode at different temperatures. The algal suspensions in steady-state (~the 30th day) was used for experiments. Figure 7 shows the chloroform formation of the filtered water with and without pre-chlorination. The effect of pre-chlorination was still prevailing, though not so apparent as described in previous sections, and the removal efficiency after treatment processes followed the similar trends as shown in previous sections. It is

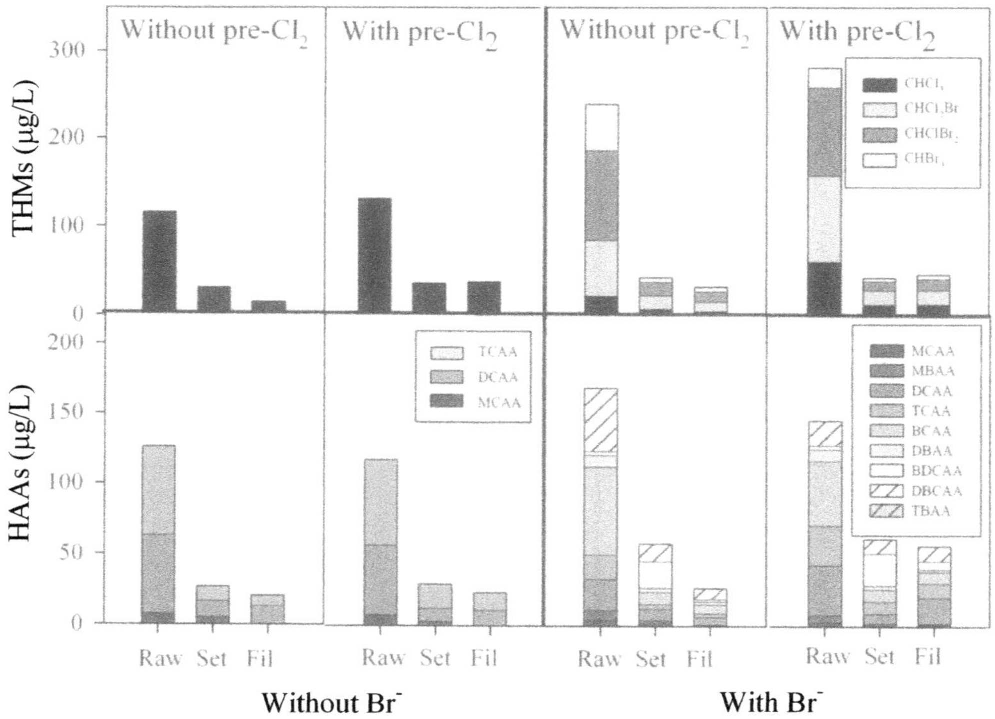

Figure 6. Effect of pre-chlorination on THMs and HAAs speciations (M. aeruginosa cultivated in chemostat for 30 days.)

noticeable that the raw water samples cultivated under 28±1 °C produced higher chloroform than those cultivated under 18±1°C. The NPDOC concentration of the samples cultivated under 28±1 °C increased from 2.1 mg/L to 2.7 mg/L, and for the samples cultivated under 18±1 °C the NPDOC increased from 1.0 mg/L to 1.1 mg/L after pre-chlorination (see Figure 1). Figure 8 shows the chloroform formation for EOM of filtered water with and without pre-chlorination. After pre-chlorination, all samples showed a 20~50% increase of chloroform formation, and higher cultivation temperature gives higher chloroform formation.

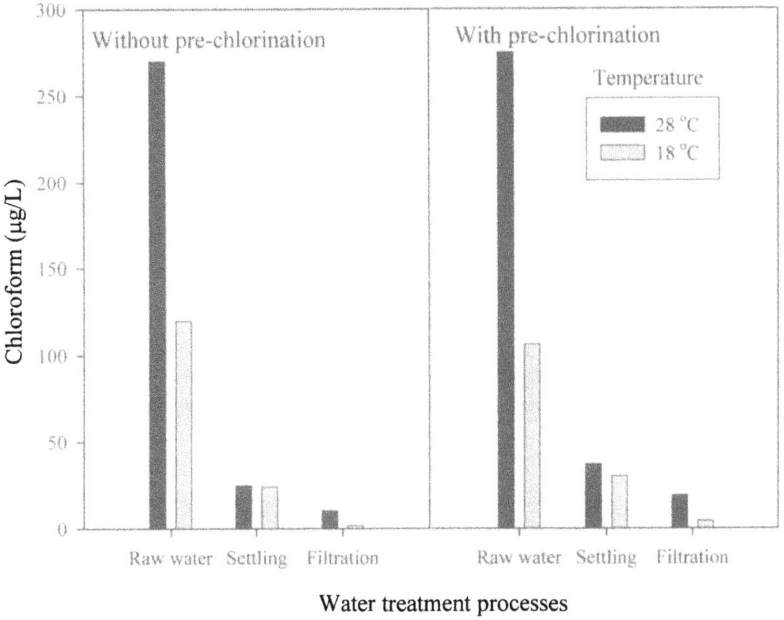

Figure 7. Effects of cultivation temperature on chloroform formation from pre-chlorination of algal suspensions cultivated in chemostat.

It has been reported that the optimal temperature for *M. aeruginosa* to reproduce is about 27 °C (*14*). The concentrations of chlorophyll a and NPDOC also indicated that the conditions of chemostat conducted at three temperatures developed similar quantity of cells (950,000 cells/mL for 28± 1 °C and 897,000 cells/mL for 18±1 °C) in steady-state. However, the relatively low chloroform formation from EOM that cultivated at 18±1 °C with and without pre-chlorination showed that effect of pre-chlorination is not apparent for EOM cultivated at low temperatures. Figure 9 shows the images of the algal cells growth under 28± 1 °C and 18±1 °C and the effect of pre-chlorination. The *M. aeruginosa* colony cultivated at 28±1 °C was more scattered while the algal cells

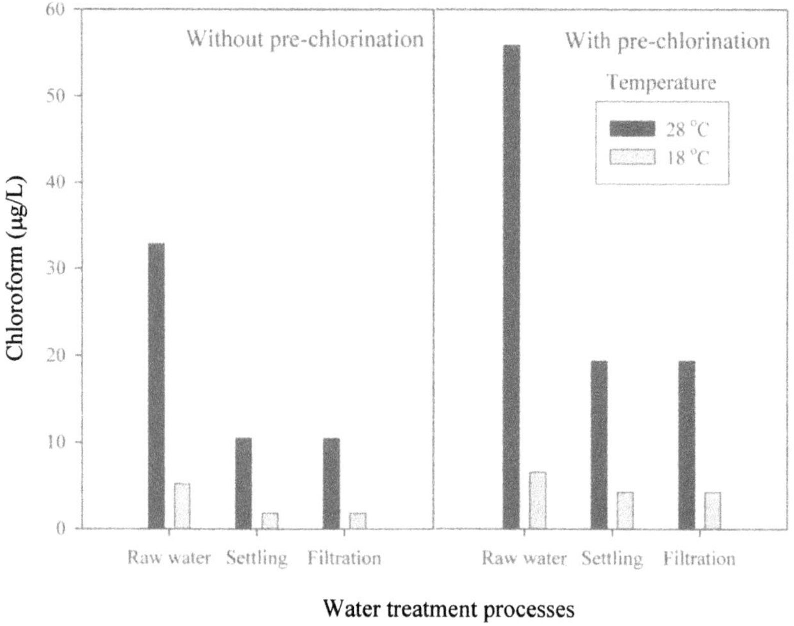

Figure 8. Effects of cultivation temperature on chloroform formation from pre-chlorination of EOM cultivated in chemostat.

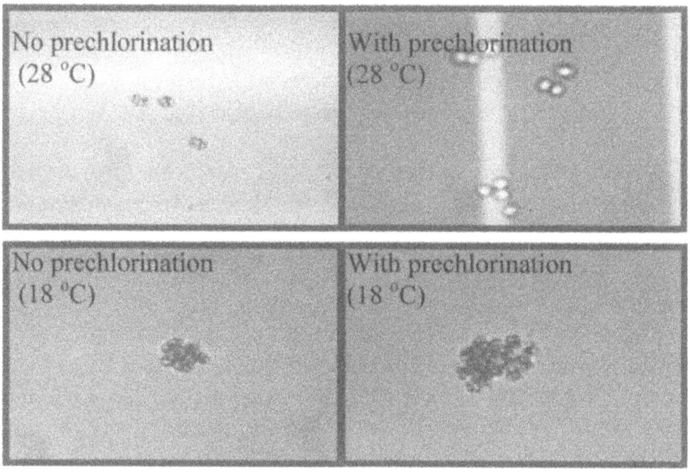

Figure 9. Effects of preuchlorination on micrograph images of M. aeruginosa cultivated in chemostate with different temperatures.

were clustered when cultivated at 18±1°C. After pre-chlorination, the algal cells cultivated at 28±1 °Cwere damaged while the algal cells cultivated at 18±1 °C maintained their original integrity. It is observed that the algae cells cultivated at 18±1 °C could resist the chlorine oxidation and thus release less DBP precursors into water.

Conclusions

The effects of pre-chlorination on *M. aeruginosa* and its impacts on DBPs formation in water treatment processes was evaluated in this study. The results showed that the algal cells cultivated at higher water temperature gives higher algal cell growth and NPDOC concentrations in water when steady-state condition is reached in chemostat mode; and a significant increase of THM precursors accompanying the algal growth was observed in batch system. It was observed that the pre-chlorination caused the cell lysing of algal cells, and the cell lysing contributes to the increase of NPDOC and hence the DBP precursors in the chlorinated water. The coagulation and sedimentation treatments reduced more than 50% of the DBPs precursors in water containing algae suspensions, and the filtration process can only enhance minor parts of THM precursors removal. Based on the results of this study, it was also observed that the primary source of DBPs precursors comes from the EOM of the algal cells, and these EOM can not be effectively removed by conventional water treatment processes. Algae cultivation at different temperatures showed that the algal cells cultivated at lower water temperature released less intracellular organic matter and less amounts of DBPs precursors after pre-chlorination than that cultivated at higher temperature. Removal of algae cells prior to oxidation will minimize DBPs formation.

References

1. Peterson, H. G.; Hrudey. S. E.; Cantin, I. A.; Perley, T. R.; Kenefick, S. L. *Wat. Res.* **1995**, *29(6)*, 1515-1523.
2. Graham, N. J. D.; Wardlaw, V. E.; Perry, R.; Jiang, J. Q. *Wat. Sci. Technol.* **1998**, *37(2)*, 83-89.
3 Wachter, J. K.; Andelman, J. B. *Environ. Sci. Technol.* **1984**, *18(11)*, 811-817.
4. Oliver, B. G. *Environ. Sci. Technol.* **1983**, *17(2)*, 80-83.
5. Karimi, A. A.; Singer, P. C. *J. Amer. Water Works Assoc.* **1991**, *83(3)*, 84-88.
6. Plummer, J. D.; Edzwald, J. K. *Environ. Sci. Technol.* **2001**, *35(18)*, 3661-3668.

7. Sugiura, N.; Yagi, O.; Sudo, R. *Environ. Technol. Letters* **1986**, *7(2)*, 77-86.
8. Environmental Analysis Laboratory, EPA, Executive Yuan, R.O.C. NIEA E508.00B.
9. APHA; AWWA; WEF. Standard Methods for the Examination of Water and Wastewater, 21st Edition. **2005**.
10. Zimba, P. V.; Dionigi, C. P.; Millie, D. F. *J. Phycol.* **1999**, *35(6)*, 1422-1429.
11. Henatsch, J. J.; Juttner, F. *Wat. Sci. Technol.* **1983**, *15(6-7)*, 259-266.
12. Boyce, S. D.; Hornig, J. F. *Environ. Sci. Technol.* **1983**, *17(4)*, 202-211.
13. Hellergrossman, L.; Manka, J.; Limonirelis, B.; Rebhun, M. *Wat. Res.* **1993**, *27(8)*, 1323-1331.
14. Yagi, O.; Ohkubo, N.; Tomioka, N.; Okada, M. *Environ. Technol.* **1994**, *15(4)*, 389-394.

Probing Formation of Disinfection By-Products

Chapter 11

Spectroscopic Studies of the Roles of Distinct Chromophores in NOM Chlorination and DBP Formation

Gregory V. Korshin and Hyun-Shik Chang

Department of Civil and Environmental Engineering, University of Washington, Seattle, WA 98195-2700

This chapter describes the results of *in situ* examination of chlorination in several waters (Lk. Washington, WA; Cincinnati, OH; Florence, NE; Sioux Fall, SD; Hillsborough River, FL) using the method of differential absorbance spectroscopy. The chlorination of NOM was observed to involve distinct groups of slowly and rapidly reacting chromophores. The fast chromophores in all water sources have identical spectroscopic characteristics while the properties of the slow chromophores exhibit some variability. The fast chromophores are consumed at the beginning of NOM chlorination, but their destruction is not associated with a significant release of total trihalomethanes (TTHM) or haloacetic acids (THAA). The engagement of the slow chromophores is closely associated with the generation of THMs and HAAs. Correlations between TTHM or THAA concentration and the differential absorbance associated with the slow chromophores were similar for all studied waters.

Since the recognition in the 1970s of chloroform and other trihalomethanes (THMs) as ubiquitous disinfection by-products (DBP) formed in drinking water, numerous volatile and hydrophilic DBP species have been found (1,2). THMs and haloacetic acids (HAAs) are by far the largest DBP groups, typically contributing ca. 50% of the total organic halogen (TOX) in chlorinated water (1,3).

Mechanistic models of DBP formation assume that the chlorination of NOM involves stepwise chlorine incorporation into predominantly aromatic reactive groups characteristic for NOM, followed by breakdown of these groups to release smaller DBPs (4-6). Such models can be employed to develop a kinetic description of NOM halogenation, but this approach is limited by uncertainties regarding the chemical and kinetic properties of the reactive NOM functionalities. Phenomenological models of DBP formation (e.g., those based on regressions relating DBP formation to DOC, chlorine dose, pH, etc.) can achieve a tolerable goodness-of-fit, but only by inclusion of a number of site-specific empirical fitting parameters for each DBP species (5,7). These complexities have necessitated the use of surrogate parameters such as absorbance of light at 254 nm (A_{254}) and the specific absorbance at that wavelength ($SUVA_{254}$) to estimate NOM reactivity towards chlorination (8,9). We previously introduced an alternative approach, in which the intensity of the differential absorbance at 272 nm (ΔA_{272}) is measured during the course of chlorination and correlated with the formation of specific DBPs and/or TOX concentrations generated in the solution (10-12).

The differential spectra of chlorinated NOM from many natural waters exhibit common features (10-12), most notably a prominent band with a maximum located between 265 to 275 nm. The intensity of this band increases at higher reaction times and chlorine doses. Its shape also exhibits subtle changes as the reaction progresses (13), but the significance of these changes have not been exhaustively examined. In our recent study (14), we presented a view that changes of the shape of the differential spectra of NOM in a representative surface water (Lk. Washington, WA) are indicative of the engagement of spectroscopically and kinetically distinct functionalities which were termed "fast" and "slow" chromophores due to their appearance in the initial phase of NOM chlorination and later in the reaction, respectively. It was also shown that the appearance of selected individual DBPs (exemplified in reference (14) by chloral hydrate and dichloroacetic acid) was closely correlated with the engagement of the slow chromophores, but not with that of the fast chromophores.

In this paper, we will examine whether these conclusions are applicable to a more diverse group of waters and individual DBP species and whether the technique of differential spectroscopy can be amended to determine yields of the most important groups for DBP species, total THMs and HAAs (TTHM and THAA) based on *in situ* measurements.

Experimental

Experiments reported in this study were carried out using water from Lake Washington (LW) located in Seattle, WA and water samples provided by Cincinnati, Florence and Sioux Falls water treatment plants (WTP). High-DOC water from Hillsborough River (Tampa, Fl) was also used in the study, after six-fold dilution with deionized water. Main characteristics of the water samples are summarized in Table I.

In all cases, chlorination was carried out at 20°C, pH 7.0 with 0.03 M phosphate buffer. All experimental procedures followed established practices that have been described in previous publications (e.g., 11). Chlorine was added as HClO and is reported here as the equivalent amount of Cl_2. The chlorine dose is presented in terms of the weight ratio of Cl_2 added to the concentration of dissolved organic carbon (Cl_2:DOC). Reaction time was varied from 5 minutes to 7 days. Absorbance spectra were measured with a dual-beam Perkin-Elmer Lambda-18 spectrophotometer using 5-cm quartz cells.

Table I. Selected Characteristics of Waters Used in the Study

Name (Location)	*Alk, mg/L as $CaCO_3$*	*DOC, mg/L*	*$SUVA_{254}$, $cm^{-1}mg^{-1}L$*
Lake Washington (Seattle, Wash.)	40	3.0	0.022
Cincinnati WTP (Cincinnati, OH)	70	2.3	0.020
Missouri R./ Florence WTP (Metropolitan Utilities District, Omaha, Neb.)	170	5.0	0.024
Big Sioux Aquifer (Sioux Falls Public Works, Sioux Fall, S.D.)	230	8.5	0.023
Hillsborough R. (City of Tampa, Tampa, Fla.)*	85	27.5	0.049

* Hillsborough R. water was diluted 6 times prior to experiments

Results and Discussion

The differential spectra of all waters examined in this study exhibited typical features observed in prior research (10-12). Specifically, the intensity of the differential absorbance at $\lambda > 250$ nm increased with increasing reaction time and Cl_2:DOC ratio, and all differential spectra had a characteristic band with a maximum around 272 nm (Figure 1). The shapes of the overall differential spectra underwent consistent changes in response to increases in reaction time and/or chlorine dose. When the differential spectra generated at varying reaction times were normalized by dividing all values by the maximum value of $-\Delta A$ in that spectrum, the characteristic band steadily broadened as the reaction progressed, as shown in Figure 1 for Missouri River water sampled at the Florence WTP. A similar trend in the width of the normalized spectrum characterized all the water sources investigated. We quantified this trend by tracking the wavelength at which the normalized differential absorbance was 0.5, as illustrated in Figure 1 for chlorinated Florence WTP water. This parameter is denoted henceforth as $\lambda^{nd}_{0.5}$.

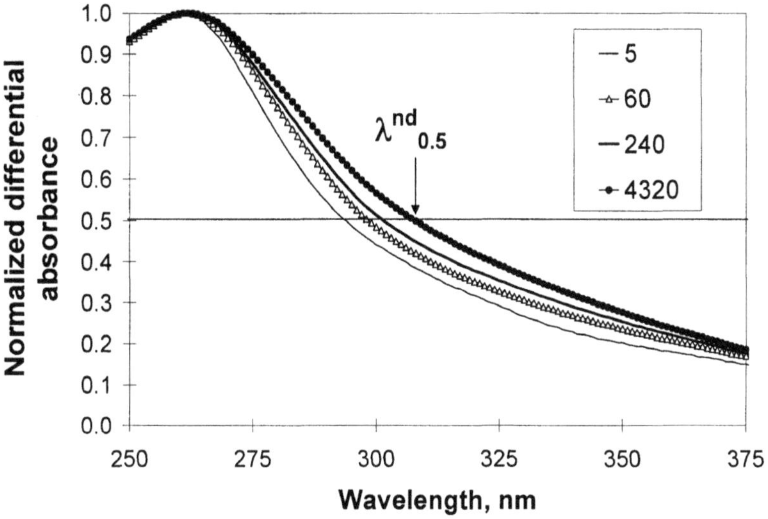

Figure 1. Changes of normalized differential absorbance spectra of chlorinated Florence WTP water at varying reaction times. Cl_2:DOC = 1.5, pH 7.0, 20°C.

For all examined waters, $\lambda^{nd}_{0.5}$ increased with reaction time and intensity of the differential absorbance,$-\Delta A_{272}$. Correlations between $\lambda^{nd}_{0.5}$ and $-\Delta A_{272}$ for four waters are shown in Figure 2. The curves are nearly linear and, although their slopes appear to be close, the intercepts exhibit some site-specificity that is prominent for Sioux Falls water. The reason that the curve for the Sioux Fall site is offset from the others is not clear, but Sioux Falls water is the only groundwater in the group, and it is possible that its NOM is therefore different from the NOM in the other waters.

The monotonic increase of $\lambda^{nd}_{0.5}$ Figure 2 indicates a steady broadening of the characteristic band in the differential spectra of NOM at increased reaction

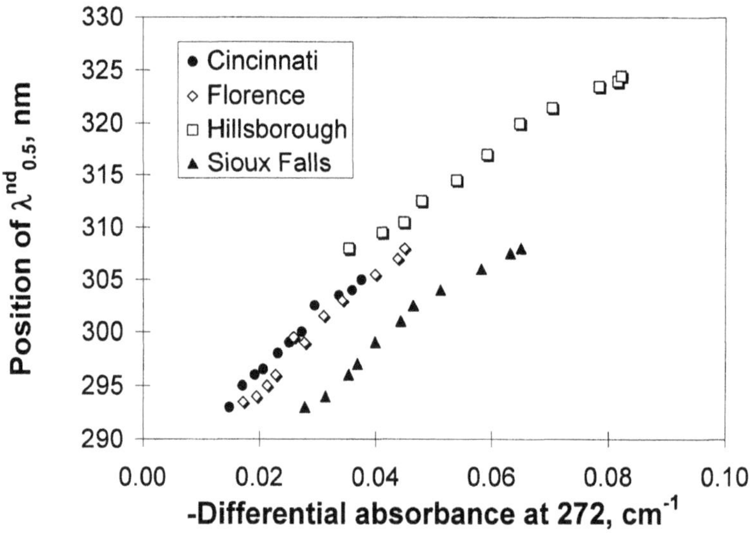

Figure 2. Comparison of changes of $\lambda^{nd}{}_{0.5}$ values as a function of overall differential absorbance values measured at 272 nm.

times and/or chlorine doses. It has been shown in our studies that this phenomenon is a manifestation of the presence of two distinct components in the differential spectrum whose contributions can be quantified via numerical deconvolution (14). Figure 3 shows the normalized spectra of these components obtained for chlorinated Lk. Washington water. In reference (14), it was also established for Lk. Washington water (and confirmed in this study, as will be described in the detail in the sections that follow) that these components are associated with chromophoric groups that either rapidly react with chlorine or emerge later in the reaction, and for that reason these components are referred to as fast and slow chromophores, respectively.

Because the differential spectra can be characterized using only two independent components, their contributions can be discerned using two wavelengths at which the relative absorbances of these components are sufficiently different. One possible such combination uses the wavelengths 272 and 350 nm. The former wavelength is close to the maxima of both components shown in Figure 3, while at the latter wavelength, the difference in the normalized absorbances of these components is pronounced. In our prior publication, we utilized the ratio of the differential absorbances at 272 and 350 nm to track the evolution of these components.

A more mathematically transparent approach can be used to examine the behavior of the slow and fast chromophores. For instance, one can use the values of relative normalized absorbances of the fast and slow chromophores at 272 and 350 nm (denoted as ε_{272}^{fast}, ε_{350}^{fast} and ε_{272}^{slow}, ε_{350}^{slow} for the fast and slow chromophores, respectively) and measurements of the differential absorbances at 272 and 350 nm to calculate, using appropriate matrix algebra, the values of differential absorbances assigned to these functionalities and denoted as ΔA_{ghast} and ΔS_{allow}, respectively. (Note that $\varepsilon_{\lambda}^{fast}$ and $\varepsilon_{\lambda}^{slow}$ values are not true molar absorbances but rather indicators of the relative shapes of the two dissimilar absorbance spectra.) Mathematically, this concept can be expressed via the following two equations that define the values of differential absorbance at 272 and 350 nm for any combination of initial chlorine dose or reaction time:

$$\varepsilon_{272}^{fast}\Delta A_{fast} + \varepsilon_{272}^{slow}\Delta A_{slow} = \Delta A_{272} \quad (1)$$

$$\varepsilon_{350}^{fast}\Delta A_{fast} + \varepsilon_{350}^{slow}\Delta A_{slow} = \Delta A_{350} \quad (2)$$

Numerical deconvolution of the differential spectra generated for chlorinated Lk. Washington water showed that the values of ε_{272}^{fast}, ε_{350}^{fast}, ε_{272}^{slow} and ε_{350}^{slow} were 1.00, 0.15, 0.99 and 0.38, respectively.

A major hypothesis explored in this study was that the $\varepsilon_{\lambda}^{fast}$, $\varepsilon_{\lambda}^{slow}$ matrices are similar for all studied waters. To test this hypothesis, it was assumed that the evolution of ΔA_{fast} and ΔA_{slow} as a function of the overall $-\Delta A_{272}$ values for all waters will have a pattern similar that observed for Lk. Washington. That is, during the initial phase of chlorination, only the fast chromophores characterized by a sharp peak close to 272 nm and a low value of ε_{350}^{fast} would be present.

However, for $-\Delta A_{272}$ values above a certain threshold value (e.g., -0.008 cm^{-1} for Lk. Washington), the contribution of the fast chromophores is expected to reach a plateau while that of the slow chromophores to increase rapidly.

When $\varepsilon_{\lambda}^{fast}$, $\varepsilon_{\lambda}^{slow}$ matrices with components practically identical to those for Lk. Washington were applied for Florence and Cincinnati WTP waters (only the value of ε_{350}^{slow} was adjusted to 0.35 and 0.37, for Florence and Cincinnati, respectively), a pattern of behavior closely matching that described above was observed (Figure 4).

Specifically, the contribution of the fast chromophores was at a plateau for all reaction periods, except the shortest reaction times for which chlorination experiments and absorbance measurements could be carried in the conventional mode (5 and 10 minutes). At the same time, the contribution of the slow chromophores emerged and increased quasi-linearly only in a $-\Delta A_{272}$ range exceeding a threshold value of ca. -0.006 and 0.010 for Cincinnati and Florence WTP water, respectively (Figure 4). (The region of very small $-\Delta A_{272}$ values was not probed in these experiments because that would require using very small chlorine doses and extremely short reaction times). A virtually instantaneous emergence and ensuing stabilization of the contribution of the fast chromophores

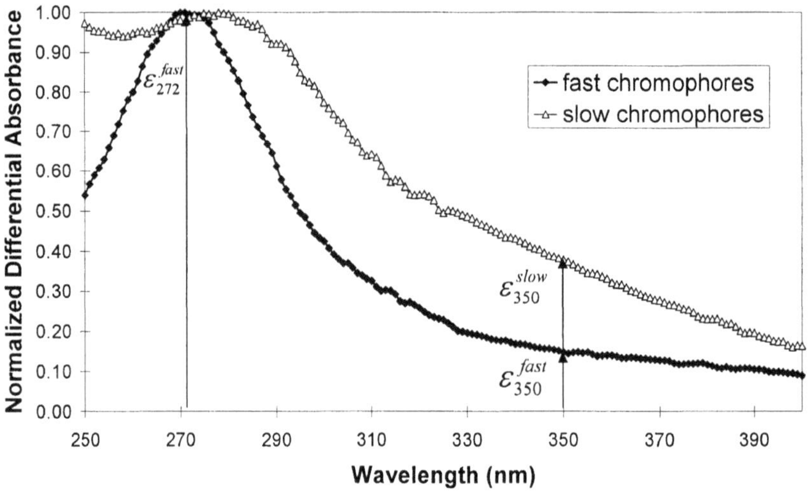

Figure 3. Normalized spectra of the fast and slow chromophores contributing to f the differential spectra of chlorinated NOM, as inferred by deconvolution of the differential spectra from the chlorination of Lk. Washington water. (Reproduced from reference 14. Copyright 2007 American Chemical Society.)

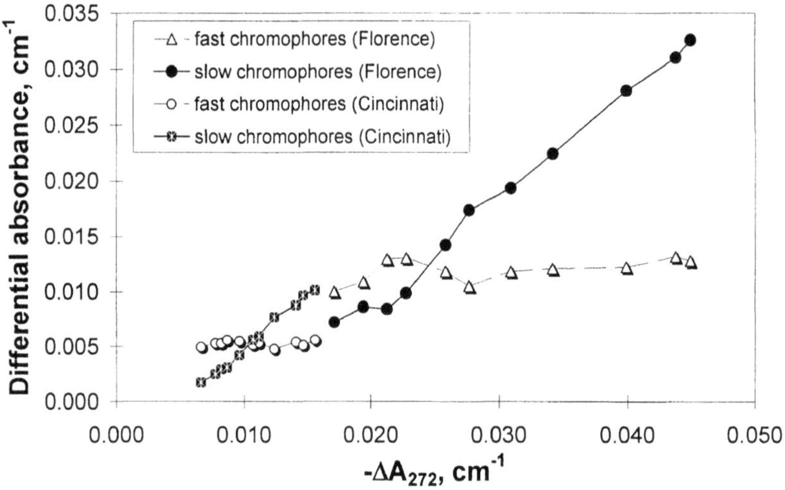

Figure 4. Development of contributions of fast and slow chromophores as a function of overall differential absorbance measured at 272 nm. Data for chlorinated Florence and Cincinnati water, pH 7.0.

and a steady development of the slow chromophores can be additionally demonstrated when the values of $-\Delta A_{272}$, ΔA_{fast} and ΔA_{slow} are represented as a function of reaction time, as exemplified in Figure 5 for Florence WTP water. Kinetic profiles represented in that figure clearly show that the former group emerges immediately following the introduction of chlorine and does not appreciably change throughout the entire reaction period. The slow chromophores also emerge quite rapidly but their contribution increases monotonically with time.

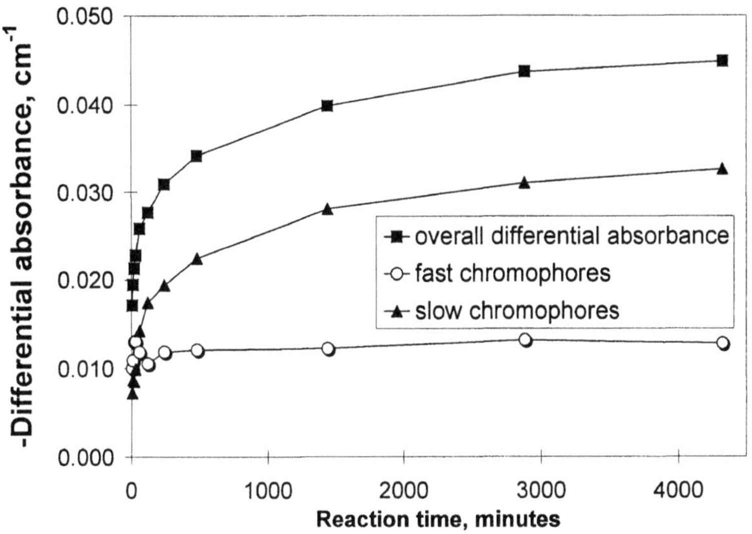

Figure 5. Time profile of the differential absorbance at 272 nm and contributions of the fast and slow chromophores in Florence WTP water chlorinated at pH 7.0, Cl_2/DOC ratio 1.5.

Data similar those shown in Figure 4 and Figure 5 were obtained for Sioux Falls WTP and Hillsborough River water. In all cases, practically identical $\varepsilon_\lambda^{fast}$, $\varepsilon_\lambda^{slow}$ matrices were used, and only the value of ε_{350}^{slow} was adjusted to 0.34 and 0.46 for Sioux Falls and Hillsborough waters, respectively.

It is unclear why the normalized relative absorbance of the slow chromophores at 350 nm exhibits some variability. It was noted, however, that the value of ε_{350}^{slow} was considerably higher in Hillsborough water that also had the highest $SUVA_{254}$ value (see Table I). Whether this is a manifestation of some sensitivity of the spectroscopic properties of the slow chromophores to the overall NOM aromaticity needs to be explored separately.

In sum, the quantitation of the behavior of the slow and fast chromophores in a variety of natural waters using measurements of differential absorbances at 272 and 350 nm and largely identical matrices of normalized absorbance coefficients indicated a similarity of the observed patterns. Given that, we proceeded to examine a related hypothesis, which was concerned with relationships between the emergence of the slow chromophores and generation of THMs and HAAs. This hypothesis assumes that that the behavior of the slow chromophores and release of individual DBPs, most importantly THMs and HAAs, should be closely correlated. Indeed, our measurements carried out for dichloroacetic acid and chloral hydrate formed in Lk. Washington water indicated the existence of such correlations, which were more linear than those for the latter DBP species and, on the other hand, overall $-\Delta A_{272}$ values (14).

Prior research established that the correlation between the concentration of virtually any stable DBPs and the overall $-\Delta A_{272}$ value is invariably very strong. At the same time, these correlations are non-linear and have a threshold $-\Delta A_{272}$ value below which little individual DBPs are generated (10-14). Major trends in the behavior of TTHM and THAA concentrations formed in several waters and plotted in the $-\Delta A_{272}$ domain (Figure 6 and figure 7; data for Hillsborough River water are not shown for simplicity) were identical to those described in prior research. To test whether the formation of TTHMs and THAAs is associated with that of the slow chromophores, the TTHM and THAA concentrations were also represented vs. their respective ΔA_{slow} values (Figure 8 and Figure 9).

Examination of these figures shows that there are three features that distinguish these plots from those shown in Figure 6 and Figure 7. Fist, in comparison with the behavior of TTHM and THAA as a function of overall $-\Delta A_{272}$ values, the TTHM or THAA vs. ΔA_{slow} relationship appear to be more linear. At the same time, the intercept of these correlations with the y-axis is non-zero. This is likely to indicate that some level of TTHM and THAA formation takes place when the fast chromophores interact with chlorine. Second, the threshold $-\Delta A_{272}$ values prominent in Figure 6 and Figure 7 below which little DBPs are formed, are absent in the ΔA_{slow} plots. Third, the slopes of THAA and THM vs. the $-\Delta A_{slow}$ correlations are comparable and generally follow one imaginary trend line. This behavior is distinct from that shown in Figure 6 and Figure 7, where the data for different waters are positioned in different areas of the graph.

The similarity of the relationships between the yield of major DBP classes and ΔA_{slow} values is additionally shown in Figure 10. In that case, sums of THAA and THM concentrations for all the studied waters are compared. To demonstrate the similarity of these correlations, a best fitting line developed for the combination of all TTHM and THAA data obtained for all waters is shown in Figure 10.

The experimental data for individual waters shown in Figure 10 do not deviate significantly from that best-fit line. This can be interpreted to signify that

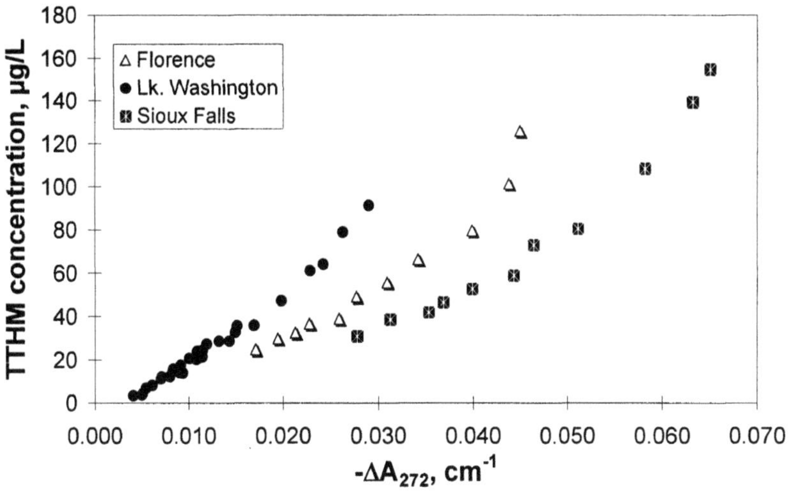

Figure 6. Relationships between the formation of total trihalomethanes (TTHM) and the intensity of differential absorbance at 272 nm in Florence, Lk. Washington and Sioux Falls water chlorinated at pH 7.0.

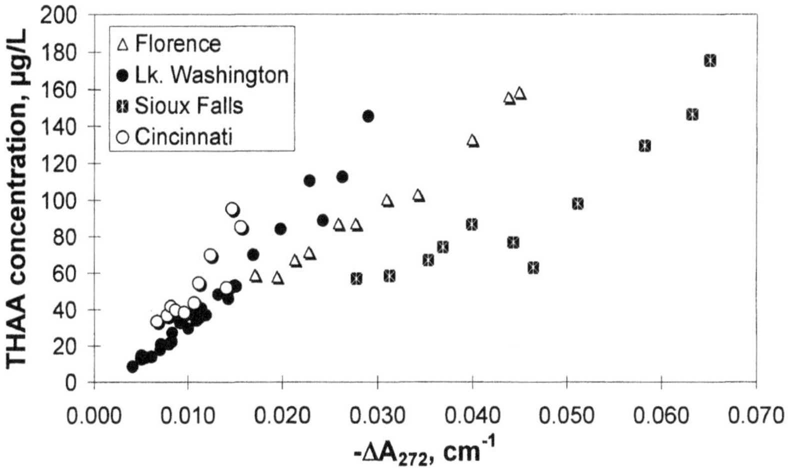

Figure 7. Relationships between the formation of total haloacetic acids (THAA) and the intensity of differential absorbance at 272 nm in Cincinnati, Florence, Lk. Washington and Sioux Falls water chlorinated at pH 7.0.

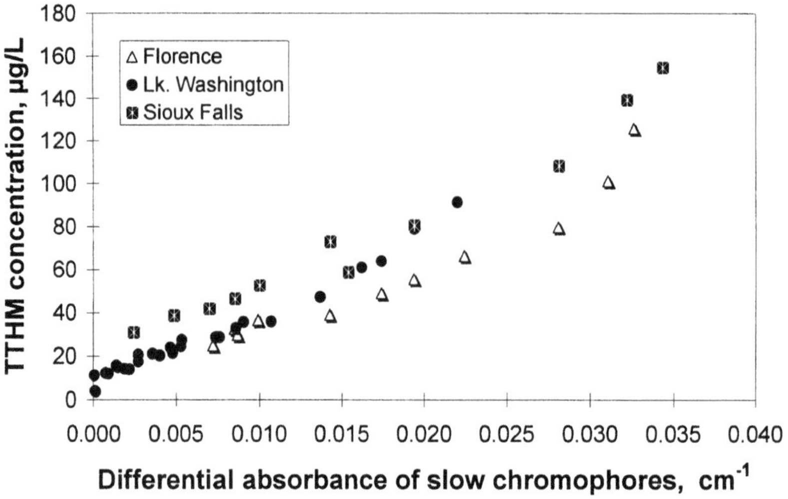

Figure 8. Comparison of relationships between the formation of total trihalomethanes and the differential absorbance associated with the slow chromophores in Florence, Lk. Washington and Sioux Falls water chlorinated at pH 7.0.

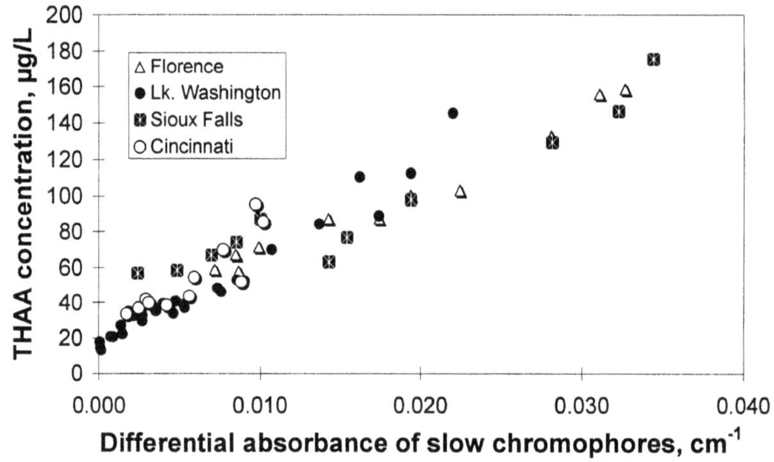

Figure 9. Comparison of relationships between the formation of total haloacetic acids (THAA) and the differential absorbance associated with the slow chromophores in Cincinnati, Florence, Lk. Washington and Sioux Falls water chlorinated at pH 7.0.

the properties of the slow chromophores which are formed as a result of the initial very rapid incorporation of chlorine into the reactive sites in NOM exhibit considerable similarity for different waters. Specifically, yields of TTHM, THAA or their sums per amount of the slow chromophores formed (and consumed) during NOM chlorination appear to be close for all sources. Further work is needed to establish whether this result is applicable to a wider variety of waters and DBP types, and whether variations of important water chemistry parameters (e.g., pH, temperature) affect this result.

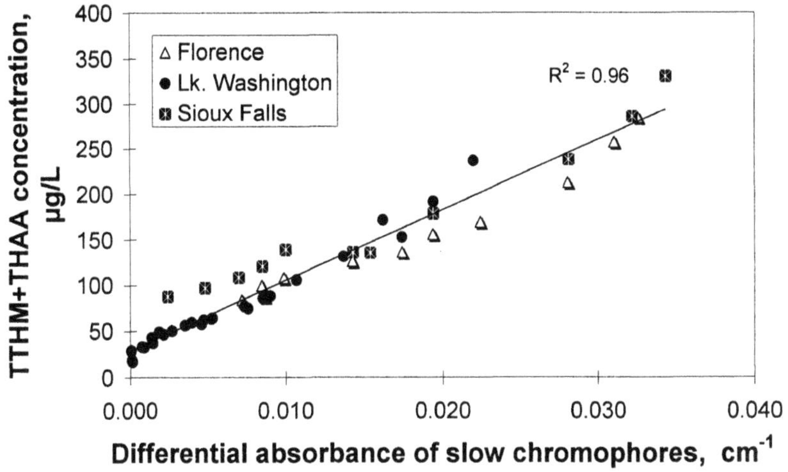

Figure 10. Comparison of relationships between the formation of the sum of total haloacetic acids (THAA) and total trihalomethanes (TTHM) and the differential absorbance associated with the slow chromophores in Cincinnati, Florence, Lk. Washington and Sioux Falls water chlorinated at pH 7.0.

Conclusions

Examination of chlorination and DBP formation in several representative waters using the method of differential absorbance spectroscopy showed that NOM chlorination in all cases proceeds via the involvement of spectroscopically distinct groups of slow and fast chromophores. The fast chromophores in the studied waters have identical spectroscopic characteristics while those of the slow chromophores exhibit some variability from one source to another.

The fast chromophores are consumed almost immediately at the beginning of NOM chlorination, but their reaction does not lead directly to the generation of THMs and HAAs. Rather, transformations of the fast chromophores largely precede the emergence of the slow chromophores. The engagement of the slow chromophores is closely associated with the generation of TTHMs and THAAs. Correlations between TTHM and THAA concentrations and, on the other hand, the magnitude of the differential absorbance specifically associated with the engagement of the slow chromophores were similar for all waters whose chlorination was probed in this study.

Acknowledgments

This study was supported by American Water Works Association Research Foundation (Project #2597). The views represented in this publication do not necessarily represent those of the funding agency.

References

1. Krasner, S.W.; McQuire, M.J.; Jacangelo, J.C.; Patania, N.L.; Reagan, K.M.; Aieta, E.M. The occurrence of disinfection by-products in US drinking water. *J. Amer. Water Works Assoc.*, **1989**, *81* (8), 41-53.
2. Richardson, S.D.; Simmons, J.E.; Rice, G. Disinfection by-products: The next generation. *Env. Sci. Technol.*, **2002**, *36* (9), 198A-205A.
3. Singer, P.C.; Chang, S.D. Correlations between trihalomethanes and total organic halides during water treatment. *J. Amer. Water Works Assoc.*, **1989**, *81* (8), 61-65
4. Cowman, G.A.; Singer, P.C. Effect of bromide ion on haloacetic acid speciation resulting from chlorination and chloramination of aquatic humic substances. *Environ. Sci. Technol.*, **1996**, *30* (1), 16-24.
5. Gallard, H.; Von Gunten, U. Chlorination of natural organic matter: Kinetics of chlorination and of THM formation. *Water Res.*, **2002**, *36* (1), 65-74.
6. Nokes, C.J.; Fenton, E.; Randall, C.J. Modeling the formation of brominated trihalomethanes in chlorinated drinking waters. *Water Res.*, **1999**, *33* (17), 3557-3568
7. Clark, R.M.; Thurnau, R.C.; Sivaganesan, M.; Ringhand, P. Predicting the formation of chlorinated and brominated by-products. *J. Env. Engineering – ASCE*, **2001**, *127*(6), 493-501
8. Edzwald, J.K.; Becker, W.C.; Wattier, K.L. Surrogate parameters for monitoring organic matter and THM precursors. *J. Amer. Water Works Assoc.*, **1985**, *77* (4), 122-132
9. Kitis, M.; Karanfil, T.; Kilduff, J.E.; Wigton, A. The reactivity of natural organic matter to disinfection by-products formation and its relation to specific ultraviolet absorbance. *Water Sci. Technol.*, **2001**, *43* (2), 9-16

10. Korshin, G.V.; Li, C.W.; Benjamin, M.M. The decrease of UV absorbance as an indicator of TOX formation. *Water Res.,* **1997**, *31* (4), 946-949.
11. Korshin, G.V.; Wu, W.; Benjamin, M.M.; Hemingway, O. Correlations between differential absorbance and the formation of individual DBP species. *Water Res.,* **2002**, *36* (13), 3273-3282.
12. Korshin, G.V.; Benjamin; M.M.; Hemingway, O.; Wu, W. 2002. Development of Differential UV Spectroscopy for On-line DBP Monitoring. AWWA Research Foundation and American Water Works Association, Denver, CO.
13. Korshin, G.V.; Benjamin, M.M.; Xiao, H.B. Interactions of chlorine with natural organic matter and formation of intermediates: Evidence by differential spectroscopy. *Acta Hydrochimica et Hydrobiologica,* **2001**, *28* (7), 378-384.
14. Korshin, G.V.; Benjamin, M.M.; Chang, H.S.; Gallard, H. Examination of NOM chlorination reactions by conventional and stop-flow differential absorbance spectroscopy. *Environmental Science and Technology,* **2007**, *41* (8), 2776-2781

Chapter 12

Formation of Volatile Disinfection Byproducts from Chlorination of Organic-N Precursors in Recreational Water

Jing Li and Ernest R. Blatchley III

School of Civil Engineering, Purdue University, West Lafayette, IN 47907

Volatile disinfection byproducts (DBPs) formed in recreational water can function as respiratory irritants, and may promote asthma. However, little information has been published regarding volatile DBP formation in recreational water settings. In this research, the formation of volatile DBPs resulting from chlorination of four organic-N precursors (creatinine, urea, L-histidine, and L-arginine) was investigated. Trichloramine, dichloromethylamine, cyanogen chloride, and dichoroacetonitrile were identified and quantified as volatile DBPs by membrane introduction mass spectrometry in bench-scale experiments involving individual organic precursors and in actual swimming pool water samples. Additional experimentation was conducted for identification of possible reaction pathways to describe the formation of these DBPs from relevant organic-N precursors.

Chlorination is the most common method used for disinfection of recreational water. Organic-N compounds are major contaminants in swimming pools, and reactions between free chlorine and organic-N compounds are common in recreational water settings. Most of these organic-N compounds originate in sweat and urine, and can result in the formation of undesirable DBPs, including some that are sufficiently volatile to be transferred to the gas phase. For example, trihalomethanes have been reported to be formed during chlorination of materials of human origin (*e.g.*, hair, lotion, saliva, skin, and urine) (*1*) and during chlorination of human urine analogs containing urea, creatinine and citric acid (*2*). Chlorination of N-containing precursors may also lead to the formation of N-chlorinated DBPs. Moreover, nitrogenous byproducts tend to be more toxic than others (*3*)

A considerable body of anecdotal evidence indicates that some of these compounds can adversely affect human health. For example, childhood asthma could be associated with exposure to chemicals in swimming pools (*4*). Trichloramine (NCl_3) has been reported to function as an irritant to the eyes and upper respiratory tract, and may contribute to acute lung injury in accidental, occupational or recreational exposures to chlorine-based disinfectants (*5, 6*). However, little information is available to describe the specific compounds responsible for volatile DBP formation in recreational waters. Additionally, most available literature related to DBPs in recreational waters focus on inorganic chloramines. Little information is available to describe the formation of other volatile DBPs in recreational water settings.

The objective of this investigation was to identify volatile DBPs resulting from chlorination of organic-N compounds that are likely to be present in recreational water settings. Four model organic-N compounds, urea, creatinine, L-histidine, and L-arginine, were selected as representative organic pollutants of recreational water (refer to Figure 1). Urea is the major nitrogenous end product of protein metabolism and is the chief nitrogenous component of urine and sweat in mammals. Creatinine is a constituent of perspiration and urine formed from the metabolism of creatine; it is found in muscle tissue and blood and is normally excreted in urine and sweat as a metabolic waste product. L-histidine and L-arginine are amino acids that are commonly found in human sweat. Urea, creatinine, and L-histidine are also the primary constituents of "body fluid analogue" (BFA), which has been used as a surrogate mixture of organic-N compounds in previous studies involving chlorination of recreational waters (*7, 8*).

Membrane introduction mass spectrometry (MIMS) was employed to monitor chlorinated DBPs because MIMS has been shown to be a suitable method for analysis of volatile compounds in aqueous samples (*9*). An important benefit of analysis by MIMS is that it yields quantitative and structural information regarding essentially all volatile DBPs present in the sample, not just a single compound. Therefore, the application of MIMS allowed identification of other volatile DBPs that were produced from chlorination reactions.

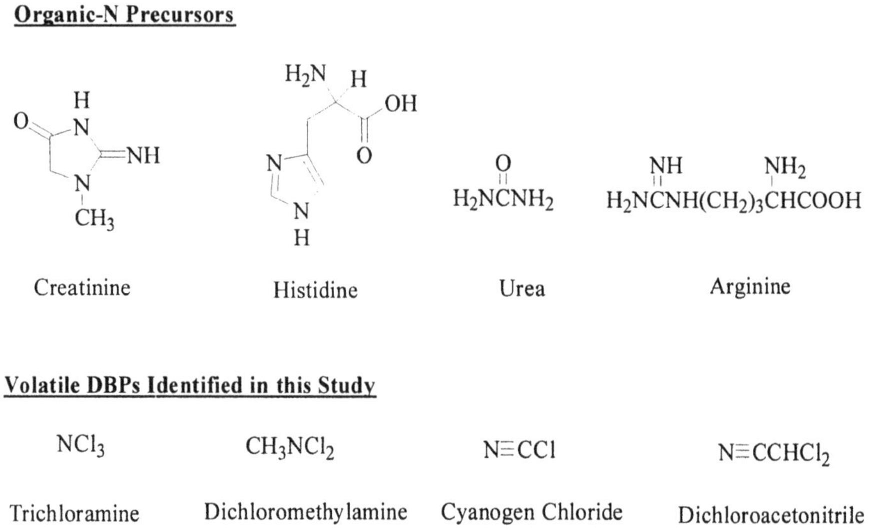

Figure 1. Illustration of the structures of model organic-N compounds and possible volatile DBPs after chlorination.

Materials and Methods

All chemicals used in this study, unless otherwise noted, were reagent-grade, purchased from Sigma-Aldrich, and used without further purification. Dilution to target aqueous-phase concentrations was accomplished with distilled, deionized water. Free chlorine stock solutions and standard solutions of inorganic chloramines were prepared in the same manner as described previously (*9*). Standard solutions of cyanogen chloride (CNCl) were prepared daily from a CNCl stock solution (2000 mg/L Protocol Analytical Supplies, Inc.). Standard solutions of chloroform ($CHCl_3$) and dichloroacetonitrile ($CNCHCl_2$) were prepared gravimetrically from pure compounds. Standard dichloromethylamine (CH_3NCl_2) solutions were prepared by chlorination of methylamine (CH_3NH_2) at Cl:N molar ratio of 2.0; the concentration of CH_3NCl_2 was estimated by DPD/FAS titration as apparent dichloramine.

All bench-scale chlorination experiments were conducted using a bicarbonate buffer system (120 mg/L as $CaCO_3$) at pH=7.5. The experiments were carried out in well-mixed, glass-stoppered, 250 mL flasks in the dark. Aliquots of a standardized sodium hypochlorite (NaOCl) stock solution were then added to the flasks. Free chlorine was added over free chlorine to precursor molar ratios (Cl:P) ranging from 1.6 to 9.6. 0.01 M HCl and 0.01 M NaOH were used to adjust the initial pH of the solution to pH 7.5. The reaction vessels were sealed and kept in the dark at room temperature. The concentrations of free and (apparent) combined chlorine were measured by DPD/FAS titration; MIMS was used to measure residual chlorine and volatile DBPs.

The MIMS system was based on a modification of an HP 5892 benchtop GC/MS comprising an HP 5972A Mass Selective Detector (MSD) equipped with electron (70 ev) ionization (EI). Mass spectrum scan mode ($49 \leq m/z \leq 200$) coupled with EI was used to identify possible DBPs, while selected ion monitoring (SIM) mode was used for quantification of volatile DBPs. Other details of the configuration and setup for MIMS system and operational conditions can be obtained from reference 9. The concentration of volatile DBPs was determined by comparison of ion abundance measurements with those developed from a series of standard solutions. Ions at m/z 61, 74, 85, 98 and 119 amu were monitored for quantification of CNCl, $CNCHCl_2$, $CHCl_3$, CH_3NCl_2, and NCl_3, respectively, and the detection limits of these compounds were 0.003 mg/L, 0.005 mg/L, 0.01 mg/L, 0.01 mg/L as Cl_2, and 0.06 mg/L as Cl_2, respectively. All compound identifications by MIMS were confirmed by analysis of standard solutions.

Swimming pool water samples were collected from indoor and outdoor swimming pools and stored in screw-capped plastic bottles with minimal headspace. The samples were immediately transported back to the Environmental Engineering Laboratories at Purdue University to allow initiation of MIMS analyses within 1 hour after collection.

Results and Discussion

Volatile DBP identification in the bench-scale chlorination experiments

The chlorination of model organic-N compounds including creatinine, urea, L-histidine and L-arginine were studied individually. At least four volatile DBPs were detected by MIMS during the chlorination of these model organic-N compounds: these included trichloramine, dichloromethylamine, cyanogen chloride, and dichloroacetonitrile (refer to Figure 1). Trichloramine, which functions as an irritant to eyes and the respiratory system, was observed to result from chlorination of all four model organic-N compounds. Also, time-course monitoring of reaction progress suggested that trichloramine may persist for several days if no measure is taken to eliminate it from the solution phase. The health effects of CH_3NCl_2 are not known; however, it is characterized by an odor similar to trichloramine. CNCl is well known to be extremely toxic to humans. $CNCHCl_2$ has been identified as an irritant of the respiratory system and skin, and has also been identified as a possible mutagen in humans (*10*). More generally, given that all of these compounds are volatile, they are expected to diminish air quality around swimming pools, especially around indoor pools.

CH_3NCl_2 and NCl_3 formation from chlorination of creatinine

Figure 2 illustrates a mixed mass spectrum of volatile DBPs after 24 hours of chlorination of creatinine at pH 7.5. The peak clusters in the spectrum indicate that two different volatile DBPs were generated. Trichloramine was identified by the existence of a peak cluster at m/z 84 ($N^{35}Cl_2^{\bullet+}$), 86 ($N^{35}Cl^{37}Cl^{\bullet+}$), 88 ($N^{37}Cl_2^{\bullet+}$) at an abundance ratio of 9:6:1, as well as a molecular ion peak at m/z 119 ($N^{35}Cl_3^{\bullet+}$). This spectral pattern was in agreement with spectra that were developed from NCl_3 standard solutions.

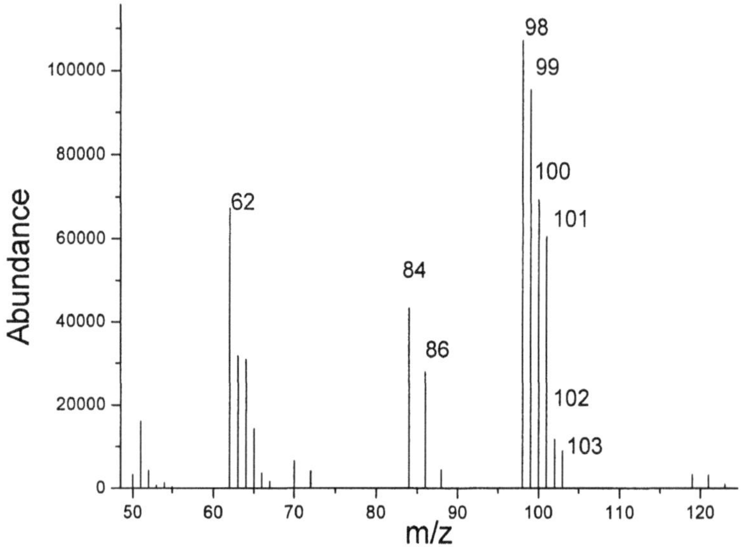

Figure 2. Mass spectrum of a chlorinated sample of an aqueous solution of creatinine after 24 hours. Initial concentration of creatinine = 1.8×10^{-4} M; initial Cl:P molar ratio = 5.0 ; initial pH = 7.5. (Reproduced from Environ. Sci. Technol. ***2007***, 41, *6732-6739. Copyright 2007 American Chemical Society.)*

In addition to the peak clusters described above that were associated with NCl_3, three additional clusters were observed. The peak clusters located at m/z 98($CH_2N^{35}Cl_2^{\bullet+}$)-100($CH_2N^{35}Cl^{37}Cl^{\bullet+}$)-102($CH_2N^{37}Cl_2^{\bullet+}$) and 99($CH_3N^{35}Cl_2^{\bullet+}$)-101($CH_3N^{35}Cl^{37}Cl^{\bullet+}$)-103($CH_3N^{37}Cl_2^{\bullet+}$), each with abundance ratios of 9:6:1, and m/z 62($CHN^{35}Cl^{\bullet+}$), 63($CH_2N^{35}Cl^{\bullet+}$), 64($CHN^{37}Cl^{\bullet+}$, $CH_3N^{35}Cl^{\bullet+}$),

65($CH_2N^{37}Cl^{\bullet+}$) and 66($CH_3N^{37}Cl^{\bullet+}$) suggested the formation of N-dichloromethylamine (dichloromethylamine). MIMS analysis of a standard solution of dichloromethylamine confirmed this finding.

The formation of trichloramine and dichloromethylamine in chlorination of creatinine was studied at different Cl:P molar ratios, ranging from 1.6 to 9.6, with an initial creatinine concentration of 1.8×10^{-5} M. In general, roughly 0.1 mg/L trichloramine (as Cl_2) was detected by MIMS after 96 hours of creatinine chlorination at Cl:P $\geq$ 3.2. The MIMS measurements indicated dichloromethylamine concentrations of less than 10 μg/L (as Cl_2) after 1 hour chlorination; however, the CH_3NCl_2 concentrations increased steadily during the first 24-hour chlorination period, ultimately yielding a concentration of approximately 1.5 mg/L as Cl_2 at chlorine to precursor molar ratio 8.0.

Time-course measurements of CH_3NCl_2 concentration resulting from chlorination of creatinine at different Cl:P molar ratios were also studied with a higher initial creatinine concentration (3.6×10^{-4} M). As shown in Figure 3, the formation of CH_3NCl_2 took approximately 20 hours to complete at Cl:P molar ratios from 1.6 to 8.0. As the free chlorine concentration increased, the yield of CH_3NCl_2 increased as well. The maximum CH_3NCl_2 yield was approximately 66% of the initial creatinine.

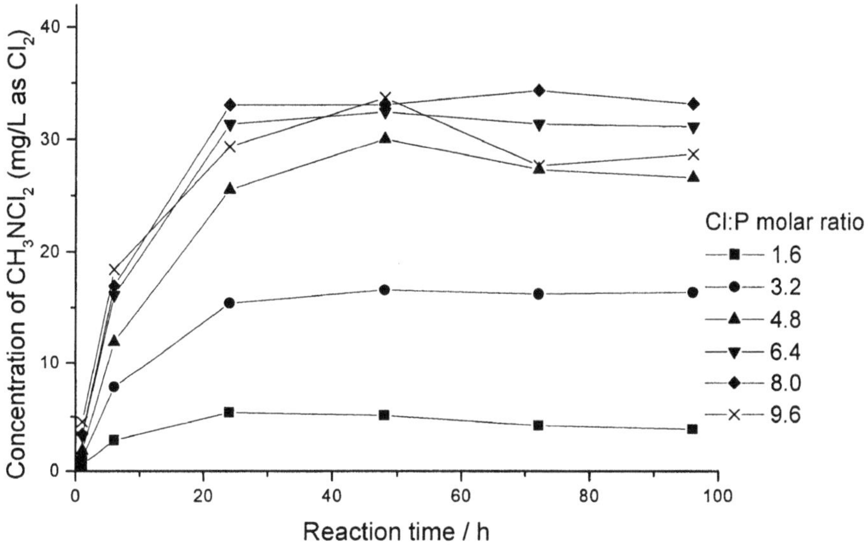

Figure 3. Time-course measurements of dichloromethylamine concentration resulting from chlorination of creatinine at different molar ratios. Experiments were conducted with a creatinine concentration of 3.6×10^{-4} M at pH 7.5. (Reproduced from Environ. Sci. Technol. ***2007****, 41, 6732-6739. Copyright 2007 American Chemical Society.)*

Additional experimentation was conducted for identification of reaction mechanisms to describe the formation of each of these DBPs from chlorination of model compounds. Some aspects of these mechanisms can be generalized to other organic-N compounds.

A proposed mechanism for formation of dichloromethylamine is illustrated in Scheme 1. The mechanism begins with chlorine substitution, followed by hydrolysis to yield urea and N-chloro-sarcosine as intermediates. Additional chlorine substitution and hydrolysis steps yield the ultimate products of trichloramine and dichloromethylamine. To test the validity of this hypothesized mechanism, an aqueous solution of sarcosine was exposed to free chlorine at Cl:P=5, and dichloromethylamine was found as the major DBP. Tachikawa *et al.* (*11*) demonstrated that urea is generated during the chlorination of creatinine. Chlorination of urea results in formation of dichlorourea; when allowed to stand, dichlorourea has been shown to decompose, yielding trichloramine as a product (*12*).

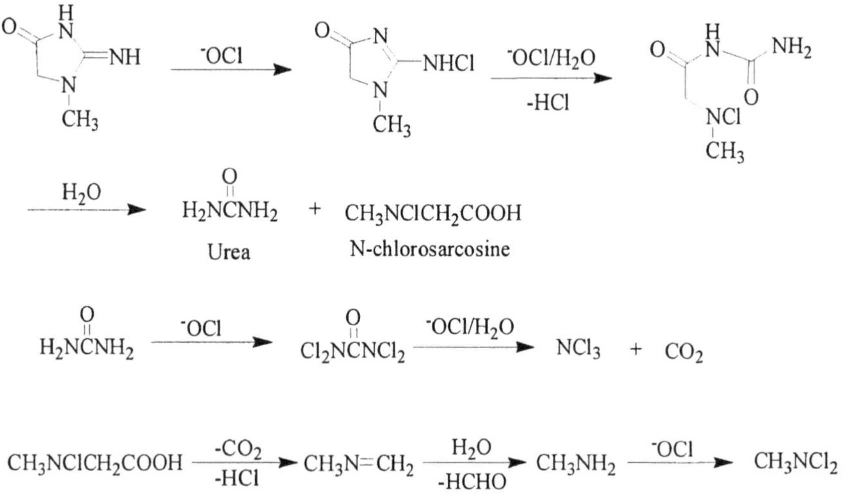

Scheme 1. Proposed mechanism for the formation of trichloramine and dichloromethlyamine from chlorination of creatinine.

$CNCHCl_2$ and CNCl formation from chlorination of L-histidine

Volatile DBP formation in chlorination of L-histidine was also investigated, with $CNCHCl_2$, CNCl, and NCl_3 all being identified by MIMS as volatile DBPs. The initial concentration of L-histidine in these experiments was 1.8×10^{-5} M.

Approximately 0.2 mg/L (as Cl_2) of $CNCHCl_2$ and 0.02 mg/L (as Cl_2) of CNCl were found to be generated at a molar ratio Cl:P = 8.0 after 96 hours of chlorination. Among the organic precursors investigated in this study, L-histidine was found to be the most reactive toward free chlorine, as defined by the rate of free chlorine consumption. The free chlorine concentration decreased steadily over the course of 96-hour chlorination and the $CNCHCl_2$ concentration increased. NCl_3 and CNCl concentrations both increased in the beginning of the reaction, reached a plateau, and then decreased. Roughly 0.22 mg/L of NCl_3 (as Cl_2) formed after 4-hour chlorination of L-histidine, followed by its decay.

In general, the reactions between free chlorine and organic-N precursors have been generalized to proceed via electrophilic substitution, in combination with hydrolysis. The reaction between free chlorine and L-histidine is assumed to proceed with an initial chlorine substitution step, followed by dechlorination and decarboxylation to yield a nitrile (*13,14*).

NCl_3 formation from chlorination of urea and L-arginine

Urea was found to yield relatively high concentrations of NCl_3, even at Cl:P as low as 1.6. For example, roughly 0.1 mg/L (as Cl_2) of NCl_3 was detected by MIMS after 1-hour chlorination of 1.8×10^{-5} M urea at Cl:P = 1.6. This observation was consistent with results from experiments involving high precursor concentrations. No other forms of residual chlorine were evident in MIMS analysis of chlorinated urea samples.

Trichloramine was the only DBP identified by MIMS in chlorination of L-arginine. No NCl_3 was detected after 1-hour chlorination of L-arginine at Cl:P molar ratios from 1.6 to 9.6, and NCl_3 was detectable only at Cl:P $\geq$ 6.4. No NCl_3 was detected at later times.

In general, chlorination of all four organic-N precursors yielded volatile DBPs, even at the low precursor concentrations that are believed to be representative of recreational waters. Inorganic chloramine, which was always observed as trichloramine (always in the presence of free chlorine), was found as a common byproduct in all cases. This finding, together with previously published results, suggests that many organic-N compounds can act as precursors to NCl_3 formation. The formation of trichloramine was strongly dependent on the Cl:P molar ratio, the structure of the precursor, and reaction time.

Analysis of actual swimming pool water samples

Volatile DBPs present in samples collected from public recreational water facilities were identified and quantified by MIMS. Table I provides a summary of volatile DBP measurements for samples collected from six public swimming pools. These data are accompanied by measurements of chlorine residuals by DPD/FAS titration.

Table I. Volatile DBP measurement in samples of recreational water.

Sample[a]	NCl_3[b] (mg/L as Cl_2)	$CHCl_3$[b] (mg/L)	$CNCHCl_2$[b] (mg/L)	Free Chlorine[c] (mg/L as Cl_2)	Combined chlorine[c] (mg/L as Cl_2)
A	0.08	0.07	0.01	1.5	1.34
B	0.07	0.13	0.03	1.95	0.25
C	0.09	0.14	0.01	0.68	1.36
D	0.16	0.08	0.02	6.52	1.76
E	0.1	0.13	0.01	5.92	1.28
F	0.07	0.08	0.01	1.72	0.76

[a]A, C, E, F: Indoor lap Swimming Pool; B: Outdoor General Use Swimming Pool; D: Outdoor Recreation Park; [b] Analysis by MIMS; [c] Analysis by DPD/FAS titration
SOURCE: Reproduced from *Environ. Sci. Technol.* **2007**, *41*, 6732-6739. Copyright 2007 American Chemical Society.

Three volatile DBPs (chloroform [$CHCl_3$], trichloramine, and dichloroacetonitrile) were detected and quantified from all six recreational water samples by MIMS. Among them, trichloramine and dichloroacetonitrile were also detected in the chlorination of model organic-N compounds. Dichloromethylamine was detected only in Sample A, and was present at a concentration of 10 μg/L. Interestingly, sample A was collected from a natatorium that is used almost exclusively for lap swimming and swimming competitions. Although no measurements of precursor concentrations of these samples were performed, the use pattern of swimming pool A is consistent with a scenario where a relatively high creatinine concentration is expected, based on excretion of sweat by swimmers. Creatinine was the only precursor in this research that yielded dichloromethylamine as a result of chlorination. However, no CNCl was detected in all swimming pool samples, which may attributed to the short half-life (about 1 hour in the presence of 0.5 mg/L free chlorine, at 25°C, pH 7) of CNCl in the presence of free chlorine (*15*). More generally, most volatile DBPs that were identified in the experiments with model organic-N compounds were also detected in municipal pool water samples.

The data presented in Table I also suggest that the concentration ranges of these volatile products in actual recreational water facilities were relatively narrow, regardless the existing concentrations of free chlorine and combined chlorine. This suggests that these volatile DBPs may be ubiquitous in chlorinated swimming pools, even in well-maintained facilities. These compounds are difficult to eliminate by simple chlorination, even shock chlorination. As a result, additional treatment may be needed in recreational water settings to improve water and air quality, relative to these volatile DBPs.

Acknowledgments

The authors are grateful to the Dupont Experimental Station and NSPF (National Swimming Pool Foundation) for financial support of this research.

References

1. Kim, H.; Shim, J.; Lee, S. *Chemosphere*, **2002**, *46*, 123-130 .
2. Judd, S. J.; Jeffrey, J. A. *Water Res.*, **1995**, *29*, 1203-1206.
3. Muellner, M. G. ; Wagner, E. D. ; McCalla, K. ; Richardson, S. D. ; Woo, Y.-T,; Plewa, M. J. *Environ. Sci. Technol.* **2007**, 41, 645-651.
4. Nickmilder M.; Bernard, A. *Occup. Environ. Med.*, **2007**, *64*, 37-46.
5. Nemery, B.; Hoet, P. H. M.; Nowak, D. *Eur Respir J.*, **2002**, *19*, 790-793.
6. Thickett, K. M.; McCoach, J. S.; Gerber, J. M.; Sadhra, S.; Burge, P. S. *Eur Respir J.*, **2002**, *19*, 827-832.
7. Judd, S. J.; Black, S. *Water Res.* **2000**, *34*, 1611-1619.
8. Judd, S. J.; Bullock, G. *Chemosphere*, **2003**, *51*, 869-879.
9. Shang, C; Blatchley, E. R. III *Environ. Sci. Technol.*, **1999**, *33*, 2218-2223.
10. Osgood, C; Sterling, D. *Mutation Research* **1991**, *261*, 85-91.
11. Tachikawa, M.; Aburada, T.; Tezuka, M.; Sawamura, R. *Water Research*, **2005**, *39*, 371-379.
12. Dowell, C. T. *J. Am. Chem. Soc.*, **1919**, *41*, 124-125.
13. Armesto, X. L.; Canle, M. L.; Santaballa, J. A. R *Tetrahedron* **1993**, *49*, 275-284.
14. Conyers, B.; Scully, F. E., Jr. *Environ. Sci. Technol.* **1993**, *27*, 261-266.
15. Na, C.; Olson, T. M. *Environ. Sci. Technol.*, **2004**, *38*, 6037-6043.

Chapter 13

Reactivity of Bromine-Substituted Haloamines in Forming Haloacetic Acids

Phillip G. Pope[1,2] and Gerald E. Speitel Jr[1,*]

[1]The University of Texas at Austin, Department of Civil, Architectural and Environmental Engineering, 1 University Station C 1700, Austin, TX 78712-0273
[2]Current Address: Carollo Engineers, 8911 Capitol of Texas Highway North, Suite 2200, Austin, TX

When drinking water is chloraminated in the presence of bromide, bromine-substituted haloacetic acids typically form. Monochloramine is the dominant haloamine species under drinking water treatment conditions. In the presence of bromide, bromine-substituted haloamines, such as mono- and dibromamine, as well as bromochloramine may also form. These bromine-substituted haloamines are typically present in much lower concentrations than monochloramine, but are significantly more reactive in forming haloacetic acids (HAAs). The reactivity of the bromine-substituted haloamine species was studied in two different surface water supplies. HAA formation from each of the bromine-substituted species (NH_2Br, $NHBr_2$, and NHBrCl) was 4 to 12 times that from monochloramine (NH_2Cl).

Background

Chloramines are widely used as a secondary disinfectant in drinking water treatment to control trihalomethane and haloacetic acid (HAA) formation; however, significant dihaloacetic acid (DXAA) concentrations have been observed to form. When bromide is present in source water, bromine-substituted haloacetic acids typically form during chloramination. Limiting the formation of bromine-substituted HAAs is of particular interest because recent studies have shown that they may pose a greater health risk than their chlorine-substituted counterparts (*1, 2, 3*). Monochloramine is the dominant haloamine species under drinking water treatment conditions. In the presence of bromide, however, bromine-substituted haloamines, such as mono- and dibromamine, as well as bromochloramine may also form (*4, 5, 6*). Even though these species are present in significantly lower concentrations than monochloramine, they may be more reactive. This research proposes that the variability in DXAA formation and speciation observed in natural waters is associated with the reactivities and relative concentrations of the different haloamine species that may be present. The objective of this research was to characterize the reactivity of bromine-substituted haloamine species in forming HAAs. Disinfection chemistry research in drinking water treatment has focused almost exclusively on disinfectant decay, and the reactivity of the predominant disinfectant present (*i.e.*, chlorine or monochloramine). Increased knowledge of the reactivity of bromine-substituted haloamines is necessary to understanding HAA formation in waters that contain bromide.

Materials and Methods

Haloamine Reactivity Experiments

The reactivity of haloamine species was studied under controlled conditions in two surface waters of differing characteristics. The individual haloamines were isolated to the extent possible and reacted with natural waters to characterize their reactivity in HAA formation. Experiments were conducted in batch reactors, and both the HAA and haloamine concentrations were measured over time.

Bromamine Reactivity

To characterize the reactivity of the bromamines in forming HAAs, natural waters were dosed with preformed bromamine stock solutions (*7*). Simultaneously, the concentrations of the individual bromamine species in the stock solutions were measured spectrophotometrically. The pH of the water matched that of the dosing solution to ensure that the bromamine speciation did not change upon dosing. The individual haloamine species were also measured spectrophotometrically (*7*) in carbonate-buffered ultra pure water (UPW) at the same experimental conditions and haloamine doses as the bromamine reactivity experiments. The dose was selected to provide a target bromamine residual concentration between 0.5 and 1 mg/L as Cl_2 at 24 hours. Samples were incubated in amber glass bottles and capped headspace free with teflon-lined septa. Samples to analyze for the nine haloacetic acids (HAA_9) were collected at 5 minutes and again at 0.5, 1, 4, 24, 48, and 72 hours. Total combined oxidant concentrations were measured alongside the HAA samples to determine total oxidant demand. Control experiments were also run with chloramines instead of bromamines to determine relative differences in reactivity. These experiments were conducted at the same pH, buffer concentration, and initial haloamine concentration as the bromamine reactivity experiments, but were dosed with preformed chloramines (*8*) instead of bromamines.

Bromochloramine Reactivity

Bromochloramine formation is known to be highly dependent on pH (*5*). Consequently, solutions at pH 9 did not promote significant bromochloramine formation. Therefore, Lake Austin (LA) source water was dosed with preformed bromochloramine (*7*) stock solutions at pH 6.3 and 7.2. Experiments were performed at pH 7.2 to provide consistency with the bromamine reactivity experiments. The pH of the dosing solution matched that of the water. Samples were incubated in amber glass bottles and capped headspace free with teflon-lined septa. Samples were taken at 5 min and again at 0.5, 1, 4, and 24 hours for HAA_9 analysis. Total combined oxidant and monochloramine concentrations were measured immediately after dosing and alongside the HAA samples to determine total oxidant demand. In addition, haloamine speciation was monitored with membrane introduction mass spectrometry (MIMS) (*7*). Because the bromochloramine stock solutions contained significant concentrations of monochloramine, several control experiments were run with monochloramine to determine relative differences in reactivity. These experiments were conducted at the same pH, buffer concentration, and initial chloramine dose as the bromochloramine reactivity experiments.

Analytical Methods

A TOC analyzer (Sievers Model 800 Series) equipped with an autosampler was used to measure the TOC and DOC (*7*). Bromide analyses were performed by ion chromatography with conductivity detection in accordance with EPA Method 300.1 (*9*), and the ultra-violet light absorbance was measured at 253.7 nm in accordance with Standard Method 5910 (*10*). Alkalinity was measured following Standard Method 2320 (*10*). HAA samples were measured in accordance with USEPA Method 552.2 (*9*) on a gas chromatograph with a micro electron capture detector and were extracted immediately to avoid complications that may arise due to sample preservation (*7*).

Total chloramine residual concentrations were measured by spectrophotometry as total chlorine using Hach DPD Total Chlorine Reagent Powder Pillows with Hach DPD Method 8021 adapted from Standard Method 4500-Cl. Monochloramine was measured by spectrometry at 655 nm using Hach Monochlor F Reagent Powder Pillows according to Hach Method 10171. Total combined oxidant residual was measured by spectrophotometry as total chlorine using Hach DPD Total Chlorine Reagent Powder Pillows with Hach DPD Method. The concentrations of the individual bromamine species were measured by spectrometry in a 10-cm pathlength quartz cell (*6*).

The Membrane Introduction Mass Spectrometry (MIMS) procedure, developed by Shang and Blatchley (*11*), was used with modifications (*7*) to detect the presence of chloro- and bromo-substituted haloamines, in particular bromochloramine. Gazda et al. (*12*) formed NHBrCl by reacting NH_2Cl with Br^- at a 12:1 Br-:NH_2Cl molar ratio and a pH of 6.5. These researchers monitored the reaction in real-time using MIMS and concluded that bromochloramine was the predominant reaction product. These reaction conditions resulted in rapid formation and decay of bromochloramine. Therefore, a lower Br^-:NH_2Cl molar ratio of 5:1 was used to slow the kinetics of bromochloramine decay. The resulting spectra of this reaction mixture (Figure 1) indicated that bromochloramine and monochloramine were the predominant species present. Ions with m/z values of 51 ($NH_2{}^{35}Cl^{\cdot+}$) and 53 ($NH_2{}^{37}Cl^{\cdot+}$) represent monochloramine, and those with values of 129 ($NH^{79}Br^{35}Cl^{\cdot+}$), 131 ($NH^{79}Br^{37}Cl^{\cdot+}$ and $NH^{81}Br^{35}Cl^{\cdot+}$) and 133 ($NH^{81}Br^{37}Cl^{\cdot+}$) represent bromochloramine. In addition, ion fragments of bromochloramine observed at 79 ($^{79}Br^{+}$), 81 ($^{81}Br^{+}$), 93 ($NH^{79}Br^{\cdot+}$), and 95 ($NH^{81}Br^{\cdot+}$) were similar to those of Gazda et al. (*12*). They also observed ions at m/z 97, but did not report observing ions clustered around m/z 160 (Br_2).

Halaomine Reactivity

The reactivity of different haloamine species was studied in Lake Austin (LA) and Metedeconk River (MR) source water. The water quality characteristics for each of these waters are shown in Table I. LA source water is a low specific ultraviolet absorbance (SUVA), high bromide, hard, high

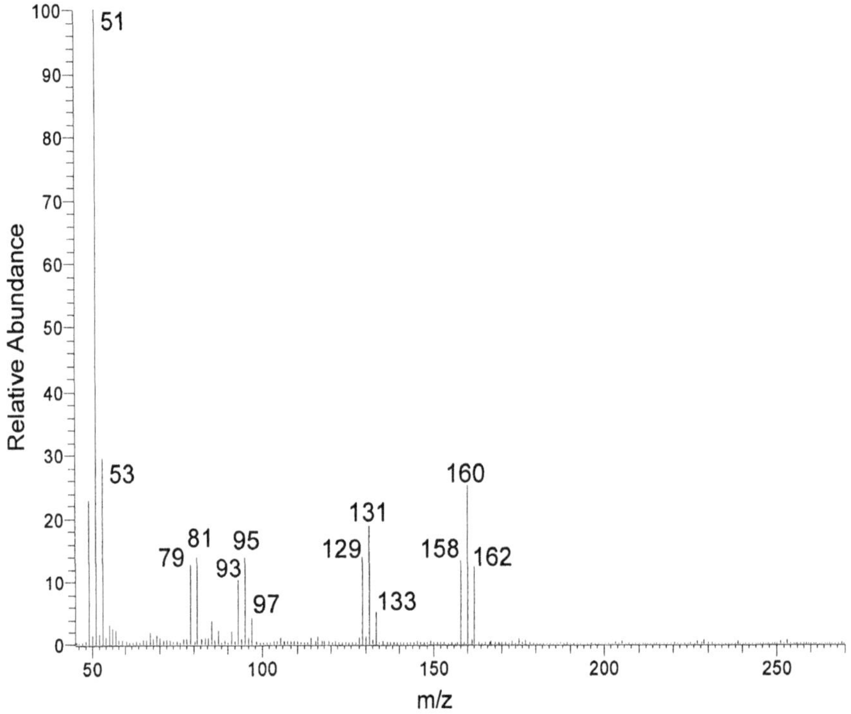

Figure 1. Mass spectrum for 0.14 mM NH_2Cl + 0.7 mM Br^- (pH 6.3 and 5 mM Carbonate buffer) 10 minutes after Br^- addition

Table I. Typical source water quality characteristics

Parameter	*LA Source*	*MRr Source*
pH	8.1	6.7
TOC (mg/L)	3.55	3.25
DOC (mg/L)	3.45	3.01
Alkalinity (mg $CaCO_3$/L)	150	14
SUVA (L/mg-m)	2.11	4.87
Bromide (µg/L)	168	27

alkalinity source water, while MR source water is characterized as a moderately high SUVA, low bromide source water.

Bromamine Reactivity

LA and MR waters were dosed with preformed bromamine stock solutions that contained predominantly mono- or dibromamine based on the pH and Br_2:N ratio selected. Bromamine chemistry is similar to that of the chloramines (*6*), and as with the chloramine system, bromamine reactions are also catalyzed by both phosphate and carbonate buffers (*6*). In light of this, only low (less than 0.5 mM) phosphate buffer concentrations were used when necessary to minimize their influence on the reactions of interest. Bromamine doses were selected to provide a target residual concentration between 0.5 and 1 mg/L as Cl_2 at 24 hours. However, both bromamine residual and HAA_9 concentrations were measured at 48 and 72 hours to confirm the reaction conditions were NOM-limited.

Lake Austin source water

The initial conditions of the bromamine reactivity experiments performed in LA water are shown in Table II. They were selected to determine the influence of small concentrations of phosphate buffer, pH, and Br_2:N ratio on HAA formation. The pH and Br_2:N ratio directly influenced which bromamine species dominated, allowing their reactivities to be compared. This comparison was made by the HAA yield, which is the ratio of HAA to DOC concentration.

Table II. Summary of bromamine reactivity experiments in LA water

pH	*Br_2:N molar ratio*	*Phosphate Buffer (mM)*	*Dose (mg Cl_2/L)*	*Dose (NH_2Br; $NHBr_2$) (mM)*
9	0.05	0	2.1	0.024; 0.003
9	0.05	0.18	2.0	0.022; 0.003
7.2	0.05	0.31	3.0	0.005; 0.019
7.2	0.667	0.33	1.9	0; 0.013

After 24 hours of incubation, each of the four experimental conditions resulted in similar HAA yield, while dibromoacetic acid (DBAA) was the predominant HAA formed (Figure 2). In addition, the results show that bromamines are significantly more reactive than chloramines. After 24 hours of contact in Lake Austin source water (Figure 3), the bromamine species formed approximately four times the HAA_9 yield as the chloramines, indicating that the bromamine species are reactive enough to play a significant role in HAA formation in waters that contain bromide. Because each experiment was dosed with approximately 2 mg/L as Cl_2 of total bromamine, the reactivity of mono- and dibromamine could be compared. Each of these conditions resulted in similar HAA yields.

Metedeconk River source water

As with LA water, MR water was dosed with different bromamine solutions formulated to produce predominantly mono- or dibromamine (Table III). However, the bromamine decay rate in MR was significantly greater than that of LA water; therefore, higher initial doses were necessary to provide the target oxidant residual at 24 hours. In addition to the higher initial doses, waters were also spiked with additional bromamines at 48 hours to ensure that the HAA formation was NOM limited. The initial and 48-hour spike dose concentrations were selected to provide a total oxidant residual between 0.5 and 1 mg Cl_2/L at 24 and 72 hours, respectively.

Similar to LA water, after 24 hours the bromamines formed predominantly DBAA in MR water (Figure 4). MR water was also more reactive than LA water, which was not unexpected since it had a greater SUVA than LA (Table II). Twenty-four hours after dosing, the bromamine species formed greater than 6 times the HAA_9 yield as the chloramines (Figure 5). Also, upon spiking the MR waters with additional bromamines, very little additional HAAs formed, indicating that these experiments were most likely NOM limited.

Table III. Summary of bromamine reactivity experiments in MR water

pH	*Br_2:N molar ratio*	*Phosphate Buffer (mM)*	*Dose (mg Cl_2/L)*	*Dose (NH_2Br; $NHBr_2$) (mM)*
9	0.05	0	4.8	0.057; 0.005
9	0.05	0.40	5.0	0.056; 0.007
7.2	0.05	0.83	9.2	0.022; 0.054
7.2	0.667	1.43	9.0	0; 0.063

Both the chloramines and bromamines formed predominantly DXAAs, and both resulted in an initial rapid period of HAA formation, followed by a slower period. Chloraminated source waters incubated at pH 7.2 exhibited greater DXAA formation than that at pH 9. During the initial 30 minutes of reaction time, similar concentrations of DXAA formed regardless of pH; however, after

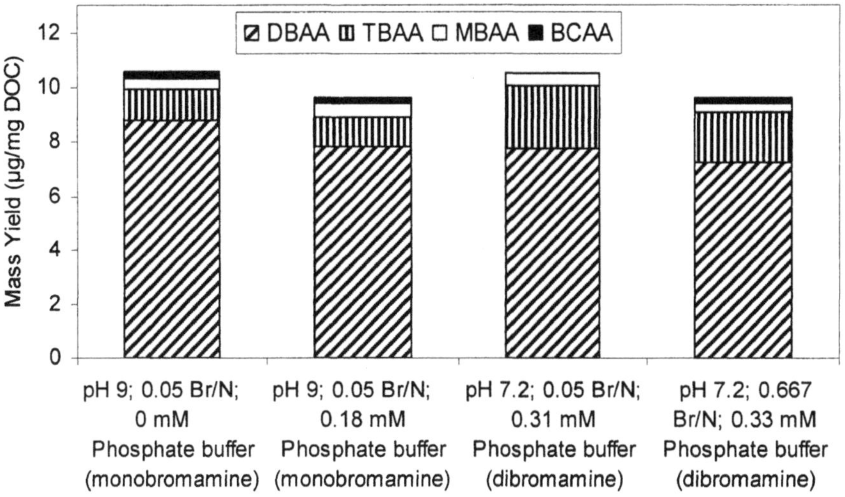

Figure 2. HAA yield and speciation in Lake Austin source water after 24 hours incubation (parentheses indicate dominant bromamine present)

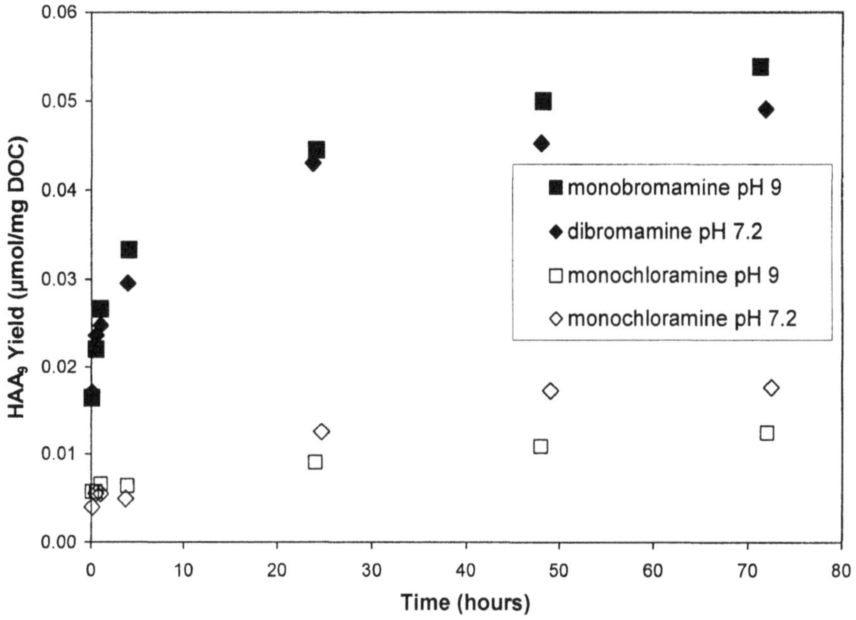

Figure 3. Haloamine reactivity in Lake Austin source water

approximately 30 minutes more rapid kinetics were observed in the water incubated at pH 7.2. The increased HAA formation at lower pH indicates that acid catalyzed chloramine reactions may be responsible. Therefore, monochloramine decay products such as NH_3Cl^+, $NHCl_2$, HOCl may be the cause of the more rapid HAA formation at low pH after the initial 30 minutes of reaction. Lei *et al.* (6) postulated that the reactions for bromamine decomposition are analogous to the chloramines. Hence, monobromamine disproportionates into dibromamine (reaction 1), and dibromamine hydrolysis (reaction 2) may result in an unidentified reaction intermediate (I), which can react with $NHBr_2$ or NH_2Br (reactions 3 and 4) to form HOBr and various other products. Therefore, NH_2Br, $NHBr_2$, or HOBr may be responsible for the observed HAA formation.

$$2NH_2Br \leftrightarrow NHBr_2 + NH_3 \quad (1)$$

$$NHBr_2 + H_2O \rightarrow I \quad (2)$$

$$I + NHBr_2 \rightarrow HOBr + \text{products} \quad (3)$$

$$I + NH_2Br \rightarrow \text{products} \quad (4)$$

Greater concentrations of HOBr may form under conditions that promote the formation and decay of $NHBr_2$. The increased tribromoacetic acid (TBAA) observed at pH 7.2 in comparison to pH 9, may be a result of either higher concentrations of $NHBr_2$ decomposition products, such as HOBr, or pH effects on NOM. At low pH, the carboxylic and phenolic functional groups of the NOM are more neutralized, decreasing the electrostatic repulsion between adjacent functional groups and causing the NOM molecule to coil (*13, 14*). This change in structure may influence the reactivity of the NOM. MR water, unlike LA, had greater HAA yields when dibromamine was the dominant bromamine species. This difference in reactivity may be a result of the NOM in MR simply being more reactive, or that dibromamine is more reactive than monobromamine. However, these data do indicate that bromamines are significantly more reactive than chloramines and, therefore, may be important contributors to HAA formation in waters that contain bromide.

Bromochloramine Reactivity

LA water was dosed with preformed bromochloramine stock solutions at pH 6.3 and 7.2. Because bromochloramine formation is highly dependent on pH (*5*), solutions at pH 9 did not promote significant bromochloramine formation. Experiments were performed at pH 7.2 to provide consistency with the

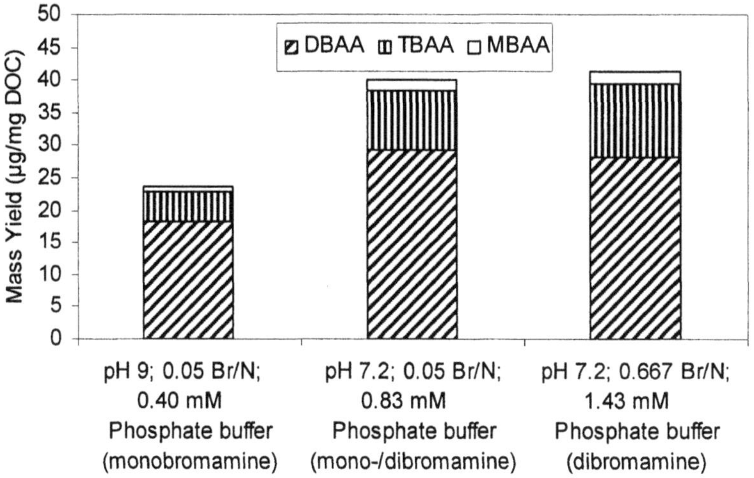

Figure 4. HAA yield and speciation in Metedeconk River source water after 24 hours incubation (parenthesis indicate dominant haloamine present)

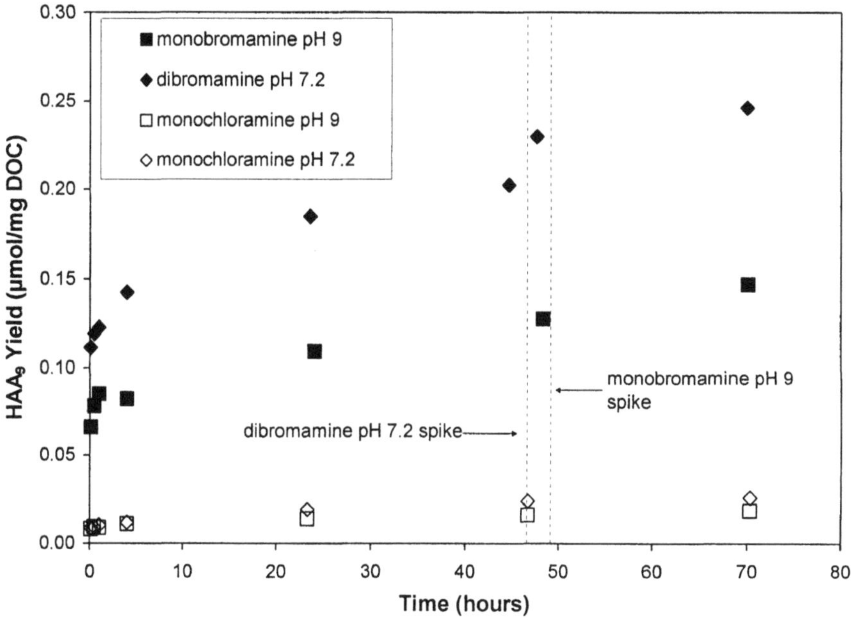

Figure 5. Haloamine reactivity in Metedeconk River source water (dashed line indicates time of additional bromamine spike)

bromamine reactivity experiments. Again, like the bromamine reactivity experiments, carbonate or phosphate buffers were used to maintain the desired pH, but only small concentrations were used to minimize any undesired reactions that may occur. A summary of the bromochloramine reactivity experimental conditions is shown in Table IV.

Total combined oxidant and monochloramine concentrations were measured immediately after dosing and at various times thereafter. The bromochloramine concentration was approximated as the difference between these two measurements because MIMS analysis indicated that a majority of the bromine-substituted haloamine was bromochloramine (7). Greater concentrations of bromochloramine were observed in the dosing solution maintained at pH 6.3. However, the haloamine residual in this solution also decayed faster than the solution incubated at pH 7.2 (Figure 6). Therefore, incubation at pH 6.3 provided greater initial concentrations of bromochloramine, but at pH 7.2, the bromochloramine residual was maintained for a longer time.

Table IV. Summary of bromochloramine reactivity experiments in LA water

pH	*Carbonate Buffer (mM)*	*Phosphate Buffer (mM)*	*Dose* (NH_2Cl; NHBrCl) (mg Cl_2/L)*	*Dose* (NH_2Cl; NHBrCl) (mM)*
6.3	3.6	0	4.2; 2.9	0.058; 0.020
7.2	2.7	0.9	5.5; 1.8	0.078; 0.013

NOTE: *NH_2Cl measured by Hach method 10171; NHBrCl determined by difference between the total oxidant concentration measured by Hach DPD method 8021 and the NH_2Cl concentration

The HAA yields with bromochloramine dosed at pH 6.3 and 7.2 (Figure 7) indicate that bromochloramine is reactive enough to play a significant role in HAA formation. A majority of the HAA formed was DBAA; however, significant quantities of TBAA and monobromoacetic acid (MBAA) were also observed. Only small amounts of bromochloracetic acid (BCAA) formed; therefore, most of the HAAs formed were bromine-substituted. Because the bromochloramine stock solutions contained significant concentrations of monochloramine, controls were run with monochloramine to determine relative differences in reactivity. These experiments were conducted at the same pH, buffer concentration, and chloramine dose as the bromochloramine reactivity experiments. Results indicated that bromochloramine was more reactive than monochloramine (Figure 8). After 24 hours, appreciably more HAA formed in the waters dosed with bromochloramine. In addition, since LA water contains

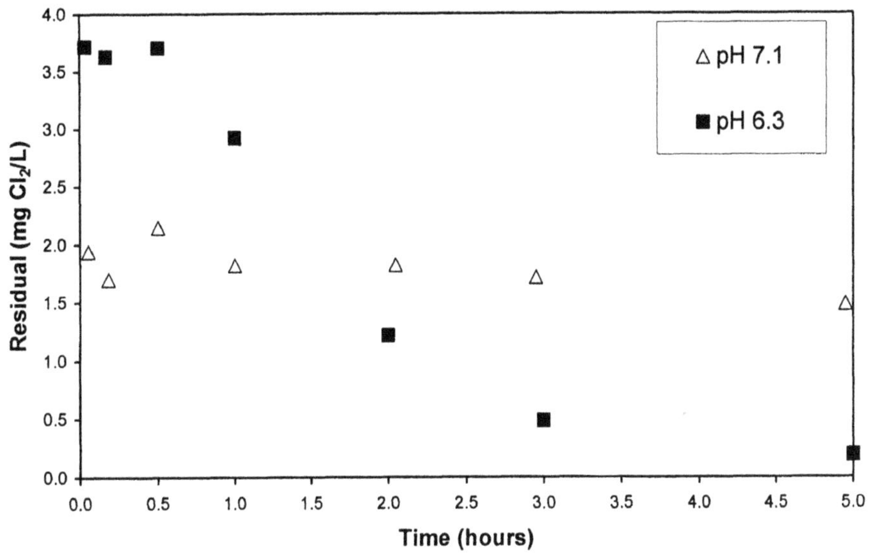

Figure 6. Bromochloramine decay kinetics in Ultrapure Water (UPW) at pH 6.3 (5:1 Br^-/NH2Cl molar ratio; 3.6 mM total carbonate) and at pH 7.1 (5:1 Br^-/NH2Cl molar ratio; 2.7 mM total carbonate; 0.9 mM total phosphate)

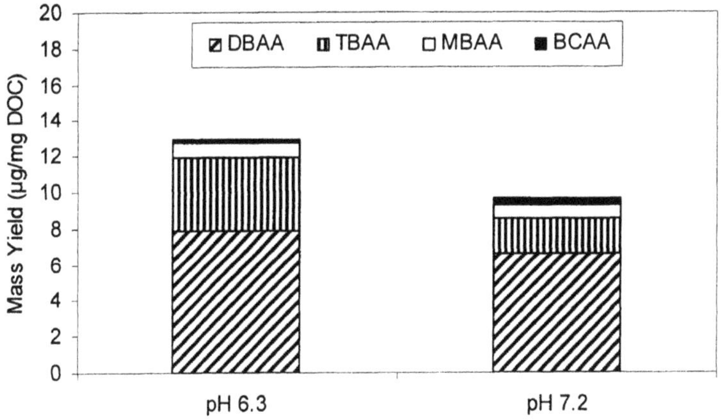

Figure 7. HAA yield and speciation in Lake Austin source water after 24 hours incubation (Predominant haloamines present - bromochloramine and monochloramine)

some bromide (168 μg/L), both bromine-and chlorine-substituted HAAs formed when the water was chloraminated. However, when the water was dosed with the bromochloramine solution, most of the HAAs were bromine-substituted. Therefore, the bromine-substituted halogen was more reactive than the chlorine-substituted halogen also present, out-competing it for available reactive NOM sites.

Bromochloramine or one of its decomposition products is responsible for the observed HAA formation. NHBrCl can form from reactions of monochloramine with bromide or HOBr. Trofe *et al.* (*5*) postulated the reaction mechanism:

$$NH_2Cl + H^+ \leftrightarrow NH_3Cl^+ \quad (4)$$

$$NH_3Cl^+ + Br^- \rightarrow NH_3Br^+ + Cl^- \quad (5)$$

$$NH_3Br^+ + NH_2Cl \rightarrow NHBrCl + NH_4^+ \quad (6)$$

The equilibrium constant for Equation 4 for is 28 M^{-1} (*15*). Therefore, under drinking water treatment conditions, only very small amounts of monochloramonium ion will be present. However, NH_3Cl^+ will react with bromide to form NH_3Br^+ (Equation 5) which rapidly reacts with monochloramine to form bromochloramine (Equation 6). This reaction mechanism demonstrates that NHBrCl formation increases as pH decreases. In addition, monochloramine reacts with HOBr to form NHBrCl (*16*). Once formed, NHBrCl decomposes in base to regenerate OBr^- as a final product by equation 7 (*16, 17*).

$$2NHBrCl + OH^- \rightarrow N_2 + 3H^+ + 2\,Cl^- + Br^- + OBr^- \quad (7)$$

Therefore, in addition to NHBrCl, its decomposition products, such as HOBr, may be responsible for the HAAs formed in the presence of NOM.

Summary and Conclusions

The bromine-substituted haloamines monobromamine, dibromamine, and bromochloramine are significantly more reactive than monochloramine in forming HAAs. Therefore, even though they are present in much lower concentrations than monochloramine under drinking water treatment conditions, they are still reactive enough to play a role in HAA formation. MR source water was more reactive than LA, forming greater HAA yields in the presence of monobromamine, dibromamine, and monochloramine (Figure 9). These data illustrate the importance of source water characteristics on the relative reactivity

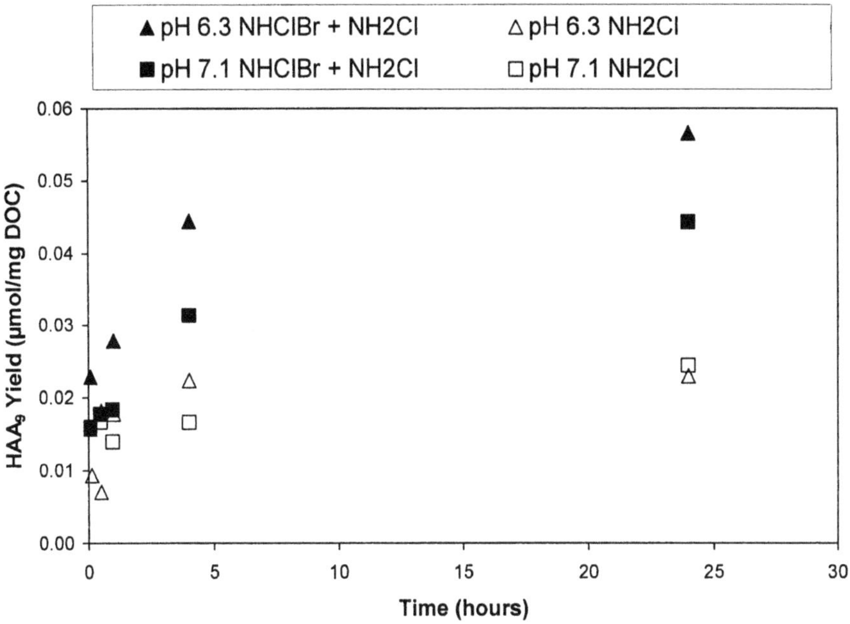

Figure 8. Bromochloramine reactivity in Lake Austin source water

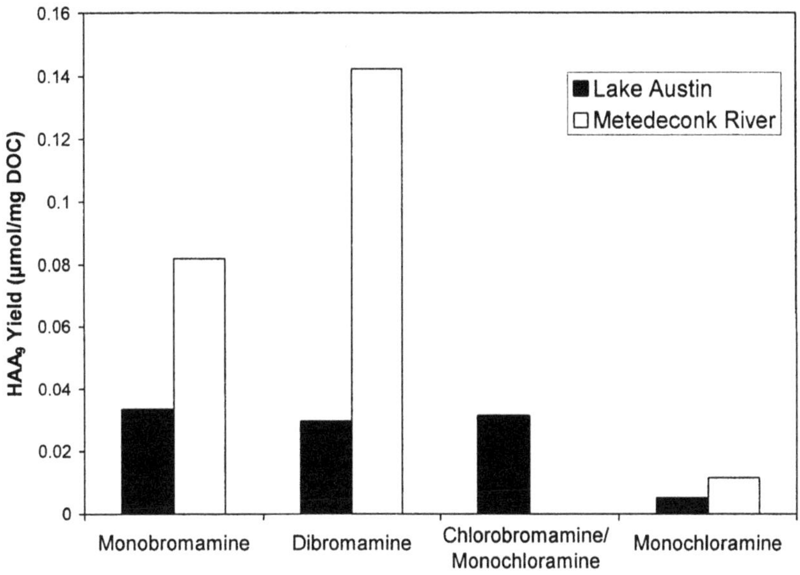

Figure 9. Different haloamine reactivities in Lake Austin and Metedeconk River source waters after 4 hour incubation at pH 7.2

of each species. The greater reactivity of MR water was expected based upon its higher SUVA (4.87) compared to Lake Austin (2.11). Therefore, the varying characteristics of NOM among source waters are significant factors in determining NOM's reactivity with the different bromine- and chlorine-substituted haloamines that may form in the presence of bromide during chloramination. The presence of bromine-substituted HAAs in chloraminated drinking water suggests that bromine-substituted haloamines are reactive enough to play a role in HAA formation even though they are present in much lower concentrations than monochloramine.

References

1. Bull, R.J., L.S. Birnbaum, K.P. Cantor; J.B. Rose, B.E. Butterworth, R. Pegram; J. Tuomisto. *Fundam Appl Toxicol.* **1995**, *28*, 155-166.
2. Kargalioglu Y., McMillan BJ, Minear RA, Plewa MJ. In *Natural organic matter and disinfection by-products, characterization and control in drinking water*; Barrett SE, Krasner SW, Amy GL, Eds.; Washington, DC. **2000**; American Chemical Society ACS Symposium series 761 pp 16–27.
3. Kargalioglu, Yahya, B.J. McMillan, R.A. Minear, and M.J. Plewa. *Teratogenesis, Carcinogenesis, and Mutagenesis.* **2002**, *22*, 113-128.
4. Haag, W. R. and M.H. Lietzke. In *Water Chlorination: Environmental Impact and Health Effects Vol.3*; R.L. Jolley, W.A. Brungs, R.B. Cumming, V.A. Jacobs, Eds.; Ann Arbor, MI; **1980**;Ann Arbor Science Publishers pp 415-425.
5. Trofe, T.W., G.W. Inman, and J.D. Johnson. *Environ. Sci. & Technol.,* **1980**, *14*, 544-549.
6. Lei, H., B.J. Marinas, and R.A. Minear. *Environ. Sci. & Technol.*, **2004**, *38*, 2111-2119.
7. Pope, P.G. Ph.D. Dissertation, The Univeristy of Texas, Austin, TX, **2006**.
8. Diehl, A.C., G.E. Speitel Jr., J.M. Symons, S.W. Krasner, C.J. Hwang, and S.E. Barrett. *Jour. AWWA,* **2002**, *92*, 76-90.
9. USEPA. Methods for the Determination of Organic Compounds in Drinking Water, Supplement III. **1995**. EPA/600/R-95/131.
10. APHA, AWWA, and WEF (American Public Health Association, American Water Works Association, and Water Environment Federation). **1998**. *Standard Methods for the Examination of Water and Wastewater.* 20th Edition, Washington, DC.
11. Shang, C. and E.R. Blatchely III. *Environ. Sci. & Technol.* **1999**, *33*, 2218-2223.
12. Gadza, M., L.E. Dejarmi, T.K. Chowdury, R.G. Cooks, and D.E. Margerum. *Environ. Sci. Technol.* **1993**, *27*, 557-561.

13. Ghosh, K and M. Schnitzer. *Soil Science*, **1980**, *129*, 266-276.
14. Murphy, E.M., J.M. Zachara, S.C. Smith, J.L. Phillips, and T.W. Wietsma. *Environ. Sci. & Technol.* **1994**, *28*, 1291-1299.
15. Gray Jr., E.T., Margerum, D. W., and R.P. Huffman; In *Organometals and Organometalloids: Occurrence and Fate in the Environment*; F.E. Brinckman and J.M. Bellama Ed.; ACS Books, Washington, DC; **1978**; pp. 264-277.
16. Gadza, M. and D.E. Margerum. *Inorg Chem.* **1994**, *33*, 118-123.
17. Valentine, R.L. Ph.D. Dissertation, UC-Berkley., Berkley, CA, **1983**.

Chapter 14

Comparison of the Performance of Spectroscopic Indices Developed to Quantify the Halogenation of Natural Organic Matter at Varying Chlorine Concentrations, Reaction Times and Temperatures

Paolo Roccaro[1], Federico G. A. Vagliasindi[1], and Gregory V. Korshin[2]

[1]Department of Civil and Environmental Engineering, University of Catania, Catania, Italy 95125
[2]Department of Civil and Environmental Engineering, University of Washington, Seattle, WA 98195-2700

The performance and significance of several absorbance and fluorescence indices introduced to probe the formation of DBPs in chlorinated water from the Ancipa reservoir (Sicily, Italy) were compared in this study. Results of the experiments confirmed the strength of correlations between the concentrations of individual DBPs formed at varying chlorine concentrations, reaction times and temperatures and differential absorbance measured at 254, 272 and 280 nm. It was also determined that strong correlations exist between the formation of individual DBPs, chlorine consumption and differential fluorescence indices that quantify changes of the shape of NOM emission spectra. These indices that include the difference between the wavelengths in the emission spectrum that correspond to 50% and 100% of the maximum intensity of fluorescence ($\Delta\lambda_{0.5}$) and the differential ratio of fluorescence intensities measured at 500 and 450 nm ($\Delta I_{500/450}$) are hypothesized to be a measure of the consumption of aromatic fluorophores in NOM, and also of the breakdown of NOM molecules. Similarly to the differential absorbance parameters, $\Delta I_{500/450}$ and $\Delta\lambda_{0.5}$ fluorescence indices can be used to probe NOM halogenation *in situ*.

Background

Chlorine and other halogen species react with natural organic matter (NOM) to produce chlorinated, brominated and, in much smaller levels, iodinated disinfection by-products (DBPs) (*1, 2*). Trihalomethanes (THMs) and haloacetic acids (HAAs) are the most investigated DBP classes but more than 500 individual DBPs have been reported in the literature (*3*).

DBP formation and speciation are affected by concentration and reactivity of NOM, level of bromide, chlorine dose, pH, temperature and reaction time. Due to the complexities of NOM halogenation's modeling, surrogate parameters have been introduced to quantify NOM reactivity and DBP formation. Among them, absorbance of light at 254 nm (A_{254}) and its DOC-normalized value ($SUVA_{254}$) are typical (*4-6*). Other indices based on the absorbance and fluorescence of NOM have been employed as well. For instance, differential absorbance at 272 nm or near (ΔA_{272}) was introduced because it shows a strong correlation with concentrations of individual DBPs and total organic halogen (TOX) values (*4, 7*). Use of absorbance of NOM at 280 nm has also been recommended because π–π* electron transitions in phenolic substances predominating NOM provide a large contribution to the absorbance in this region. Strong relationships between the molar absorbance of organic carbon at 280 nm (ε_{280}) and NOM aromaticity or its apparent number- and weight-averaged molecular weights have been found (*5, 8, 9-11*). Absorbance of NOM at 254 and 272 nm is also a very good indicator of NOM aromaticity (*8, 9, 12*). Strong correlations between differential absorbance at 280 nm and THM or HAA formation potentials (FP) have been reported to exist thus indicating that aromatic structures in NOM are the primary sites responsible for THMs and HAAs formation (*13*).

Relationships between NOM fluorescence and its properties have been explored in some extent (*5, 14*). Fluorescence spectra of NOM exhibit some level of pH dependence that is associated with protonation/deprotonation of mostly aromatic fluorophores, conformational changes of NOM molecules and effects caused with their interactions with metals (*15, 16*). In addition to aromatic groups, quinones can affect the fluorescence of NOM (*17, 18*). Fluorescence excitation-emission matrices (EEMs) of NOM revealed the existence of distinct fluorophore groups that are deemed to be related to allochtonous or autochtonous NOM sources. Based on that, a fluorescence index (FR) defined as the ratio of NOM emission intensities at 450 nm and 500 nm and obtained with an excitation of 370 nm was developed. In principle, FR can help ascertain contributions of dissimilar sources of NOM, and it is correlated with NOM aromaticity and molecular weight (*14, 19*). Only a few studies have been carried out to probe NOM halogenation and DBPs formation using fluorescence (*20-22*). Although limited, these studies have demonstrated the existence of strong correlations between DBP concentration and changes in the position of wavelength that corresponds to 50% of the maximum intensity of fluorescence ($\lambda_{0.5}$) (*22*). Such relationships need to be explored in more detail to determine whether they can be universally used to predict the formation of DBPs. In accord with that, this study explored the performance of selected absorbance and fluorescence indices in probing DBP formation.

Experimental

Chlorination experiments were conducted using untreated water from the Ancipa reservoir (Sicily, Italy). It is one of the main water sources in central Sicily (about 250,000 inhabitants currently served, with a potential of up to 350,000 inhabitants). A sample of Ancipa water was taken in December 2005, transferred to the research laboratory of Department of Civil and Environmental Engineering of the University of Washington, where it was filtered through a 0.45 μm nylon membrane filter (Polycap™ AS Capsule Filter) and stored at 4°C. The DOC concentration in the filtered sample was 2.9 mg/L, $SUVA_{254}$ was 2.78 $L \cdot mg^{-1} \cdot m^{-1}$, pH 7.9, while the alkalinity and hardness were 170 and 140 mg/L as $CaCO_3$, respectively.

Chlorination was carried out with free chlorine at pH 7.0 in the presence of 0.03 mol/L phosphate buffer and temperatures 3±1, 20±1 and 34±1°C in headspace-free 1.6 L PTFE sampling bags, which were used to prevent the loss of volatile DPBs when samples were taken at different reaction times (10 minutes to 7 days). Initial chlorine doses were 0.25, 0.5, 0.75, 1.0, 1.5 and 2.0 mg Cl_2 per mg DOC. Chlorinated samples were analyzed for DBPs only if chlorine residual was found. Na_2SO_3 or, for HAA analyses, NH_4Cl were used to quench residual chlorine. Chlorine concentrations were determined using the DPD colorimetric method. Absorbance spectra were obtained with a Perkin-Elmer Lamda 18 spectrophotometer using 5 cm quartz cells. All reported spectra were normalized to the cell length of 1 cm. A Perkin-Elmer LS-50B fluorometer was used for fluorescence measurements, using excitation at 320 nm. TOC was analyzed using an O.I. Analytical 1010 Total Organic Carbon Analyzer. Concentrations of THMs, other volatile and semi-volatile DBPs and 9 HAAs were determined using standard analytical procedures (EPA methods 551.1 and 552.2) and a Perkin-Elmer AutoSystem gas chromatograph equipped with an electron capture detector. Other analytical conditions were similar to those reported in (*7*).

Results and Discussion

Kinetics of DBP Formation in Chlorinated Ancipa Water

In agreement with the data of previous studies (e.g., *1, 5-7, 22*), concentrations of all major DBP species increased monotonically with time at given chlorine dose (Figure 1). Chloroform ($CHCl_3$) was the predominant THM compound, while trichloroacetic acid (TCAA) and dichloroacetic acid (DCAA) were two major HAA species. A significant fraction of THMs and HAAs was comprised by brominated compounds. The halogenation of NOM present in Ancipa water and release of HAAs and THMs was relatively rapid. At chlorine doses higher than 1 mg per mg of DOC, total trihalomethanes (TTHM) concentrations reached 35-44% of their maximum levels observed for a 7-day reaction period within 2 to 4 hours of chlorination. Comparable yields of HAAs (38-53%) were observed in the same conditions. Higher chlorine doses caused higher levels of THMs and HAAs species to be produced. Experiments carried out at 3°C, 20°C and 34°C showed a notable acceleration of DBP formation at higher temperatures (Figure 2).

Changes in UV-Absorbance and Fluorescence Spectra Observed During NOM Chlorination

The absorbance spectra of chlorinated Ancipa water were characterized by a monotonic decrease of the absorbance intensity with increasing reaction time. The shape of the corresponding differential absorbance spectra (computed as $\Delta A_\lambda = A_\lambda^t - A_\lambda^{t=0}$) exhibited slight but consistent changes at increased chlorine doses, reaction times and temperatures. Despite the presence of such changes, the differential spectra of chlorinated Ancipa water in all cases had a prominent peak near 272 nm and their intensity ΔA_λ increased with increasing reaction time (Figure 3), similarly to observations for other surface waters (e.g., *7*).

Unlike UV absorbance, the intensity of fluorescence of chlorinated Ancipa water underwent non-monotonic changes at varying reaction times, similarly to observations reported in (*22*). Despite the complex behavior of the emission intensity, values of $\lambda_{0.5}$, which denotes the position of the normalized emission band (by the maximum intensity) at its half-intensity for wavelengths > λ_{max} exhibited a consistent blue shift (Figure 4). Its magnitude gradually increased at higher reaction times and/or chlorine doses. To quantify this phenomenon, the difference between the values of $\lambda_{0.5}$ and λ_{max} ($\Delta\lambda_{0.5} = \lambda_{0.5} - \lambda_{max}$) was calculated. It represents a measure of the width of the NOM emission band.

Because prior research has indicated that the ratio of fluorescence intensities of NOM at 450 and 500 nm, or FR value, is correlated with its aromaticity and

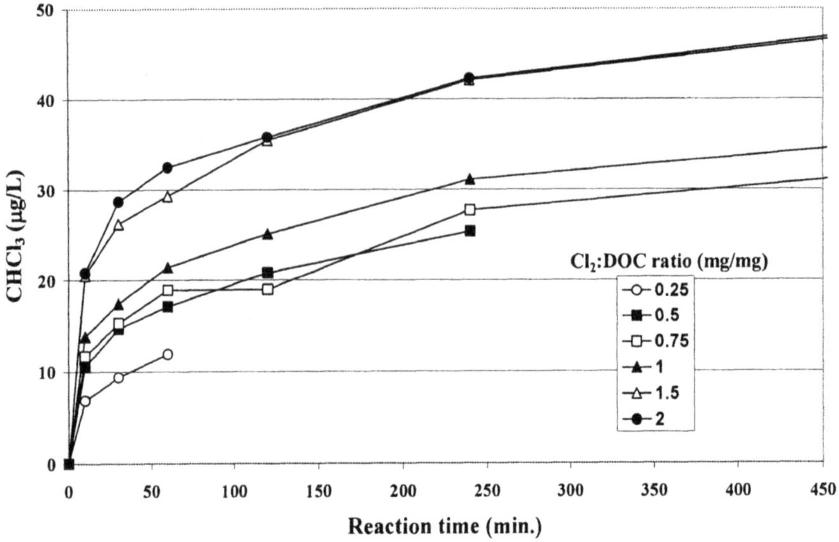

Figure 1. Kinetics of chloroform formation at varying Cl_2 to DOC ratios. Chlorinated Ancipa water, pH 7, temperature 20°C.

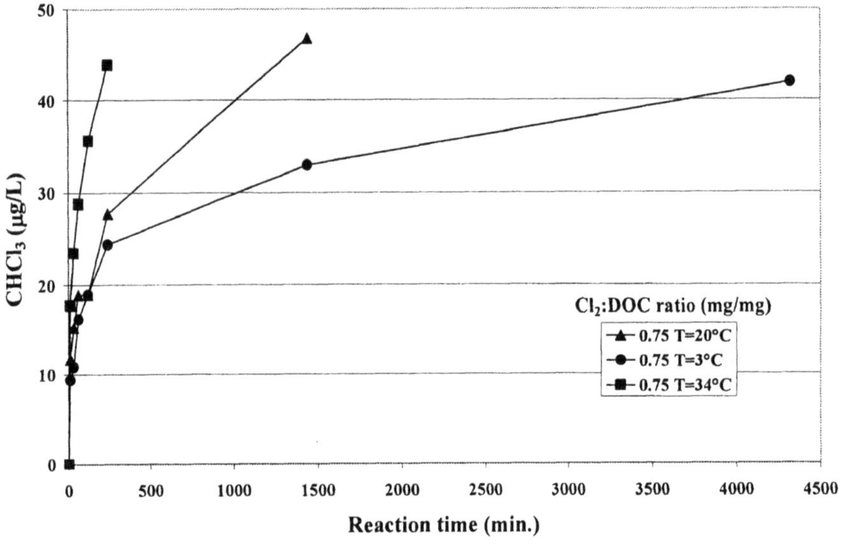

Figure 2. Effects of temperature on the kinetics of chloroform formation in Ancipa water chlorinated at pH 7 and Cl_2 to DOC weight ratio 0.75.

molecular weight (*14, 17, 18*), changes of $I_{500/450}$ for chlorinated Ancipa water were measured. $I_{500/450}$ ratios for chlorinated Ancipa water decreased monotonically at increased chlorine doses and reaction times. This may be hypothesized to be a manifestation of chlorination-induced breakdown of relatively large aromatic molecules of NOM, which is intrinsically linked to the release of individual DBP species (*20*). To explore this hypothesis in more detail, another differential fluorescence index denoted as $\Delta I_{500/450}$ was introduced. It was defined as shown below

$$\Delta I_{500/450} = \left(\frac{I_{500}}{I_{450}}\right)^{t} - \left(\frac{I_{500}}{I_{450}}\right)^{t=0} \quad (1)$$

Like the differential absorbance, $\Delta I_{500/450}$ was always negative (this corresponds to relatively more pronounced changes of the fluorescence intensity at 500 nm compared to those at 450 nm at any given chlorination conditions) and its intensity increased with reaction time.

Relationships Between DBP Formation and Differential Absorbance Measured at Selected Wavelengths

In agreement with previous studies of NOM halogenation (*7*), strong correlations between differential absorbance values at 272 nm (ΔA_{272}) and concentrations of chlorinated and brominated DBPs were found to exist, as shown in Figure 5 for the major THM and HAA species ($CHCl_3$, $CHBrCl_2$, TCAA and DCAA). A strong and unambiguous correlation also existed between changes of absorbance and chlorine consumption (Figure 6). Similarly strong correlations between ΔA_{254} and ΔA_{280} values determined for varying chlorine dose and reaction times and, on the other hand, chlorine consumption or concentrations of individual chlorinated and brominated DBPs were observed.

Those relationships were applicable for all chlorine doses and reaction times, although markedly different kinetics of DBPs formation and chlorine consumption were observed for different chlorine doses. DBP and ΔCl_2 data generated at highly variable experimental conditions and plotted vs. the ΔA_{272} coordinate could be fitted with a single fitting function. This confirms the common observation generated in prior studies of NOM chlorination by the method of differential absorbance spectroscopy that ΔA_{272} (or, alternatively ΔA_{254} or ΔA_{280}) is a fundamentally important system parameter that reflects that engagement and transformation of the halogen attack sites in NOM.

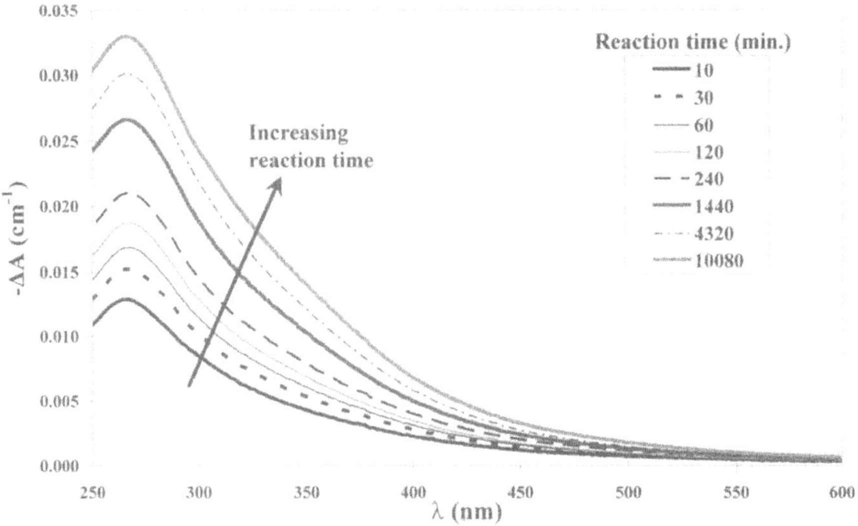

Figure 3. Differential absorbance spectra. Chlorinated Ancipa water, pH 7.0, 20°C, Cl_2 to DOC ratio 1.5.

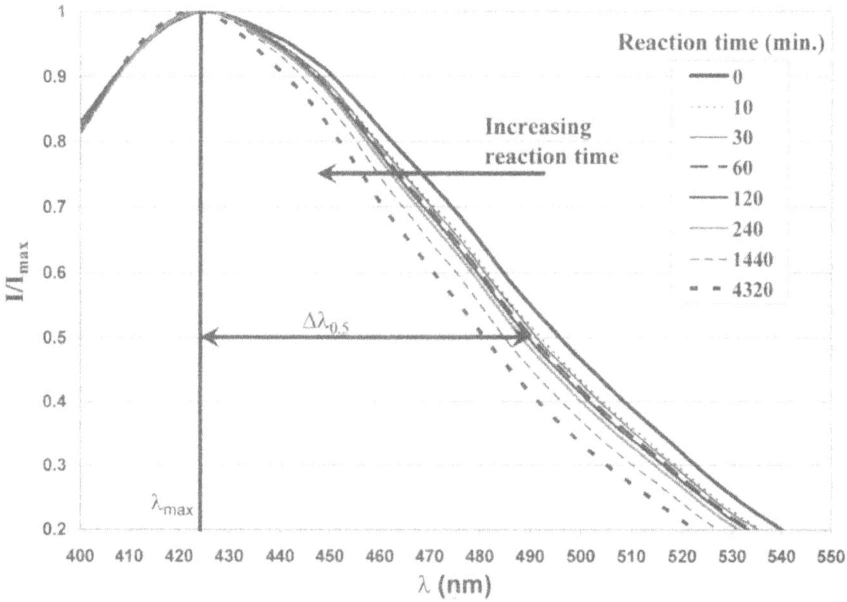

Figure 4. Normalized fluorescence spectra. Chlorinated Ancipa water, pH 7.0, 20°C, Cl_2 to DOC ratio 1.0.

Results generated in this study indicate the above conclusion can be expanded to include temperature as another NOM halogenation parameter whose variations cause different patterns of DBP release in the time domain but they collapse into the same function in the A_λ domain. Indeed, markedly different kinetics of DBPs formation and chlorine consumption were observed for different temperatures (Figure 2), but similar relationship between halogenated DBP and ΔA_{272} continued to exists for all temperatures, as illustrated in Figure 7 for $CHCl_3$. This result enhances the point that the differential absorbance is an intrinsic system parameter that enables *in situ* monitoring DBPs formation in distribution systems. Its use is all the more appealing for systems experiencing considerable short term fluctuations of temperature, as well as those with high variability of chlorine consumption and residence time. However, possible effects of seasonal and short-terms variations of NOM properties on ΔA_{272} vs. DBP correlations need to be examined in more detail.

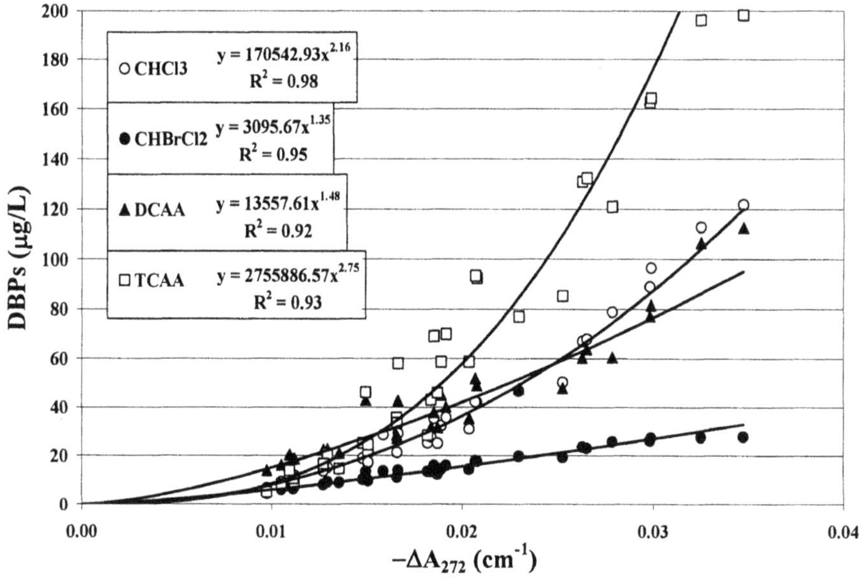

Figure 5. Correlations between selected individual chlorinated and brominated DBPs and ΔA_{272} values. Chlorinated Ancipa water, pH 7.0, 20°C, Cl_2 to DOC ratios from 0.25 to 2.0, reaction times 10 minutes to 7 days.

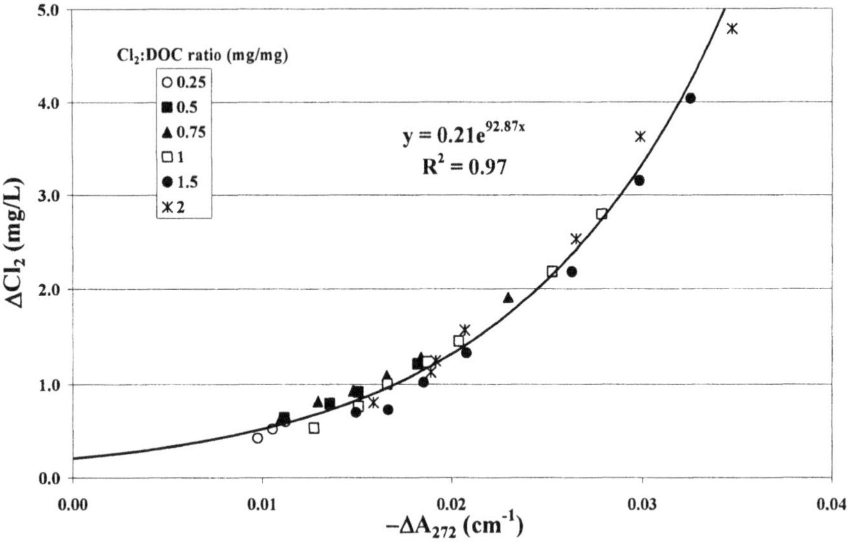

Figure 6. Correlations between chlorine consumption and ΔA_{272} values. Chlorinated Ancipa water, pH 7.0, 20°C, Cl_2 to DOC ratios from 0.25 to 2.0, reaction times 10 minutes to 7 days.

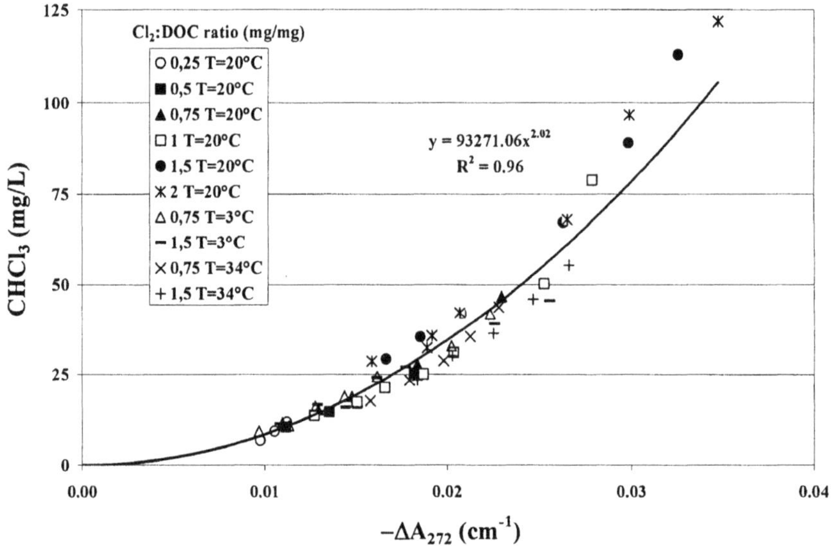

Figure 7. Correlations between chloroform and ΔA_{272}. Chlorinated Ancipa water, pH7, 3°C, 20°C, 34°C, Cl_2 to DOC ratios 0.25 to 2.0, reaction times 10 minutes to 7 days.

Examination of NOM Halogenation and DBP Release Using Fluorescence Indices

Analysis of the behavior of an alternative set of spectroscopic parameters, namely that $\Delta\lambda_{0.5}$ or $\Delta I_{500/450}$ fluorescence indexes, showed that they too can be employed for NOM halogenation monitoring. Figure 8 shows such correlations for $\Delta\lambda_{0.5}$ values and major THM and HAA species. The fact that the experimental data for each single individual DBP compound collapsed, similarly to the behavior of these compounds in the ΔA_{λ} domain, into a single dataset when represented vs. the respective $\Delta\lambda_{0.5}$ values supports a conclusion that the changes of $\Delta\lambda_{0.5}$ are indicative of microscopic details of NOM halogenation processes. Strong relationships between $\Delta I_{500/450}$ values and DBP concentrations were observed (Figure 9). Finally, $\Delta\lambda_{0.5}$ and $\Delta I_{500/450}$ values exhibited an unambiguous relationship with chlorine consumption, as exemplified for $\Delta\lambda_{0.5}$ in Figure 10.

All spectroscopic indices used in this study were stongly correlated with the concentrations of various DBPs. For instance, $CHCl_2Br$, bromochloroacetic acid, bromodichloroacetic acid were correlated with ΔA_{272}, with regression coefficients of 0.96, 0.94 and 0.92, respectively. The best fit functions that correspond to these regressions were mostly power functions, although in other cases linear or exponential relationships were employed (Figures 5, 7, 8 and 9). These best-fit functions were selected based on the maximum attainable R^2 values. As such, they are not necessarily indicative of the intrinsic chemistry of NOM halogenation. The fundamental functional type of the ΔA_{272} vs. DBP correlations remains to be ascertained, but evidence concerning the contributions of slow and fast NOM reactive sites (*23*) indicates that these correlations are likely to be polynomial.

The above results prove that the $\Delta\lambda_{0.5}$ and $\Delta I_{500/450}$ fluorescence indices are powerful parameters for *in situ* monitoring of the formation of DBP species. The behavior of $\Delta\lambda_{0.5}$ and $\Delta I_{500/450}$ is in many respects similar to that of ΔA_{272} (Figure 11). This similarity indicates that $\Delta\lambda_{0.5}$ and $\Delta I_{500/450}$ indices are likely to be indicative of the consumption of the reactive aromatic groups in NOM. To explore this issue, we decided to examine the relationship between the aromaticity of NOM that is correlated with the specific absorbance of NOM at 280 nm, and the spectroscopic fluorescence indexes introduced or additionally explored in this study. According to (*9*), the aromaticity of NOM (calculated as the percentage of organic carbon in NOM associated with aromatic functional groups) is the following linear function of ε_{280}:

$$\Phi_{280}(\%) = 0.057\varepsilon_{280} + 3.0 \tag{2}$$

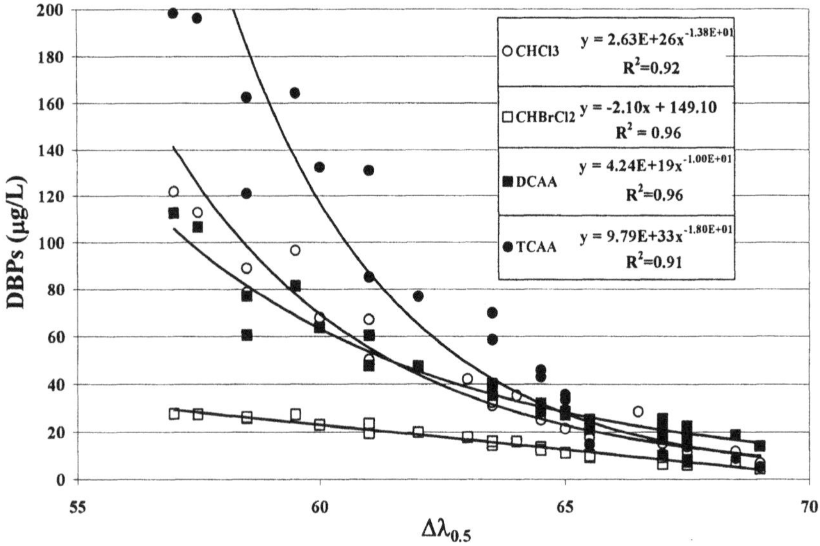

Figure 8. Correlations between selected individual chlorinated and brominated DBPs and $\Delta\lambda_{0.5}$ values. Chlorinated Ancipa water, pH 7.0, 20°C, Cl_2 to DOC ratios 0.25 to 2.0, reaction times 10 minutes to 7 days.

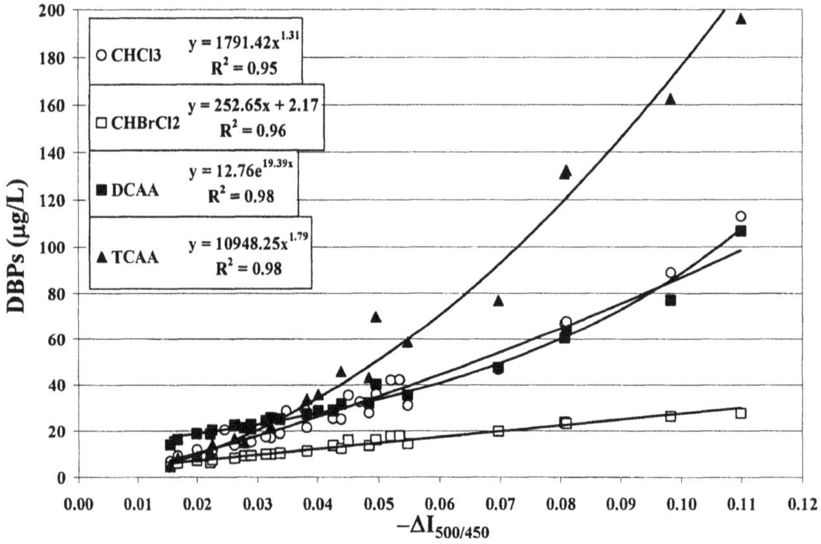

Figure 9. Correlations between selected individual chlorinated and brominated DBPs and $\Delta I_{500/450}$ values. Chlorinated Ancipa water, pH 7.0, 20°C, Cl_2 to DOC ratios 0.25 to 2.0, reaction times 10 minutes to 7 days.

Based on the relationships shown in Figure 11 , and applying absorbance data for chlorinated Ancipa water generated at 280 nm, we determined that there are convincing reasons to view changes $I_{500/400}$ or $\lambda_{0.5}$ values as a manifestation of the behavior of NOM aromaticity during halogenation. Analytically, the relationships between $I_{500/400}$ or $\lambda_{0.5}$ and NOM aromaticity can be described by the following best-fit formulas:

$$I_{500/400} = 0.02\Phi_{280} + 0.23 \qquad (R^2=0.98) \tag{3}$$

$$\lambda_{0.5} = 2.16\Phi_{280} + 460 \qquad (R^2=0.97) \tag{4}$$

The existence of strong $I_{500/400}$ or $\lambda_{0.5}$ vs. A_{280} correlations that constitute the basis for the derivation of equations (3) and (4) does not disprove an alternative viewpoint according to which transformations of the fluorescence spectra of NOM are caused not only by the changes of the aromatic fluorophores, but also of conformations and molecular weights of NOM molecules. In NOM halogenation reactions, the degradation of the aromatic fluorophores and breakdown of NOM molecules is accompanied by the release of products of their fragmentation, notably individual DBPs (*20, 23*). The chlorine attack on and degradation of, the aromatic sites and ensuing breakdown of NOM molecules processes must occur quasi-simultaneously. The near synchronicity of these two different processes is likely to be the fundamental reason of the existence of the correlations between any of the several examined spectroscopic indexes (ΔA_λ, $\Delta I_{500/400}$ and $\Delta\lambda_{0.5}$) and individual DBPs species.

Conclusions

This study explored the performance of a well established (ΔA_{272}) and alternative spectroscopic (ΔA_{280}, ΔA_{254}, $\Delta\lambda_{0.5}$ and $\Delta I_{500/450}$) indices, all which were found to be unambiguously correlated with DBPs formation and chlorine consumption in Ancipa reservoir water. Strong relationships between the behavior of differential absorbance (e.g., ΔA_{272}) and fluorescence indices ($\Delta\lambda_{0.5}$ and $\Delta I_{500/450}$) were also shown to exist. This observation reinforces the concept that both absorbance and fluorescence indices are complementary indicators of the engagement of aromatic groups and DBPs formation.

Based on the observed data, it can be concluded that any of the examined spectroscopic indices, or their combination, can be used to predict and ultimately control the formation of individual DBP species in practically important situations. One prominent aspect of such an approach is that applications of differential absorbance and/or fluorescence spectroscopy allow directly estimating the response of NOM to halogenation in a wide range of water

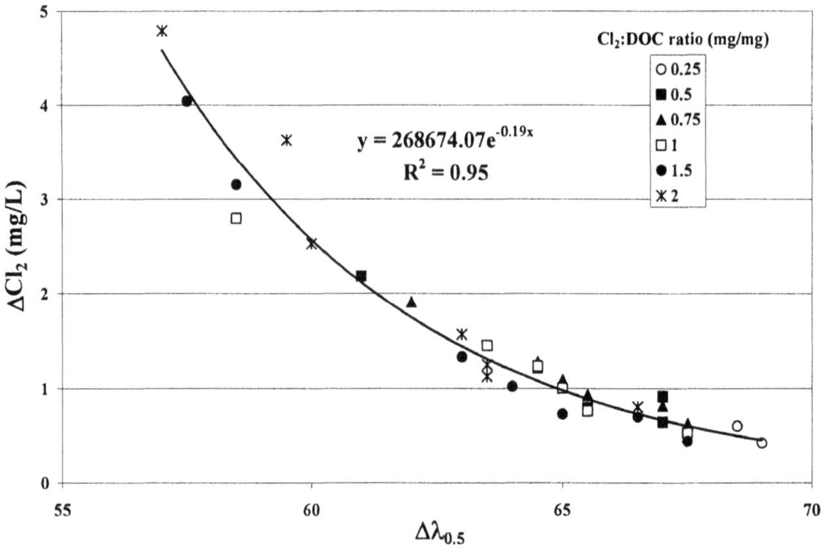

Figure 10. Correlations between chlorine consumption and $\Delta\lambda_{0.5}$ values. Chlorinated Ancipa water, pH 7.0, 20°C, Cl_2 to DOC ratios 0.25 to 2.0, reaction times 10 minutes to 7 days.

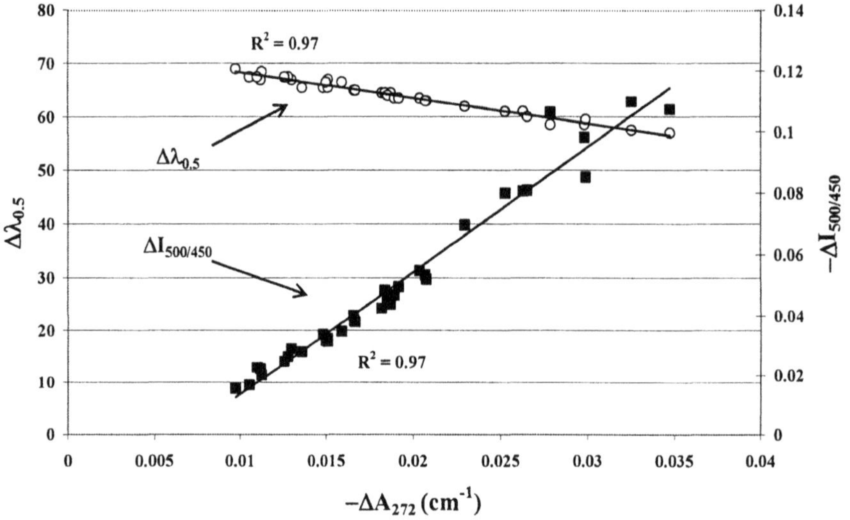

Figure 11. Correlations between developed fluorescence indices and ΔA_{272} values. Chlorinated Ancipa water, pH 7.0, 20°C, Cl_2 to DOC ratios 0.25 to 2.0, reaction times 10 minutes to 7 days.

treatment (chlorine dose) and distribution (reaction time and temperature) conditions. This approach also allows carrying out rapid assessments of the performance of alternative water treatment processes.

Because of their strong relationships with the intrinsic chemistry of NOM, the spectroscopic indices discussed in this study can be potentially used to develop a complete mechanistic model of DBP formation. Further investigations that need to be performed to determine all relevant details of such a model for representative surface waters, for instance Ancipa Reservoir water, include a more detailed exploration of the effects of pH, temperature, bromide concentration and NOM properties (hydrophobic, transphilic and hydrophilic fractions) in DBP formation.

Acknowledgments

Authors are thankful to Professor Mark M. Benjamin (University of Washington) and to Professor Rodolfo M. A. Napoli (University of Napoli "Parthenope") for their interest in and support of this study. This study was partially supported by the United States EPA/Cadmus (grant 069-UW-1) and the Italian Ministry of Instruction, Universities and Research (MIUR), through the Program of Research Projects of National Interest "Control and Monitoring of Drinking Water Quality". Views expressed in this paper do not necessarily reflect those of the funding agencies.

References

1. Reckhow, D.A.; Singer, P.C.; Malcolm, R.L. *Environ. Sci. Technol.* **1990**, *24*(11), 1655-1664.
2. Richardson, S.D.; Thruston, A.D., Jr.; Rav-Acha, C.; Groisman, L.; Popilevsky, I.; Juraev, O.; Glezer, V.; McKague, A.B.; Plewa, M.J. and Wagner, E.D. *Environ. Sci. Technol.* **2003**, *37*(17), 3782-3793.
3. Krasner, S.W.; Weinberg, H S.; Richardson, S.D.; Pastor, S.J.; Chinn, R.; Sclimenti, M.J.; Onstad, G.D.; Thruston, A.D. *Environ. Sci. Technol.* **2006**, *40*(23), 7175-7185.
4. Korshin, G.V.; Li, C.-W.; Benjamin, M.M. *Water Res.* **1997**, *31*(4), 946-949.
5. Croué J.-P.; Korshin, G.V.; Benjamin, M.M. *Characterization of Natural Organic Matter in Drinking Water*; American Water Works Association Research Foundation: Denver, CO. **2000**.
6. Kitis, M.; Karanfil, T.; Wigton, A.; Kilduff, J.M. *Water Res.* **2002**, *36*(15), 3834-3848.
7. Korshin, G.V.; Wu, W.W.; Benjamin, M.M.; Hemingway, O. *Water Res.* **2002**, *36*(13), 3273-3282.

8. Chin Y.P.; Alken G. and O'Loughlin E. *Environ. Sci. Technol.* **1994**, *28*(11), 1853-1858.
9. Peuravuori J. and Pihlaja K. *Anal. Chim. Acta* **1997**, *337*(2), 133-149.
10. Peuravuori J. and Pihlaja K. *Environ. Sci. Technol.* **2004**, *38*(22), 5958-5967.
11. Peuravuori J.; Monteiro A.; Eglite L.; Pihlaja K.C. *Talanta* **2005**, *65*(2), 408–422.
12. Wu, W.W.; Chadik, P.A.; Davis, W.M.; Delfino, J.J.; Powel, D.H. In *Natural Organic Matter and Disinfection By-Products: Characterization and Control in Drinking Water*, Amy L.G.; Krasner S.W.; Barrett S.E. Eds. ACS symposium series 761; American Chemical Society: Washington, DC, **2000**; pp 109-121.
13. Karanfil, T.; Kitis, M.; Kilduff, J.M.; Wigton, A. In *Natural Organic Matter and Disinfection By-Products: Characterization and Control in Drinking Water*, Amy L.G.; Krasner S.W.; Barrett S.E. Eds. ACS symposium series 761; American Chemical Society: Washington, DC, **2000**; pp 190-205.
14. McKnight, D.M.; Boyer, E.W.; Westerhoff, P.K.; Doran, P.T.; Kulbe, T.; Andersen, D.T. *Limnol. Ocean.* **2001**, *46*(1), 38-48.
15. Cabaniss, S.E. *Environ. Sci. Technol.* **1992**, *26*(6), 1133-1139.
16. Pullin, M.J. and Cabaniss, S.E. *Environ. Sci. Technol.* **1995**, *29*(6), 1460-1467.
17. Klapper, L.; McKnight, D.M.; Fulton, J.R.; Blunt-Harris, E.L.; Nevin, K.P.; Lovley, D.R.; Hatcher P.G. *Environ. Sci. Technol.* **2002**, *36*(14), 3170-3175.
18. Cory R.M. and McKnight D.M. *Environ. Sci. Technol.* **2005**, *39*(21), 8142-8149.
19. Schwede-Thomas, S.B.; Chin, Y.-P.; Dria, K.J.; Hatcher, P.; Kaiser E. and Sulzberger, B. *Aquat. Sci.* **2005**, *67*(1), 61-71.
20. Korshin, G.V.; Kumke, M.U.; Li, C.W. and Frimmel, F.H. *Environ. Sci. Technol.* **1999**, *33*(8), 1207-1212.
21. Li, C.W. and Korshin, G.V. *Chemosphere* **2002**, *49*(6), 629-636.
22. Korshin, G.V. and Fabbricino M. *Water Science and Technology: Water Supply* **2005**, *4*(4), 227-233.
23. Korshin, G.V.; Benjamin, M.M.; Chang, H.-S.; Gallard, H. *Environ. Sci. Technol.* **2007**, *41*(8), 2776-2781.

Formation and Control of Disinfection By-Products

Chapter 15

Combined Treatments for Enhanced Reduction of Trihalomethane Precursors

Rolando Fabris, Christopher W.K. Chow and Mary Drikas

CRC for Water Quality and Treatment, Australian Water Quality Centre, SA Water Corp.
Private Mail Bag 3, Salisbury, South Australia 5108

Research has determined that there is a portion of natural organic matter (NOM) that cannot be removed by coagulation processes. As such, alternative treatment technologies must be applied to achieve improvements in drinking water quality. Three different powdered activated carbons (PACs) were applied in combination with coagulation to treat a high dissolved organic carbon (DOC) source water with a focus on reducing trihalomethane formation potential (THMFP). A steam-activated, coal-based carbon was shown to most effectively reduce DOC and THMFP. In addition, a combined treatment protocol utilising adsorbent technologies (MIEX® and PAC) with coagulation was effective in reducing DOC by up to 96% and THMFP by up to 97%. Due to differing mechanisms of NOM removal, the technologies applied were complimentary in removing DOC of various character, including material that is typically recalcitrant to coagulation.

Introduction

Many published studies have determined that there is a portion of the natural organic matter (NOM) that cannot be removed by coagulation processes *(1-4)*. This material can be termed recalcitrant NOM and is primarily composed of low molecular weight hydrophilic neutral organics such as polysaccharides, proteins and amino sugars *(5,6)*. Previous investigations have also established that coagulants alone were incapable of producing required reductions in chlorine reactivity and disinfection by-product (DBP) formation in some water sources *(7)*. As such, alternative treatment technologies must be considered if further improvements in drinking water quality are to be achieved.

Alternative treatment technologies for NOM removal fall into four main categories: biological degradation; oxidative processes such as UV irradiation and ozonation; adsorbent technologies such as ion-exchange resins, metal oxides and activated carbon; and membrane filtration, specifically nanofiltration and reverse osmosis.

In terms of practical application, perhaps the simplest of the alternative technologies to implement into existing plant infrastructure is activated carbon and more specifically, powdered activated carbon (PAC). The ability to dose periodically into the treatment plant with minimal changes to operation makes this a popular choice for removal of algal metabolites or micro-pollutants, however the use of PAC specifically for NOM removal is less prevalent. This concept is explored further through the work presented in the following section.

Of greater complexity to implement, but developed specifically for NOM removal, MIEX®, a magnetic ion-exchange resin, has shown effectiveness in enhancing DOC removal, while reducing the need for chemical based treatments, such as coagulation (*8,9*). The enhanced aggregation and settling of the resin due to magnetic properties allows application in stirred contacters that can cause operational issues for many other ion-exchange resins. It has shown effectiveness for removal of a wide range of organic compounds of various molecular weights and can also reduce the dosing requirement for disinfectants and hence reduces DBP formation *(10-12)*. In practical applications, adsorbents will also typically require a clarification step to remove or reduce turbidity, such as traditional coagulation/media-filtration or microfiltration (MF).

As most of the described techniques are not practical to apply in isolation, it is common to combine several technologies which are complimentary to each other in removing selected contaminant materials efficiently and economically. In such a way, both particulates and dissolved species, such as NOM, algal metabolites and micro-pollutants can be targeted within the treatment strategy. This is conducive to the 'multi-barrier' approach to drinking water treatment that has gained considerable interest and approval in recent years *(13)*.

Evaluation of Activated Carbon

Activated carbon has long been used as a treatment technology for the removal of algal toxins, tastes and odours and micropollutants, such as pesticides and pharmaceuticals. It is also one of the most effective methods for removal of recalcitrant NOM and this can be optimised through careful choice of carbon material based on selection of the pore volume distribution and adsorption kinetics *(14-17)*. While granular activated carbon (GAC) is preferred for long term application of activated carbon, PAC, through virtue of greater surface area, requires lower doses and less contact time to produce equivalent removals of target compounds *(18)* and is more easily applied in a laboratory based investigation.

Three different powdered activated carbons were applied in combination with alum to treat a high DOC source water with a focus on improving NOM removal, and specifically recalcitrant THM precursors.

PAC evaluation protocol

The source water used for this investigation was Myponga Reservoir, located about 50km south of Adelaide, Australia. The water is sourced via surrounding catchment and is generally considered a high colour and DOC source (115 Hazen units and 14.5mg/L), respectively, at the time of the investigation.

Aluminium sulphate stock solution (20,000mg/L as $Al_2(SO_4)_3.18H_2O$) was prepared by dilution of liquid aluminium sulphate solution (7.5% as Al_2O_3) with ultrapure water. Preliminary jar testing indicated that a dose of 100mg/L of aluminium sulphate (alum) was optimal for DOC removal. The jar testing procedure employed has been previously described *(3)* All jar tests were pH controlled to between 6.2 and 6.4.

The three PACs used were chosen to represent a variety of source materials and activation methods. Carbon A is a steam activated, coconut based carbon, which is primarily microporous (<2nm pore width) in structure. Carbon B is a chemically activated, wood based carbon, which is primarily mesoporous (2-50nm pore width) *(19)*. Carbon C is a steam activated, coal based carbon which has a broad cross section of pore diameters. Realistic carbon doses of 20, 40 and 60mg/L were chosen. PAC was dosed at the same time as the alum and contact times were effectively 15 minutes, after which the PAC settled with the floc.

Analysed parameters included absorbance at 254nm (UV_{254}), DOC, trihalomethane formation potential (THMFP) and rapid fractionation. Samples for UV_{254} and DOC were filtered through 0.45μm membranes. Absorbance was measured through a 1cm quartz cell and DOC was obtained using a Sievers 820

Portable TOC analyser (GE Analytical Instruments, USA). THMFP was determined by addition of 20mg/L Cl_2 to 100mL of pre-warmed (30°C) sample in a brown glass bottle with no headspace. Reaction was allowed for 4 hours at 30°C in a covered water bath before quenching the residual chlorine with excess ascorbic acid. THM concentrations were determined by headspace GC (Clarus 500, Perkin Elmer, USA) with electron capture detection. Rapid fractionation technique uses selected ion-exchange resins to separate DOC into four fractions based on character and molecular weight. A 500mL sample was acidified and flowed through packed columns of Supelite™ DAX-8 and then Amberlite® XAD-4 resin with intermediate samples taken for DOC analysis. Samples were then adjusted to pH8 and flowed through a strong anion exchange resin (Amberlite® IRA-958). All resins were obtained from Supelco (Sigma Aldrich, USA). Fraction concentrations were obtained by calculation of DOC concentration measured before and after each resin. Fractions produced are defined as very hydrophobic acids (VHA), slightly hydrophobic acids (SHA), charged hydrophilics (CHA) and neutral hydrophilics (NEU). More specific information on the technique have been described elsewhere *(10)*.

Characterisation of PAC performance

The removal of UV_{254} following alum jar tests with PAC is shown in Figure 1. While UV absorbance was reduced by more than 80% for all carbons and doses applied, there were variations in the trend with increasing dose that indicate different effects for each carbon. Carbon A showed marginal improvement with increasing dose indicating that either the pore sizes were too small for the target compounds (NOM), or the kinetics of adsorption of NOM was too slow for the selected contact time. Carbon B showed maximum UV_{254} and DOC reduction at 20mg/L, the lowest applied dose, and then decreased with increasing PAC dose. It is postulated that this is due to the competing effects between increased NOM adsorption with higher carbon doses, and more significantly, disadvantageous interactions between the carbon and the coagulant. Carbon B, like many chemically activated carbons, have acidic surfaces giving a negative charge at pH conditions near neutral. Newcombe and Drikas (*20*) showed carbon B to have a relative surface charge of approximately -0.15 mmol/g at the applied coagulation pH range (6.2-6.4). In contrast, a steam activated coal based carbon, similar to Carbon C, exhibited a positive charge (+0.10 mmol/g) in the same pH range. For this reason, there is the possibility that an electrostatic attraction exists between the positively charged alumino-hydroxy complexes of the coagulant and the surface of carbon B. Essentially, Carbon B appeared to exert a coagulant demand that decreased the overall efficiency of the treatment. The applied experimental protocol, with PAC and alum dosed simultaneously, would have served to highlight this effect. This phenomenon would not normally be observed in traditional PAC applications for drinking water treatment as sufficient contact time is usually allowed for contaminant adsorption, prior to addition of the coagulant.

Carbon C exhibited increasing removal with dose, with the greatest removal attained at 60mg/L, the highest dose applied in the investigation. The importance of characterizing UV_{254} reduction lies in the fact that it shows strong correlation with disinfectant demand *(21)*. A relationship between UV_{254} and THMFP also exists, however it is weaker due to the fact that other parameters such as pH, temperature and bromide concentration have a similar or stronger influence on THM formation than NOM character.

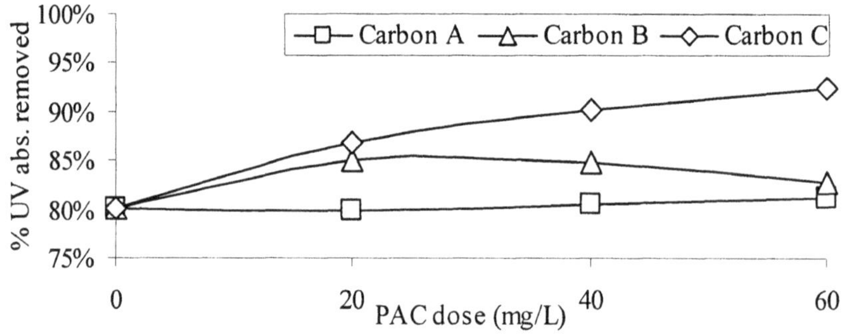

Figure 1. UV absorbance removed with100mg/L alum and increasing PAC dose in Myponga Reservoir water

DOC removal efficiencies varied for the three carbons with 100mg/L alum (within 2.7mg/L DOC), however with the exception of carbon C, the optical character of the residual NOM was not noticeably different to alum alone (Table 1). While the difference in residual DOC after carbon A and B was 0.4 mg/L, the specific UV absorbance (SUVA) showed no variation indicating that the additional material removed was of very similar spectral character as the bulk sample. However when comparing Carbon B and C treated water, the 2.3mg/L difference in DOC reduction was clearly reflected in the SUVA also (Δ=0.4). This demonstrated the different selectivities of the three activated carbons for DOC removal in this water source and highlights the need for evaluating activated carbons in the source water matrix against the target compounds to establish their suitability.

Rapid fractionation was used to determine the effects of the carbon adsorption on the character of the organic matter and was expressed as percentage of the raw water fraction concentration remaining after treatment

Table I. SUVA, DOC and THMFP reduction maxima for Myponga Reservior.

	SUVA *(cm^{-1})*	*DOC* *(mg/L)*	*% DOC* *removal*	*THMFP* *(μg/L)*	*%THMFP* *reduction*
Raw	4.1	14.5	-	470	-
Alum	2.1	5.6	62%	108	77%
Alum + Carbon A	2.1	5.3	63%	106	77%
Alum + Carbon B	2.1	4.9	66%	72	85%
Alum + Carbon C	1.7	2.6	82%	74	84%

(Figure 2). Being primarily microporous, it was observed that carbon A was not very capable of adsorbing the larger MW fractions (VHA and SHA) and mostly removed additional CHA material over alum treatment alone. However, as CHA material made up only 5% of the total Myponga Reservoir NOM at the time of the investigation, the reduction of overall DOC amounted to 0.3mg/L and THMFP was not improved over alum alone. Carbon B showed improvement in VHA removal over alum treatment alone and therefore adsorbed an additional 0.7mg/L DOC, with the removal of other fractions being inconsistent. Regardless, the THMFP reduced by 8% compared to alum alone. Carbon C removed an additional 3.0mg/L DOC over alum treatment, through improved removal of all character fractions, showing the advantage of a broad pore volume distribution for NOM reduction. This translated to a reduction in SUVA also, which was not achieved by the other PACs. THMFP was reduced by a similar level to carbon B which indicates that the relationship of character fraction concentration and THMFP is not significant enough to be used for describing DBP formation in isolation. This is consistant with literature findings *(21)*.

The literature has shown how carbons properties such as pore size and adsorption kinetics affect their performance for removal of other problematic compounds as well as DOC. It is however the heterogeneous and varied nature of NOM that makes it the most difficult group of adsorbates to characterise. For this reason, when a carbon is evaluated for DOC removal, it is difficult to make generalisations about its performance, as the result may not translate to different water sources or differing water quality (seasonal variation). Carbon C, a steam activated coal-based carbon, was known to exhibit a wide cross section of pore sizes and this appeared to favour its performance in this instance but may also indicate greater adaptability for application in other water sources. Carbon C was the most effective carbon for treatment of the source water in combination with alum, as it showed more effective DOC removal than carbons A and B for the given contact time and resulted in equal lowest THMFP.

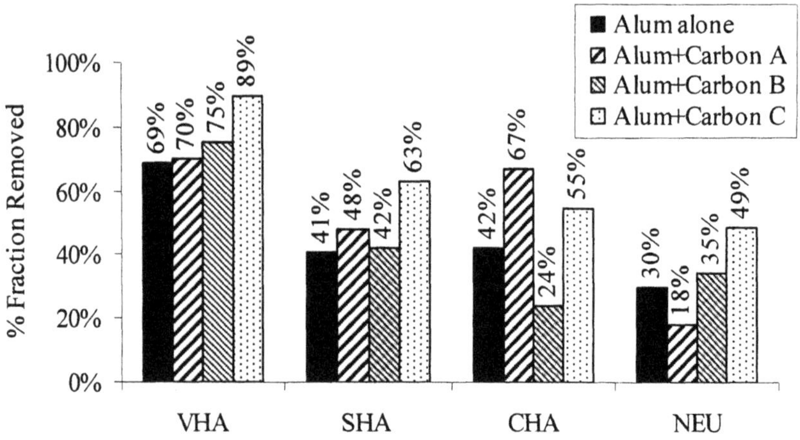

Figure 2. Percentage fraction removed for alum and carbon treatments of Myponga Reservoir water. VHA= Very hydrophobic acids, SHA= Slightly hydrophobic acids, CHA= Charged hydrophilics and NEU= Neutral hydrophilics

Combined Treatments

Subsequent to the carbon selection, a laboratory evaluation of two adsorbent technologies, MIEX® resin and PAC were applied in combination with a coagulation process to further improve NOM removal. Reductions in secondary water quality parameters such as trihalomethane formation potential (THMFP), used as a surrogate of general DBP formation, were also targeted. The capacity to retrofit the process into an existing coagulation treatment plant was also a significant consideration.

Methodology of combined treatment evaluation

The order of treatment followed a logical sequence that could be practically applied in a treatment plant. MIEX® was added as a slurry at a dose of 10mL/L in 2L gator jars (B-KER[2], Phipps and Bird, USA) with stirring at 120rpm for 20 minutes using a gang paddle stirrer. After settling occurred (approximately 10 minutes), the supernatant was decanted and PAC at concentrations of 20, 40 and 60mg/L was dosed. Following 30 minutes contact at 200rpm, 20mg/L alum (as $Al_2(SO_4)_3.18H_2O$) was added and flash mixed for 1 minute at 200rpm, then slow mixed for 14 minutes at 20rpm. Stirring was then ceased and flocs (including PAC) were allowed to settle for 15 minutes prior to filtration through an 11μm pore-size paper filter. In this investigation, aluminium sulphate (alum) was chosen to remove both the natural water turbidity and to coagulate the PAC. Appropriate dose ranges and conditions for all treatment steps were predetermined by jar testing. Control of pH was undertaken at both the MIEX® contact and coagulation steps. The PAC chosen was the same steam activated, coal based carbon identified in the first part of the investigation.

Treated water was analysed for filtered turbidity, true colour at 456nm by the method of Bennett and Drikas *(22)*, UV_{254}, DOC, THMFP and fractionation, as previously described. High performance size exclusion chromatography (HPSEC) was used to derive the molecular weight (MW) distribution. HPSEC was performed using a Waters Alliance 2690 separations module and 996 photodiode array detector (PDA) at 260nm (Waters Corporation, USA). Phosphate buffer (0.1M) with 1.0M NaCl was flowed through a Shodex KW802.5 packed silica column (Showa Denko, Japan) at 1.0mL/min. This column provides an effective separation range from approximately 100 to an exclusion limit of 50,000 Da. Apparent MW was derived by calibration with polystyrene sulphonate (PSS) molecular weight standards.

Combined treatments water quality

Water from Myponga Reservoir was fractionated using the rapid fractionation technique to determine the baseline character of the water source with regards to treatability (Figure 3). Seventy two percent of Myponga Reservoir NOM was determined to be VHA, with less than 7 percent NEU material, indicating a strong humic character and greater influence from allochthonous source NOM. As such, reasonable ease of treatment was expected.

Water quality parameters for Myponga Reservoir treated water are presented in Table II. MIEX®, being an adsorbent resin, is not generally

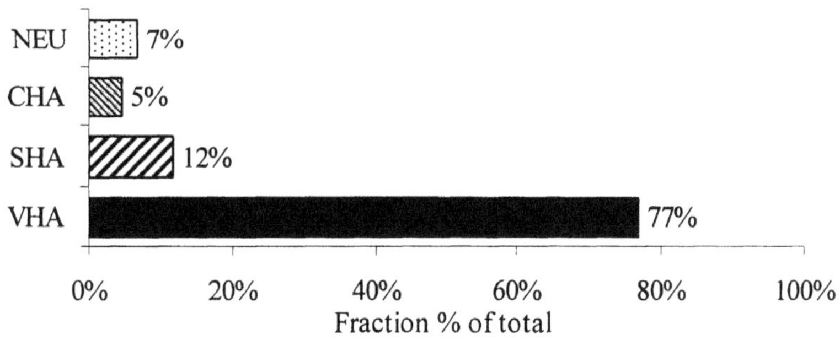

Figure 3. Character fraction composition of Myponga Reservoir water.

effective for turbidity removal and the small reduction observed was due to concurrent settling with the resin. Following the subsequent coagulation stage however, both the PAC particles and natural water turbidity were efficiently reduced (>96%). Colour and UV_{254} were reduced by 86 and 88%, respectively by MIEX® alone, while the subsequent PAC dosing and coagulation improved removal to greater than 98% for both parameters. For DOC removal, 79% was removed by MIEX® contact alone and up to 96% after combined treatment. This demonstrates that the applied treatment protocol is very effective for reduction of traditional water quality parameters to levels previously unachievable by inorganic coagulants alone *(23)*. To examine the effect of the treatments on organic character, molecular weight distribution of the treated waters at intermediate treatment steps was determined.

Table II. Water quality parameters for Myponga Reservoir combined treatment

	Turbidity (NTU)	Colour (HU)	UV_{254} (/cm)	DOC (mg/L)
Myponga Raw	1.88	85	0.492	12.8
MIEX® alone	1.70	12	0.060	2.7
+20mg/L PAC*	0.07	0	0.008	1.0
+40mg/L PAC*	0.08	0	0.003	0.6
+60mg/L PAC*	0.08	0	0.002	0.5

*NOTE: PAC treatment includes 20mg/L alum coagulation.

MIEX® proved very effective for removal of DOC in this application, leaving only low MW recalcitrant NOM and very high MW (>50,000 Da) colloidal material. This is illustrated in the MW distribution in Figure 4a. The coagulant dose applied was sufficient for removal of these colloidal materials. As these compounds are considered to have high specific UV absorbance (SUVA), but little contribution to the bulk water DOC, the effect of their removal is negligible in absolute DOC terms but significant in terms of UV_{254} and colour removal (0.060 to 0.008cm^{-1} and 12 to 0HU, respectively). As such, the removal of the colloidal component is an important consideration as part of the combined treatment. In cases where minimal chemical addition is desired, the coagulation step could potentially be replaced with membrane filtration, conditional on the ability of the filtration to remove the organic colloids.

The benefit of the PAC dosing was in the adsorption of the low MW neutral hydrophilic compounds that are recalcitrant to both coagulation and also the ion-

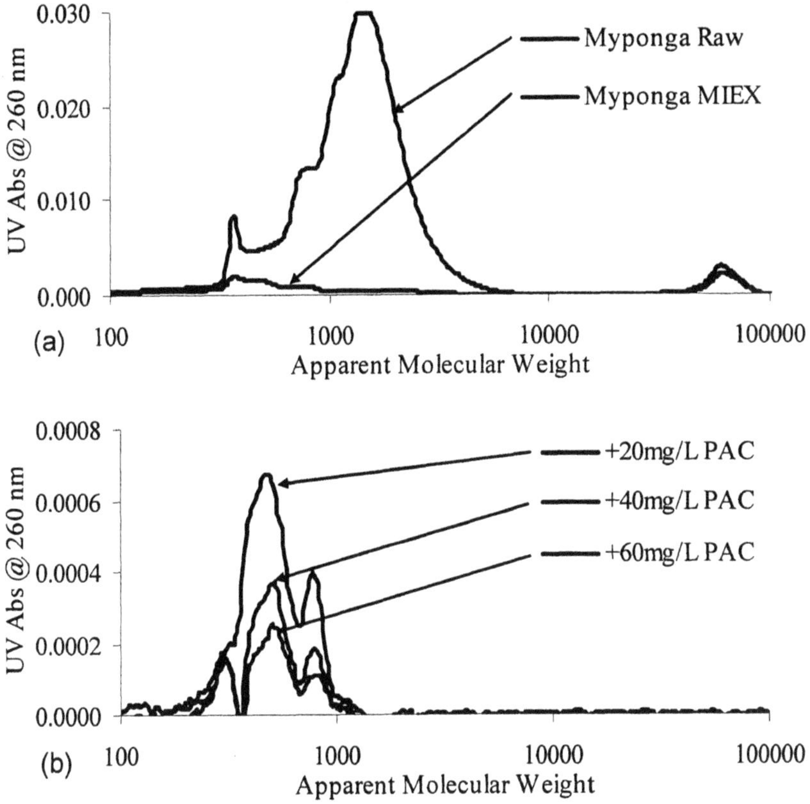

Figure 4. Molecular weight distribution for (a) Myponga Raw and Myponga MIEX treated; and (b) Myponga combined treated.

exchange mechanisms that are prevalent in MIEX® treatment. However, it is important to note that the efficient removal of recalcitrant NOM by the PAC is enhanced by the removal of competitive and more abundant NOM fractions by the MIEX®, prior to activated carbon contact. Recalcitrant components of NOM are typically colourless and have low SUVA *(24)*. For this reason, the reduction of DOC with increasing carbon dose is not strongly reflected in UV_{254} and colour (Table II).THMFP was used as a surrogate parameter to indicate DBP formation. While it is acknowledged that different classes of DBPs have varied pathways of formation, when used for comparative purposes, THM formation can give an indication of the susceptibility of the treated water organic matter to form DBPs. MIEX® treatment reduced Myponga Reservoir THMFP from 581µg/L to 129µg/L, a reduction of 78%. Treatment with PAC and alum showed progressively improved THM reduction with increasing PAC dose (Figure 5). The remaining THM precursors present following MIEX® treatment were primarily small MW recalcitrant organic compounds and were removed by the PAC in a linear decrease observed with increasing carbon dose (Figure 4b). It is also expected that the removal of the >50,000Da colloidal materials in the Myponga water by the coagulation step was beneficial in reducing THM formation at all applied carbon doses. Despite the considerable THMFP reductions achieved, specific THMFP (THMFP per unit DOC) remained largely unaffected by any treatment (Figure 5), indicating that THM reductions were a product of bulk DOC reduction, rather than removal of any particular organic character fractions.

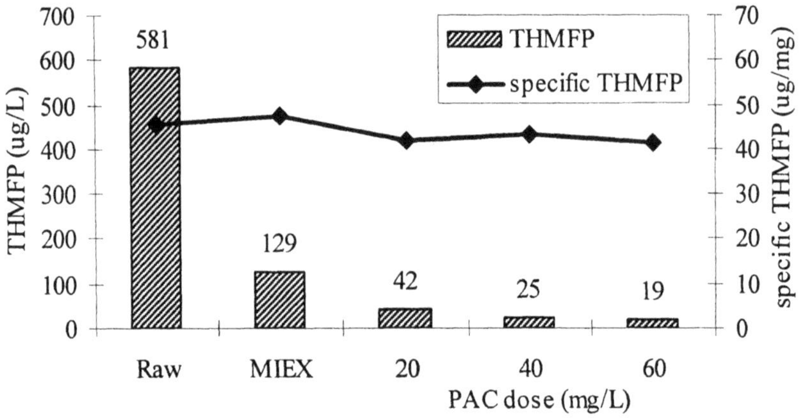

Figure 5. Trihalomethane formation potential(THMFP) and specific THMFP for combined treated Myponga Reservoir

Summary of findings – Combined treatments

In the second part of this investigation, a combined treatment protocol utilising MIEX® and PAC with coagulation was effective in greater than 96% reduction of traditional water quality parameters such as turbidity, colour and UV absorbance in Myponga Reservoir. Concurrently, DOC reduced by up to 96% at the highest applied PAC and alum doses. This resulted in reductions of THMFP, a surrogate for DBP formation, of between 92 and 97%. Due to differing mechanisms of NOM removal, the technologies applied were complimentary to each other in removing DOC of varied character, including material that is typically recalcitrant to traditional water treatment processes.

References

1. Randtke, S.J. Organic Contaminant Removal by Coagulation and Related Process Combinations. *Journal of American Water Works Association* **1988**, 80(5), 40-56.
2. Edwards, M. Predicting DOC removal during enhanced coagulation. *Journal of American Water Works Association* **1997**, *89*(5), 78-89.
3. Chow, C.W.K.; van Leeuwen, J.A.; Drikas, M.; Fabris, R.; Spark, K.M. and Page, D.W. The Impact of the Character of Natural Organic Matter in Conventional Treatment with Alum. *Water Science & Technology* **1999**, *40*(9), 97-104.
4. van Leeuwen, J.; Chow, C.; Fabris, R.; Withers, N.; Page, D. and Drikas, M. Application of a fractionation technique for better understanding of the removal of natural organic matter by alum coagulation. *Water Science & Technology: Water Supply* **2002**, *2*(5), 427-433.
5. Leenheer, J.A. Comprehensive assessment of precursors, diagenesis, and reactivity to water treatment of dissolved and colloidal organic matter. *Water Science and Technology: Water Supply* **2004**, *4*(4), 1-9.
6. Allpike, B.P.; Heitz, A.; Joll, C.A.; Kagi, R.I.; Abbt-Braun, G.; Frimmel, F.H.; Brinkmann, T.; Her, N. and Amy, G. Size Exclusion Chromatography to Characterize DOC Removal in Drinking Water Treatment. *Environmental Science and Technology* **2005**, *39*(7), 2334-2342.
7. Fabris, R.; Chow, C.W.K. and Drikas, M. The Impact of Coagulant Type on NOM Removal. *Proceedings of Ozwater Convention & Exhibition*, Perth, Western Australia, April **2003**, CD-ROM.
8. Chow, C.; Cook, D. and Drikas, M. Evaluation of magnetic ion exchange resin (MIEX®) and alum treatment for formation of disinfection by-products and bacterial regrowth. *Water Science and Technology: Water Supply* **2002**, *2*(3), 267-274.
9. Singer, P.C. and Bilyk, K. Enhanced Coagulation using a Magnetic Ion Exchange Resin. *Water Research* **2002**, *36*(16), 4009-4022.
10. Chow, C.W.K.; Fabris, R. and Drikas, M. A New Rapid Fractionation Technique to Characterise Natural Organic Matter for the Optimisation of Water Treatment Processes. *Journal of Water Supply: Research and Technology – AQUA* **2004**, *53*(2), 85-92.
11. Fearing, D.A.; Banks, J.; Guyetand, S.; Eroles, C.M.; Jefferson, B.; Wilson, D.; Hillis, P.; Campbell, A.T. and Parsons, S.A Combination of Ferric and MIEX for the treatment of a humic rich water. *Water Research* **2004**, *38*(10), 2551-2558.
12. Morran, J.Y.; Drikas, M.; Cook, D. and Bursill, D.B. Comparison of MIEX treatment and coagulation on NOM character. *Water Science and Technology: Water Supply* **2004**, *4*(4), 129-137.

13. *Australian Drinking Water Guidelines 6th Edition*, National Health and Medical Research Council, Australia **2004**.
14. Newcombe, G.; Drikas, M.; Assemi, S. and Beckett, R. Influence of Characterised Natural Organic Material on Activated Carbon Adsorption: I. Characterisation of Concentrated Reservoir Water. *Water Research* **1997a**, *31*(5), 965-972.
15. Lebo, J.A.; Huckins, J.N.; Petty, J.D.; Cranor, W.L. and Ho, K.T. Comparisons of coarse and fine versions of two carbons for reducing the bioavailabilities of sediment-bound hydrophobic organic contaminants. *Chemosphere* **2003**, *50*(10), 1309-1317.
16. Schreiber, B.; Brinkmann, T.; Schmalz, V. and Worch, E. Adsorption of dissolved organic matter onto activated carbon - the influence of temperature, absorption wavelength, and molecular size. *Water Research* **2005**, *39*(15), 3449-3456.
17. Treguer, R.; Couvert, A.; Wolbert, D.; Suty, H. and Randon, G. Influence of porosity and surface chemistry of commercially available powdered activated carbons for the removal of dissolved organic carbon. *Water Science & Technology: Water Supply* **2006**, *6*(3), 27-34.
18. Matsui, Y.; Murase, R.; Sanogawa, T.; Aoki, N.; Mima, S.; Inoue, T. and Matsushita, T. Rapid adsorption pretreatment with submicrometre powdered activated carbon particles before microfiltration. *Water Science and Technology* **2005**, *51*(6-7), 249-256.
19. Newcombe, G.; Drikas, M. and Hayes, R. Influence of Characterised Natural Organic Material on Activated Carbon Adsorption: II. Effect of pore volume distribution and adsorption of 2-Methylisoborneol *Water Research* **1997b**, *31*(5), 1065-1073.
20. Newcombe, G. and Drikas, M. Adsorption of NOM onto Activated Carbon: Electrostatic and Non-electrostatic Effects. *Carbon* **1997**, *35*(9), 1239-1250.
21. Fitzgerald, F.; Chow, W.K. and Holmes, M. Disinfectant demand prediction using surrogate parameters - a tool to improve disinfection control. *Journal of Water Supply: Research and Technology – AQUA* **2006**, *55*(6), 391-400.
22. Bennett L.E. and Drikas M. The evaluation of colour in natural waters. *Water Research* **1993**, *27*(7), 1209-1218.
23. van Leeuwen, J.A.; Chow, C.; Fabris, R.; Drikas, M. and Spark, K. Enhanced coagulation for dissolved organic carbon removal in conventional treatment with alum. *18th Federal Convention of the AWWA*. Proceedings. 11-14th April, **1999**, Adelaide, Australia.
24. Drikas, M.; Chow, C.W.K. and Cook, D. The impact of recalcitrant organic character on disifection stability, trihalomethane formation and bacterial regrowth: An evaluation of magnetic ion exchange resin (MIEX) and alum coagulation. *Journal of Water Supply: Research and Technology – AQUA* **2003**, *57*(2), 475-487.

Chapter 16

Removal of Trihalomethane Precursors Using the MIEX® Dissolved Organic Carbon Process in Combination with Granular Activated Carbon

Mary Drikas, Mike Dixon, and Jim Y. Morran

CRC for Water Quality and Treatment, Australian Water Quality Centre, SA Water Corporation, Private Mail Bag 3, Salisbury, South Australia 5108

A pilot plant study comparing the removal of trihalomethane (THM) precursors by granular activated carbon (GAC) following conventional treatment with and without MIEX pre-treatment was completed. The quality of the treated water following GAC filtration was found to be substantially better when incorporating MIEX. This difference was reduced when the empty bed contact time (EBCT) of the GAC filters was decreased from 20 mins to 5 mins and the GAC filter aged, however the removal of dissolved organic carbon (DOC) after GAC filtration remained marginally more effective with MIEX treated water. The additional DOC removal with MIEX pre-treatment also resulted in reduced chlorine demand and trihalomethane (THM) formation. THM formation following MIEX pre-treatment was considerably lower than could be achieved with conventional coagulation alone, becoming almost negligible following GAC filtration when the EBCT was 20 minutes. MIEX pre-treatment was shown to preferentially remove THM precursors, both by direct removal and by improving the effectiveness of GAC filtration.

The important role of natural organic matter (NOM) in the disinfection process has been well established (*1-3*). Studies have shown that removal of NOM reduces disinfectant demand and results in decreased formation of disinfection by-products (DBP) confirming that NOM provides the major source of precursors in formation of DBPs (*1-7*). Consequently removal of NOM has been adopted as one of the main mechanisms to reduce DBP formation. This has led to the optimisation of existing treatment processes and the development of new processes which focus on NOM removal. The MIEX DOC Process was developed specifically to remove NOM from raw water sources utilising a contact process rather than column filtration (*6,8*). The application of this process in operating treatment plants is increasing (*9-13*). As the MIEX DOC process only removes dissolved organics it is necessary to link the process with another technique to remove suspended matter.

The Mt Pleasant Water Treatment Plant (WTP) in South Australia encompasses the MIEX DOC process and also enables comparison of two subsequent turbidity removal processes – conventional treatment (comprising coagulation, flocculation, sedimentation, and filtration) and submerged microfiltration (MF) (*13*). This has proven the effectiveness of operating MIEX in two possible scenarios – retrofitting into a conventional treatment plant or in a greenfield operation utilising MF to remove particulates. The successful operation of this plant has shown that a high level of removal of dissolved organic carbon (DOC) can be achieved using MIEX combined with a second step for removal of turbidity. This results in high quality water requiring low doses of chlorine to achieve stable disinfectant residuals with associated low levels of trihalomethanes. Whilst the use of MIEX significantly improves NOM removal, there are other water quality contaminants such as pesticides or algal metabolites that require additional processes for effective removal. Granular activated carbon (GAC) filters may be included in the treatment train for such a purpose. Moreover as GAC is a broad adsorbent it will also result in further NOM removal (*14-16*). The effect of combining these two NOM removal processes and the impact on disinfectant demand and the formation of disinfection by-products (DBP) was evaluated in a pilot plant study. The direct comparison of the efficiency of NOM removal by GAC following coagulation with and without MIEX pretreatment is discussed.

Methodology

The Mt Pleasant WTP consists of two separate treatment trains; both utilise the MIEX DOC Process as the first treatment process but are followed by two different particulate removal streams – either conventional treatment (comprising coagulation, flocculation, sedimentation, rapid filtration) or microfiltration. Stream 1 at the Mt Pleasant WTP incorporates MIEX followed by conventional treatment. During the period July 2005 to January 2007, MIEX was applied at an average resin dose of 8mL/L for 10 minute contact followed by sedimentation and removal of the resin before entering the conventional treatment stage, Coagulation during this period utilised an average alum dose of 6.5 mg/L (as $Al_2(SO_4)_3.18H_2O$) and 0.2 mg/L poly DADMAC. A conventional pilot plant without MIEX pre-treatment was established on site at the Mt Pleasant WTP as a control stream to enable direct comparison of the water quality produced with the full scale operation of the MIEX pre-treated water. The conventional pilot plant consists of coagulation, flocculation, sedimentation and rapid filtration. The alum dose was chosen both using a model (*17*) and jar tests to achieve the optimum DOC removal (defined as the point of diminishing return, where an additional 10 mg/L alum produces <0.1 mg/L DOC reduction). The pH was not optimised but was between 6.5-6.8 throughout the study. The alum dose over the study period averaged 40 mg/L (as $Al_2(SO_4)_3.18H_2O$).

Two pilot filters utilizing coal based GAC (Calgon F400, Calgon Carbon Corporation. Pittsburgh, USA) were established in July 2005 with 20 minutes empty bed contact time (EBCT) to treat the two waters, comprising conventional treatment, both with and without MIEX. In June 2006, after 312 days operation of the GAC filters, the EBCT was reduced to 5 minutes to accelerate the reduction in adsorption capacity. This was to accentuate differences between the GAC filters in removal of algal metabolites undertaken in a simultaneous study. Details of the treatment trains supplying feed waters to the GAC columns are summarized in Table I.

Table I. Description of Treatment Trains Supplying GAC filters

Conv	Conventional treatment - pilot plant	GAC 1
MIEX Coag	WTP Stream 1 – MIEX treatment followed by conventional treatment	GAC 2

Samples were taken on a regular basis from (a) the raw water, (b) after MIEX pre-treatment, (c) after the filtration stage of both the conventional pilot plant and the WTP Stream 1 (MIEX treatment followed by conventional treatment) and (d) following GAC filtration. Monitoring of the water quality included NOM characterization by determining molecular weight distribution and rapid fractionation, chlorine demand and trihalomethane formation potential (THMFP). Samples for DOC and molecular weight determination were filtered through 0.45 um pre-rinsed membranes. DOC was measured using a Sievers 820 Portable TOC analyser (GE Analytical Instruments, USA). Molecular weight was determined using high performance size-exclusion chromatography (HPSEC). HPSEC was undertaken using a Waters Alliance 2690 separations module and 996 photodiode array detector at 260 nm (Waters Corporation, USA). Phosphate buffer (0.1 M) with 1.0 M sodium chloride was flowed through a Shodex KW802.5 packed silica column (Showa Denko, Japan) at 1.0 mL/min. Apparent molecular weight was derived by calibration with poly-styrene sulphonate molecular weight standards of 35, 18, 8 and 4.6 kDa.

Chlorine demand was determined by addition of 20-30 mg/L chlorine to 150 mL of sample and storage at room temperature (20°C) in the dark. Residual chlorine was titrated after 72 hours by the DPD ferrous titrimetric method (*18*) and the demand calculated by the difference. THMFP was determined by addition of 20 mg/L chlorine to 100 mL of pre-warmed (30°C) sample in a brown glass bottle with no headspace. Reaction was allowed for 4 hours at 30°C in a covered water bath before quenching the residual chlorine with excess ascorbic acid. THM concentrations were determined by headspace gas chromatography with electron capture detection (Clarus 500 with Turbomatrix 110 (Perkin Elmer Corporation, USA)). The rapid fractionation analysis separates DOC into four fractions by adsorption onto different adsorbent resins in a sequential process based on character and molecular weight. A 500mL sample was acidified and flowed through packed columns of Supelite™ DAX-8 and then Amberlite® XAD-4 resin with intermediate samples taken for DOC analysis. Samples were then adjusted to pH 8 and flowed through a strong anion exchange resin (Amberlite® IRA-958). All resins were obtained from Supelco (Sigma Aldrich, USA). Fraction concentrations were obtained by calculation of DOC concentration measured before and after each resin. Fractions produced are defined as very hydrophobic acids (VHA), slightly hydrophobic acids (SHA), charged hydrophilics (CHA) and neutral hydrophilics (NEU). Specifics of the technique and definitions have been described elsewhere (*19*).

DOC Removal

Raw water DOC prior to any treatment varied between 2.8 and 5.8 mg/L (average 4.1 mg/L) over the 505 day study period between July 2005 and January 2007. The average DOC arising from the various treatment processes and the calculated standard deviation is summarised in Table II. The data following GAC filtration has been averaged for the two different periods encompassing EBCT of 20 mins and 5 mins.

The DOC removal following conventional coagulation has averaged 34% whilst MIEX alone averaged 46% with a further average 10% removal by the additional coagulation step. Table II illustrates that the MIEX treated water contained less DOC and therefore also produced water with lower DOC after GAC filtration. This difference between the two treatment trains was maintained when the EBCT was reduced from 20 mins to 5 mins. However the amount of DOC removed by each GAC was reduced due both to lower EBCT and loss of carbon adsorption capacity with time.

Table II. Average DOC (July 05 – January 07)

	Raw (n=130)	*MIEX* (n=130)	*Pre GAC* (n=130)	*Post GAC EBCT 20mins (n=80)*	*Post GAC EBCT 5min) (n=50)*
Conv	4.1 ± 0.8	-	2.7 ± 0.5	1.5 ± 0.6	2.2 ± 0.3
MIEX Coag	4.1 ± 0.8	2.2 ± 0.5	1.8 ± 0.3	0.6 ± 0.3	1.3 ± 0.1

NOTE: Units are mg/L

A comparison of the DOC removal by the two treatment trains over the study period is shown in Figure 1. The change in GAC EBCT at 312 days is shown by a dashed line. In terms of DOC removal, GAC filtration following MIEX and coagulation (Post GAC 2) showed the best performance with DOC below 0.5 mg/L for the first 240 days. The DOC after GAC 2 filtration also appeared more stable most likely because the DOC entering the GAC filter was more consistent when utilizing MIEX treated water. The MIEX treated water maintained consistently lower DOC levels than the conventional train, with DOC after MIEX treatment alone being lower than conventional treatement after GAC filtration (post GAC 1) after 200 days operation of the GAC filters. The impact of the decrease in EBCT after 312 days is also clearly visible in this figure.

The effect of the MIEX pre-treatment on GAC performance can be seen more clearly in Figure 2 which summarises the actual DOC removed by the GACs fed from the two treatment trains. Initially DOC removal was slightly higher in GAC 1, sourced from the conventional train (average 1.9 compared with 1.6 mg/L), but within 60 days this stabilized to achieve similar removals with both GAC 1 and 2 (average 1.2 compared with 1.3 mg/L) with an EBCT of 20 mins. It is interesting to note that both GAC filters removed approximately the same amount of DOC regardless of the inlet DOC concentration, although due to the lower inlet DOC concentration, the percentage removal by GAC filtration was substantially higher with the MIEX treated water (GAC 2) than the non-MIEX treated water (GAC 1). After 312 days operation when the EBCT was reduced to 5 minutes, both GAC filters were only removing minimal DOC, however, GAC 2 (MIEX treated) continued to marginally outperform GAC 1(non-MIEX treated) (average 0.3 mg/L) compared with 0.2mg/L).

Impact on Disinfection Demand and By-Product Formation

The 3 day chlorine demands of the two treatment trains over the study period is shown in Figure 3. The trending of the chlorine demand is similar to the DOC trends, with the MIEX treatment (before and after coagulation) providing lower demands than the conventional treatment. The stability of the

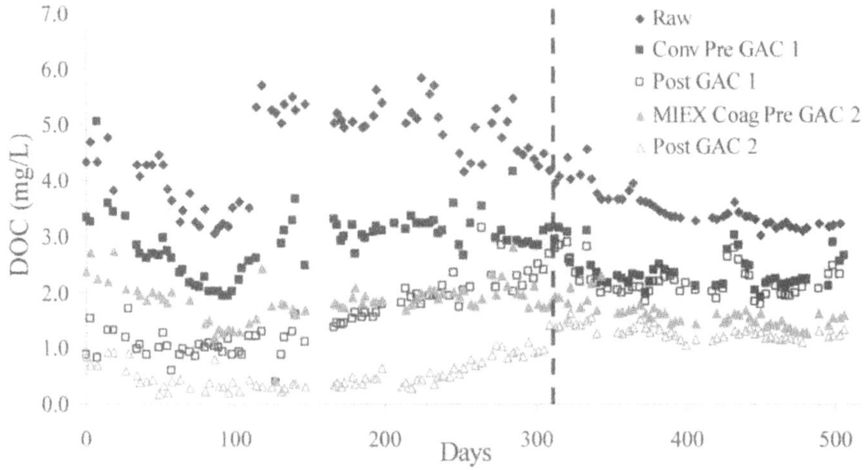

Figure 1. DOC before and after GAC filtration in both treatment trains

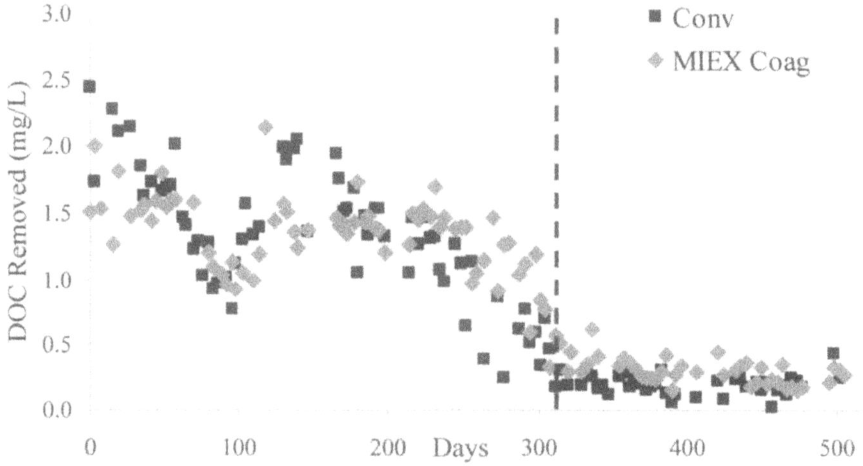

Figure 2. DOC removed by GAC filtration in both treatment trains

water quality after MIEX treatment is also apparent particularly for the period between 100 to 300 days where the chlorine demand of the conventional train was impacted more by raw water quality changes than the MIEX treatment train. GAC filtration following MIEX and coagulation (Post GAC 2) reduced chlorine demand to minimal levels initially whilst GAC filtration enabled conventional treatment (Post GAC 1) to achieve similar chlorine demands to that achieved by MIEX coagulation alone. Once the EBCT was reduced, only marginal improvement in chlorine demand following GAC filtration was observed.

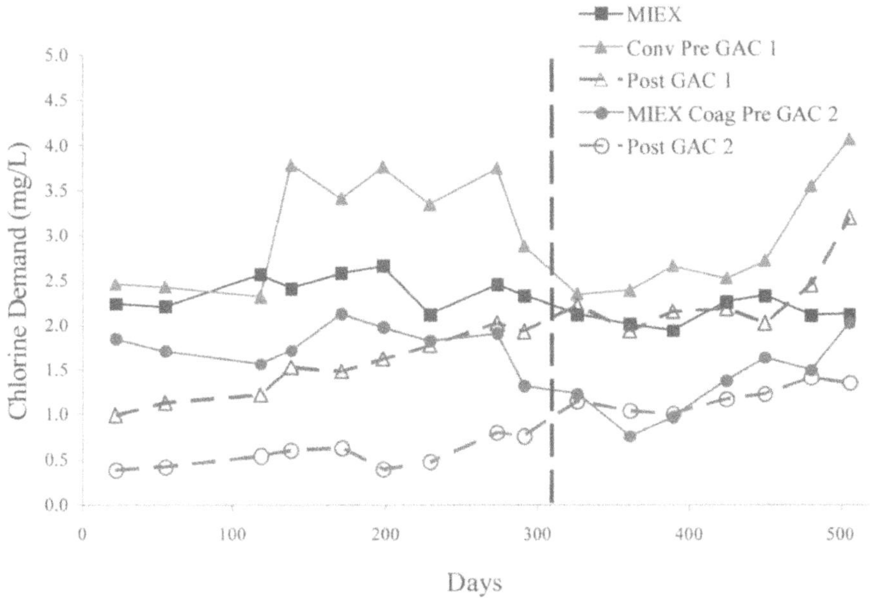

Figure 3. Chlorine demand before and after GAC in both treatment trains

The THMFP of the same process streams are plotted in Figure 4. The lower THMFP resulting after MIEX treatment (before and after coagulation) is apparent. The reduction of the EBCT after 312 days had the greatest impact on the reduction in THMFP following GAC filtration of the conventional stream (Post GAC 1) with very little change in THMFP after this period. The GAC following MIEX treatment (Post GAC 2) continued to reduce the THMFP even after the EBCT was reduced to 5 minutes, providing better attenuation of THMFP as determined in this study than chlorine demand. This suggests that MIEX treatment assists in the preferential removal of THM precursors. Table III contains data extracted from Day 198, including specific chlorine demand and THMFP (normalized per mg/L DOC), to support this.

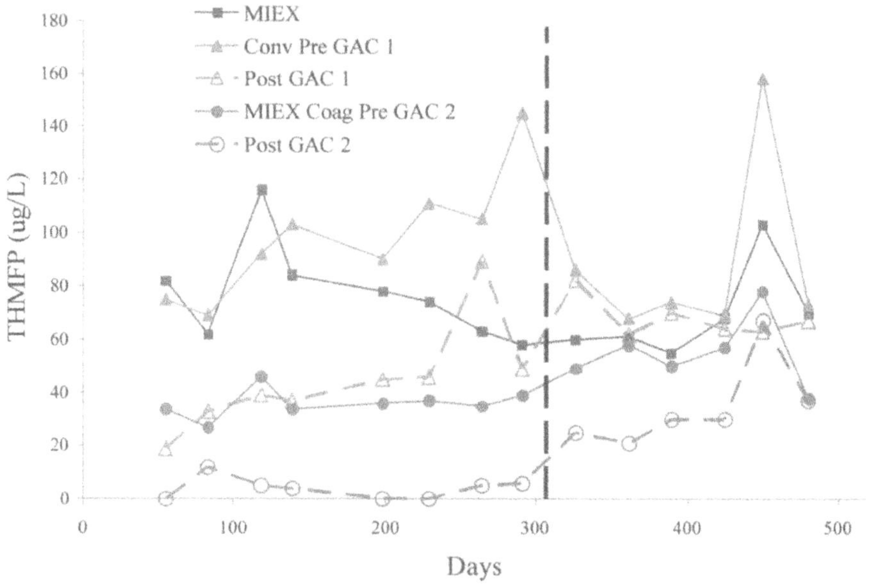

Figure 4. THMFP before and after GAC in both treatment trains

Table III. Data for both treatment streams before and after GAC filtration (Day 198)

	Raw	*Conv*	*Post GAC 1*	*MIEX*	*MIEX Coag*	*Post GAC 2*
DOC (mg/L)	5.4	3.1	1.8	3.4	1.9	0.6
Chlorine Demand (mg/L)	5.8*	3.8	1.6	2.7	2.0	0.4
Specific Chlorine Demand (ug/mg)	1.1	1.3	0.9	0.8	1.1	0.7
THMFP (ug/L)	184	90	45	78	36	4
Specific THMFP (ug/mg)	34	29	25	23	19	6

*Filtered through 0.45 um

The 3 day chlorine demand and the THMFP of the raw water were reduced substantially by the MIEX pre-treatment. The data also confirms a correlation between DOC and chlorine demand, as seen from the specific chlorine dmand and identified by other studies (*20*). There was a reduction in specific THMFP following MIEX treatment alone or MIEX followed by coagulation which did not occur with conventional treatment suggesting that the MIEX process assisted in the preferential removal of THM precursors. In addition, there was a substantial decrease in specific THMFP following GAC filtration of the MIEX treated water (post GAC 2) suggesting that the MIEX treatment train also improved the effectiveness of GAC filtration for removal of THMFP beyond that possible by conventional treatment alone.

Character of DOC Removed

The types of organics removed by the different processes were investigated using molecular weight determined by HPSEC chromatography using UV absorbance at 260nm. An extract of this data for the same date, Day 198, is summarised in Figure 5 for the organic character before GAC filtration. This illustrates that the conventional process removed predominantly higher molecular weight organics whereas the MIEX treatment removed organics across the whole molecular weight range with the additional coagulation step not showing any further removal of organics. However other studies have shown that the additional coagulation step can result in further removal of high molecular weight organics (*7, 21*).

Comparing Figure 5 with the THMFP data listed in Table III illustrates that reduction of THMFP from 184 ug/L to 90 ug/L by conventional treatment corresponded with removal of the high molecular weight fraction. MIEX treatment alone removed organics across a broader molecular weight range than observed with coagulation but surprisingly only resulted in a slightly reduced THMFP (78 ug/L), compared with conventional treatment. This was almost halved to 36 ug/L after the MIEX treated water was coagulated. The molecular weight scans of the MIEX treated water before and after coagulation are almost identical except that the MIEX has not removed the "colloidal" fraction observed as a low intensity peak at >50,000Da. This fraction has been shown to exert a significant chlorine demand, produce THMs and to be effectively removed by coagulation processes (and was also removed by the conventional treatment in this study) (*22- 23*). The high THMFP after MIEX treatment alone and the improvement noted after coagulation of the MIEX treated water is attributed to the presence and subsequent removal of this fraction.

Molecular weight distribution after GAC filtration is shown in Figure 6 for the samples taken for Day 198 and includes the raw and MIEX treated waters for comparison. Similar to MIEX treatment, GAC filtration also removed organics

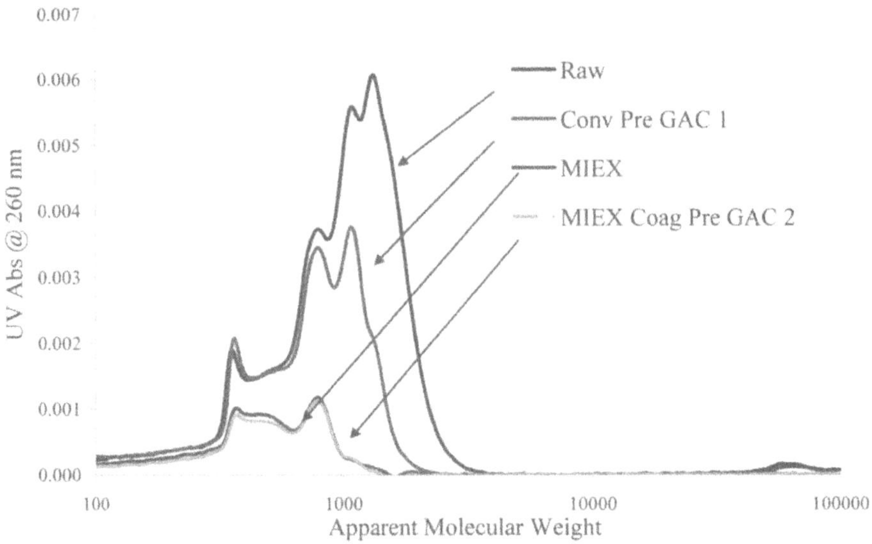

Figure 5. Molecular weight distributions before GAC filtration (Day 198)

across the whole molecular weight range. However, as previously illustrated (refer Figure 1), the organics removal remained constant with the MIEX process whereas the efficiency of DOC removal with the GAC filtration decreased with time to the extent that after 300 days minimal reduction in organics was detectable. The decrease in adsorption capacity of GAC is well documented (*16,24*) and, whilst biodegradation may continue to occur after adsorption is exhausted, the extent of DOC removal is substantially lower at around 10-20% DOC removal.

GAC filtration after conventional treatment (post GAC 1) reduced the organics across the complete molecular weight range and particularly the lower molecular weight range. This reduced the THMFP to 45 ug/L, similar to that achieved by MIEX followed by coagulation. Comparing the molecular weight scans in Figure 6, this suggests that removal of molecular weight below 600 Da and above 1000 Da did not significantly impact THM formation. The key difference in molecular weight distribution between the conventional and MIEX coagulated processes after GAC filtration is the removal of the range between 600-1000 Da. This suggests that in addition to higher molecular weight compounds previously identified as responsible for THM formation (*3-5*), and easily removed by coagulation processes, compounds in this molecular weight range also produce notable contributions to THM formation (*25*).

The low level of DOC remaining after GAC filtration following MIEX coagulation treatment and the absence of organics in the molecular weight scans

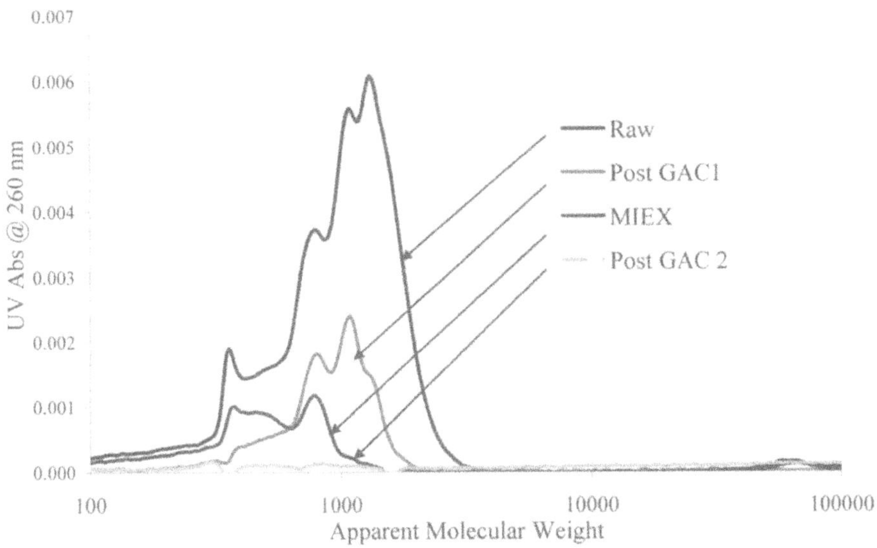

Figure 6. Molecular weight distributions after GAC filtration (Day 198)

support the conclusion that MIEX pretreatment improves GAC effectiveness. This can be attributed to the removal of a broad range of organics both by reduction of the quantity reaching the carbon adsorption pores and by removal of larger organics that may cause pore blockage. This improvement is greater than can be achieved by conventional treatment alone but similar improvement could be achieved by other processes which removed a broad range of organics such as nanofiltration (*26*). Whether there is any impact on biogical growth on the carbon and resulting biodegradation is not clear at this time.

Rapid fractionation was also used to determine whether the various processes were removing organics of different character. The fractionation data summarised in Figure 7 was obtained for the same samples (Day 198) as the molecular weight determination illustrated in Figures 5 and 6. This indicates that whilst conventional treatment decreased all fractions to some extent, MIEX (either alone or in conjunction with coagulation) removed all the charged fraction and more of the hydrophobic (VHA and SHA) fractions than coagulation. GAC filtration after conventional treatment only removed small additional quantities of all fractions confirming the residual presence of organic compounds of varying character. GAC filtration after MIEX and coagulation pre-treatment (post GAC 2), removed the remaining hydrophobic fractions, leaving only a proportion of the neutral fraction. As the neutral fraction has been identified as consisting generally of non-UV absorbing organics this confirms the lack of any observable peaks in the molecular weight scans which were detected using by UV absorbance.

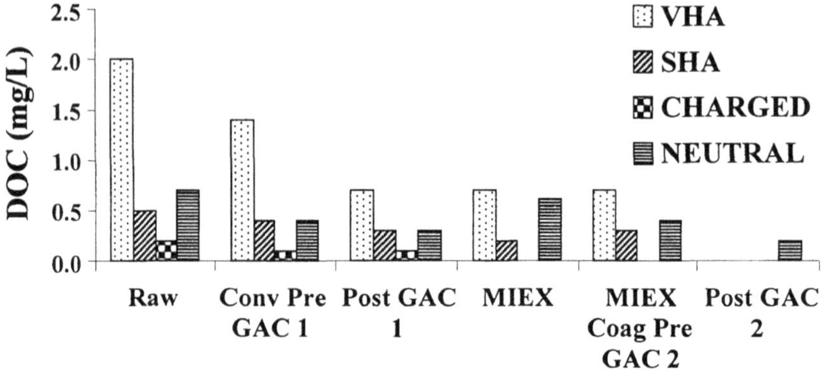

Figure 7 Rapid fractionation of the coagulation processes before and after GAC filtration (Day 198)

Conclusions

This study compared the effectiveness of GAC filtration for removal of DOC and the subsequent formation of THMs following coagulation with and without MIEX pre-treatment over a 505 day period. The EBCT of the GAC filters was initially 20 mins but this was reduced to 5 mins after 312 days. The study has shown that the removal of DOC following GAC filtration was significantly better when the treatment train incorporated MIEX. After 200 days operation DOC after MIEX treatment alone was lower than conventional treatement after GAC filtration. This difference was reduced when the EBCT of the GAC fitlers was decreased from 20 mins to 5 mins and the GAC filter aged, however the removal of DOC after GAC filtration remained marginally more effective with MIEX treated water for the entire study period. The additional DOC removal with MIEX pre-treatment also resulted in reduced chlorine demand and THM formation. THM formation incorporating MIEX pre-treatment was considerably lower than could be achieved with conventional coagulation alone, becoming almost negligible following GAC filtration when the EBCT was 20 minutes. MIEX pre-treatment was shown to preferentially remove THM precursors, both by direct removal and by improving the effectiveness of GAC filtration. This was attributed to removal of a higher proportion and a broader range of organics (as defined by molecular weight and fractionation) by MIEX than possible with conventional treatment. This enhanced the ability of GAC to remove further organics resulting in lower THM formation potential. The lower THM formation was maintained for the length of the study even after the EBCT of the GAC filter was reduced.

References

1. Edzwald, J.K.; Becker, W.C.; Wattier, K.L. Surrogate parameters for monitoring organic matter and THM precursors. *Journal of American Water Works Association* **1985**, *77*(4), 122-132.
2. Reckhow, D. A.; Singer, P.C. Chlorination by-products in drinking waters: from formation potentials to finished water concentrations. *Journal of American Water Works Association* **1990**, *82*(4), 173-180.
3. *Disinfection By-products in Water Treatment: The Chemistry of Their Formation and Control* Minear, R.A.; Amy, G.L., Eds; Lewis Publishers: **1996**.
4. Martin-Mousset, B.; Croue, J.P.; Lefebvre, E.; Legube, B. Distribution and characterization of dissolved organic matter of surface waters. *Water Research* **1997**, *31*(3), 541-553.
5. Kitis, M.; Karanfil, T.; Kilduff, J.E.; Wigton, A. The reactivity of natural organic matter to disinfection byproducts formation and its relation to specific ultraviolet absorbance. *Water Science and Technology* **2001**, *43*(2), 9-16.
6. Bursill, D.; van Leeuwen , J.; Drikas, M.; Problems related to particulate and dissolved components in water: the importance of organic matter. *Water Science & Technology: Water Supply* **2002**, *2*(1) 155-162.
7. Drikas, M.; Chow, C.W.K.; Cook, D.; The impact of recalcitrant organic character on disifection stability, trihalomethane formation and bacterial regrowth: An evaluation of magnetic ion exchange resin (MIEX) and alum coagulation. *Journal of Water Supply: Research and Technology – AQUA* **2003**, *57*(2), 475-487.
8. Morran, J.Y.; Bursill, D.B.; Drikas, M.; Nguyen, H.; A new technique for the removal of natural organic matter, *Proceedings of the AWWA WaterTECH Conference*, Sydney, Australia, May **1996**.
9. Smith, P.; Botica, C,; Long, B,; Allender, B.; Design, construction, commissioning, and operation of the world's first large scale MIEX® water treatment plant, *Proceedings of the AWWA Annual Conference*, New Orleans, LA, June **2002**.
10. Hammann, D.; Bourke, M.; Topham, Evaluation of a magnetic ion exchange resin and ozone to achieve EPA DBP standards at the village of Palm Springs. *Proceedings of the AWWA Conference*, Orlando, Florida **2003**.
11. Nestlerode, F.; Bourke, M.; Teply, M.; Installation of a magnetic ion exchange resin treatment process to improve TOC removal and meet EPA Stage 1 DBP standards at the Green Valley WTP., *Proceedings of the CA/NV AWWA Conference*, Reno, NV, **2005**.
12. Bourke, M.F.; Full-scale impact of a magnetic ion exchange process on downstream coagulation performance and distribution system DBP levels. *Proceedings of the Water Quality Technology Conference*, Denver, CO, November **2006**.

13. Drikas, M.; Morran, J.Y.; Cook, D.; Bursill, D.B.; Operating the MIEX® process with microfiltration or coagulation. *Proceedings of the AWWA Water Quality Technology Conference*, Philadelphia, USA, November **2003**.
14. Summers, R.S.; Hopper, S.M.; Solarik, G.; Owen, D.M.; Hong, S.; Bench scale evaluation of GAC for THM Control. *Journal of American Water Works Association* **1995**, *87*(8), 69-80.
15. Newcombe, G.; Drikas, M.; Assemi, S.; Beckett, R.; Influence of characterised natural organic material on activated carbon adsorption: I. Characterisation of concentrated reservoir water. *Water Research* **1997a**, 31(5), 965-972.
16. Metz, D.H,.; DeMarco, J.; Pohlman, R.; Canon, F.S.; Moore, B.C.; Effect of multiple GAC reactivations on disinfection byproduct precursor removal. *Water Science and Technology: Water Supply* **2004**, *4*(4), 71-78.
17. van Leeuwen, J.; Daly, R.; Holmes, M.; Modeling the treatment of drinking water to maximize dissolved organic matter removal and minimize disinfection by-product formation, *Desalination* **2005**, *177*, 81-89.
18. *Standard Methods for the Examination of Water and Waste Water*; Method 4500-Cl (F); APHA, AWWA and WEF: Washington, DC, **1998**; 20th Edition.
19. Chow, C.W.K.; Fabris, R.; Drikas, M.; A rapid fractionation technique to characterise natural organic matter for the optimisation of water treatment processes. *Journal of Water Supply: Research and Technology – AQUA* **2004**, *53*(2), 85-92.
20. Fitzgerald, F.; Chow, C.W.K.; Holmes, M.; Disinfectant demand prediction using surrogate parameters – a tool to improve disinfection control. *Journal of Water Supply: Research and Technology – AQUA* **2004**, *55*(6), 391-400.
21. Morran, J.Y.; Drikas, M.; Cook, D.; Bursill, D.B.; Comparison of MIEX treatment and coagulation on NOM character, *Water Science and Technology: Water Supply* **2004**, *4* (4), 129-137.
22. Lin, C.F.; Huang, Y.J.; Hao, O.J.; Ultrafiltration processes for removing humic substances: effect of molecular weight fractions and PAC treatment *Water Research* **1999**, *33*(5), 1252-1264.
23. Allpike, B.P.; Heitz, A.; Joll, C.; Kagi, R. Abbt-Braun, G.; Frimmel, F.:Brinkmann, T.; Her, N.: Amy, G.; Size Exclusion Chromatography To Characterize DOC Removal in Drinking Water Treatment *Environmental Science and Technol*ogy **2005**, *39*, 2334-2342
24. Newcombe, G.; Collett, A.; Drikas, M.; Roberts, B.;Granular Activated Carbon Pilot Plant Studies *Water* 1996, 23(3), 29-31.

25. Fabris, R.; Chow, C.W.K.; Drikas, M; Practical application of a combined treatment process for removal of recalcitrant NOM – alum and PAC, *Water Science and Technology: Water Supply* **2004**, *4*(4), 89-94
26. Her, N.; Amy, G.; Park, H. R; Song, M; Characterising algogenic organic matter (AOM) and evaluating associated NF membrane fouling, *Water Research* **2004**, 38(6), 1427-1438

Chapter 17

Natural Dissolved Organic Matter Removal and Subsequent Disinfection By-Product Formation: A Comparison of Ion Exchange and Activated Carbon

Yongrui Tan[1], James E. Kilduff[1,*], and Tanju Karanfil[2]

[1]Civil and Environmental Engineering, Rensselaer Polytechnic Institute, 317 MRC Building, 110 8th Street, Troy, NY 12180
[2]Department of Environmental Engineering and Science, Clemson University, 342 Computer Court, Anderson, SC 29625

A weak base ion exchange resin and a granular activated carbon (GAC) were compared for their ability to sorb natural dissolved organic matter (DOM) and reduce specific ultra violet absorbance (SUVA). Disinfection by-product formation of water having high bromide content after separate ion exchange and carbon adsorption and subsequent chlorination was compared. The ion exchange resin showed superior dissolved organic carbon (DOC) uptake, although SUVA reduction by the two sorbents was comparable. The concentrations of total trihalomethanes (THM4) and the sum of six haloacetic acids (HAA6) were reduced by both sorbents, although the resin was capable of meeting drinking water regulations at much lower doses. The effect of ion exchange and carbon adsorption on THM4 reactivity (yield) was markedly different, and the results demonstrate that ion exchange removes reactive DOM species that GAC does not. Bromine incorporation factor (BIF) values for both sorbents increased initially and were similar at low dosages; however, as dosage increased further the BIF after resin treatment decreased rapidly and remained significantly lower than the BIF after GAC treatment.

Introduction

During chlorination of natural water, chlorine reacts with naturally present dissolved organic matter (DOM) to form potentially hazardous chlorinated hydrocarbons, so-called disinfection byproducts (DBPs) *(1)*. In the presence of bromide ion, brominated DBPs are formed in parallel with chlorinated species. Research has suggested that DBP toxicity increases with increasing bromine substitution *(2)*. Controlling DBP formation is increasingly important as the allowable concentrations of such DBPs as trihalomethanes (THMs) and haloacetic acids (HAAs) in finished water are lowered *(3)* and the regulation of individual DBP species, especially the brominated ones, is considered.

Bromide is naturally present in drinking water sources, with concentrations ranging from <5 to 429 μg/L in the United States *(4)*. Bromide can be rapidly oxidized by chlorine to form hypobromous acid ($HOBr/OBr^-$) and subsequently react with DOM to form brominated DBPs. Bromine reacts faster with DOM and substitutes more efficiently than chlorine *(5)*. The yields of bromine substituted DBP species relative to total DBP formation generally increase with higher Br^-/DOC values, where DOC represents dissolved organic carbon concentration *(6,7)*.

DOM is a complex and heterogeneous mixture of organic compounds consisting of aromatic and aliphatic moieties with carboxyl, phenolic, carbonyl, alcoholic hydroxyl, and methoxyl functional groups with various physical and chemical properties (e.g. structure, functionality, molecular weight, ultra violet (UV) absorbance, and fluorescence). As a result of its heterogeneity, DOM components exhibit different adsorption characteristics, depending on the mechanism of separation. Under the near-neutral pH conditions typical of most natural waters, carboxyl groups on aquatic DOM are mostly in their ionized form, while most phenolic groups are unionized.

The removal of DOM prior to chlorination has been identified as one strategy to reduce the DBP formation. Alum coagulation and activated carbon adsorption were shown to be effective in reducing DBP formation potentials by removing more hydrophobic and aromatic DOM components *(8,9)*. These processes generally remove components having high specific ultra violet absorbance (SUVA, often measured at $\lambda = 254$ nm, i.e., $SUVA_{254}$) preferentially. While high $SUVA_{254}$ components are major precursors of THMs and HAAs, hydrophilic components with lower $SUVA_{254}$ values can also yield significant DBPs during chlorination *(9-13)*. More importantly, some research has indicated that hydrophilic fractions were more reactive with bromine than their corresponding hydrophobic fractions from the same water source *(8,9)*. In addition, bromide concentration is generally conservative in conventional treatment, while DOC is greatly reduced, resulting in a shift of Br^-/DOC ratio to higher values. This can increase the bromine incorporation in DBP formation and the relative fraction of brominated species *(5,6,14)*. Therefore, even though the total DBP concentration may be reduced in the treatment process, the health risk may not be lowered concomitantly because of the shift to more toxic brominated DBPs.

Anion exchange processes have received increasing attention in recent years as an efficient technique for removing DOM and controlling DBP formation *(11; 15-18)*. Anion exchange removes DOM molecules primarily by ionic group attachment (e.g. carboxyl group) *(19)*; therefore, it has the potential to target more hydrophilic, charged DOM components that are not readily removed by conventional treatment techniques *(11)*. Used in combination with conventional processes, it can provide the means to remove a broader spectrum of DBP precursors.

As an added advantage, anion exchange is able to remove bromide ion while reducing DOC, although the removal efficiency depends on the concentration of other anions *(16,18)*. Removal of bromide with DOC reduction is potentially beneficial if the Br^-/DOC ratio can be maintained at low values after treatment, resulting in lower bromine incorporation as compared to conventional processes. However, relatively little information is available regarding the effect of anion exchange treatment on DBP speciation and bromine incorporation during subsequent chlorination. Furthermore, discrepancies exist among the limited published research regarding the treatment efficiency and the preference for DOM components by resins *(6,16,18)*.

Activated carbon adsorption is effective in reducing DBP formation by removing organic precursors *(20)*. A positive relationship was observed between DBP formation potential and $SUVA_{254}$ values after activated carbon adsorption *(8)*. However, it was also found that low SUVA components of DOM have higher bromine incorporation. More importantly, activated carbon adsorption does not remove bromide, resulting in a shift of Br^-/DOC ratio to higher values.

Some studies have been conducted to directly compare the treatment efficiency of activated carbon adsorption and anion exchange in terms of DOM removal. Afcharian et al. *(20)* reported comparable removal of DOC and UV_{254} in Seine River water after passing through anion exchange resin (S6328A) and GAC (F400) columns. Heijman et al. *(21)* observed a greater than 70% reduction in UV_{254} absorbance in a German groundwater, after adsorption using GAC (ROW 0.8S) with a dose of 154 mg/L. This was comparable to the result using an anion exchange resin (A860) at a dose of 40 mg/L. The removal efficiency of GAC, however, decreased significantly after regeneration with 0.1 N NaOH. Chen *(22)* obtained a relatively low removal of TOC (40%) in batch adsorption of a drinking water in Taiwan, using a GAC (G-840) dose of 2 g/L after a short contact time of 24 hrs. In contrast, at the same dosage, 65% removal was achieved by a strong base anion exchange resin (Amberlite IRA-458).

Because different mechanisms control DOM removal by activated carbon adsorption and anion exchange in water treatment, they potentially remove different DBP precursors. However, there is a lack of information in literature providing a direct and systematic comparison of these two treatment techniques in DBP precursor removal and DBP formation reduction. In this study, we used a weak base anion exchange resin and an activated carbon to treat bromide-spiked RO-isolated reservoir water. Water samples obtained from resin and carbon treatment at various dosages were subsequently chlorinated, and the resulting DBP concentrations were measured. The objectives of our research were to: 1) Compare anion exchange and carbon adsorption in terms of their sorption of DOM, preferential sorption on the basis of molecular weight, and reduction of SUVA; and, 2) Compare the reduction of DBP formation after anion exchange and carbon adsorption, and their effect on DBP speciation and bromine incorporation.

Materials and Methods

Natural Water

Water from the Tomhannock Reservoir, located in Troy, New York (TMK), was filtered, softened, and concentrated in the field using a reverse osmosis unit as described elsewhere *(23,24)*. After transport to the lab, RO concentrate was adjusted to pH 4, and refrigerated at 4°C in the dark. The characteristics of TMK water have been reported previously *(12)*. Prior to isotherm experiments, DOM solutions were microfiltered (using pre-cleaned 0.45-μm Gelman Supor membrane filters), diluted with reagent-grade I water (Milli-Q system, Millipore Corp., Billerica, MA, USA) to DOC concentrations of about 4 mg/L, and spiked to yield a high bromide concentration of 1.14 mg/L (3.44 meq/g DOC). This water is referred to as TMK-Br.

DOM Adsorption

The adsorbents used in this study included an anion-exchange resin (M43, Dow, Inc.), and a bituminous coal-based activated carbon (TOG, Calgon, Inc.). The M43 resin is a weak base, tertiary amine, macroporous resin with an exchange capacity of 1.55 eq/L and a water content of 40 to 50%. The TOG carbon is similar to F400 but designed for point-of-use water treatment applications. Both adsorbents had a geometric mean particle size near 500 μm. Adsorbents were cleaned and conditioned using procedures reported previously *(13,25)*. Isotherm experiments were conducted using the bottle-point method employing batch reactors (amber bottles with Teflon-lined rubber caps) kept well mixed by continuous end-over-end tumbling. DOM solutions were microfiltered (using pre-cleaned 0.45-μm Gelman Supor membrane filters) and initial pH was adjusted to pH 7.5 using a dilute reagent grade sodium hydroxide solution; no buffer was employed. Based on preliminary rate studies, contact times of 10 and 14 days were used for the resin and the GAC, respectively. After equilibration, aliquots of the solution phase were withdrawn from the reactors, and analyzed for total organic carbon (TOC-V_{CSH}, Shimadzu) and UV absorbance at $\lambda = 254$ nm (Cary 100, Varian). Bromide concentrations were measured using a reagent-free ion chromatography system (ICS-2000, Dionex). The batch approach used in this research should apply directly to mixed reactors, which are being used increasingly for disinfection byproduct formation control via NOM removal by ion exchange resins. However, the results will generally not apply directly to column results, because of different mass transfer rates among components, resulting in chromatographic effects that we do not account for here.

Molecular Weight Distribution

The molecular weight distribution (MWD) of DOM solutions was measured before and after ion exchange sorption using an HPSEC technique using UV detection at 254 nm *(25)*. Poly(styrene sulfonate) (PSS) was used as an external standard to calibrate a Waters Protein Pak 125 column with silica based gel stationary phase operated on a HP 1100 HPLC system. An ionic strength of 0.1 M and pH 6.8 was maintained in all standard and sample solutions.

Chlorination and DBP formation

Supernatants obtained from DOM isotherms were chlorinated using the Uniform Formation Conditions (UFC) protocol, as proposed by Summers et al. *(26)*. After the chlorination period, THMs and haloacetonitriles (HANs) were extracted with hexane according to Standard Method 6232 B. HAAs were reacted with acidic methanol to yield esters, and extracted with methyl-tert-butyl-ether (MTBE) according to EPA Method 552.2. Four THM (chloroform (TCM), bromodichloromethane (BDCM), chlorodibromomethane (DBCM), and bromoform (TBM)), four HAN (trichloro-, dichloro-, bromochloro-, and dibromoactonitrile) and six HAA species (chloro- (MCAA), bromo- (MBAA), dichloro- (DCAA), trichloro- (TCAA), bromochloro- (BCAA), and dibromoacetic acid (DBAA)) were identified and quantified using a gas chromatography system (Agilent 6890) equipped with micro-ECD detector. THM and HAA standards were purchased (Supelco) and spiked into reagent-grade I water to prepare external standards that were then extracted and analyzed in the same manner as samples.

Results and Discussion

DOM Uptake and Effects on MWD and SUVA

The uptake of TMK DOM by the weak base resin and the GAC is shown in Figure 1. The equilibrium liquid phase concentration was normalized by dosage (*D*) to control for any effects of initial concentration on the DOC isotherm *(see, e.g., 25)*. It is clear that the resin exhibits higher DOM uptake from the TMK-Br water than the GAC. There was little effect of bromine on the uptake by either of the sorbents (data not shown).

The MWD of DOM remaining in solution after sorption by the resin and the GAC was measured using HPSEC, to investigate which DOM components were adsorbed preferentially. As shown in Figure 2, with an increase in GAC dose, sorption reduces the total mass of DOM in solution, and adsorption preferentially removes DOM components with *lower* MW, especially at low carbon doses, when adsorption sites are more limited and the competition among DOM components is high. This is consistent with the results observed in previous research for activated carbon adsorption *(25)*, and is explained by size exclusion effects. In contrast, with an increase in M43 *resin* dose, components having *intermediate* molecular weights are removed preferentially at low dosages; with an increase in dosage, *higher* molecular weights DOM species are then removed in preference to smaller ones. Clearly, GAC and ion exchange target different DOM components, although both are effective in reducing DOC individually. Therefore, a combination of GAC adsorption and anion exchange could cover a broad spectrum of DOM components with different properties.

Selective uptake of DOM components having different SUVA values was evaluated by measuring the change in SUVA (λ = 254 nm) after sorption, as shown in Figure 3. For both the resin and the GAC, the $SUVA_{254}$ values decreased with increasing sorbent dose, indicating a preferential removal of high $SUVA_{254}$ components from solution. However, the pattern of SUVA removal was different. The rate of decrease in SUVA with adsorbent dose was greater for the M43 resin, up to dosages of about 200 mg/L. At higher dosages, SUVA removal by the resin was minimal, whereas removal by the GAC continued.

Taken in combination with the results from Figure 2, this suggests that the smallest DOM components in TMK water do not have highest SUVA values. This is consistent with the findings of Kitis et al. *(22)*. Because they target different DOM components, the generally similar behavior of these two sorbents with regard to SUVA removal is rather surprising, although it appears that GAC can reduce SUVA to lower levels, albeit at high dosages.

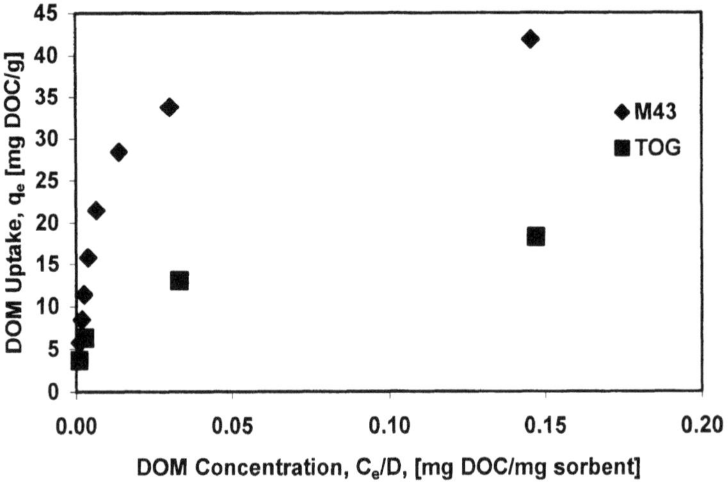

Figure 1. Uptake of DOM from TMK-Br bromide spiked water by M43 weak base resin (C_0 = 4.2 mg/L) and TOG GAC (C_0 = 4.7 mg/L).

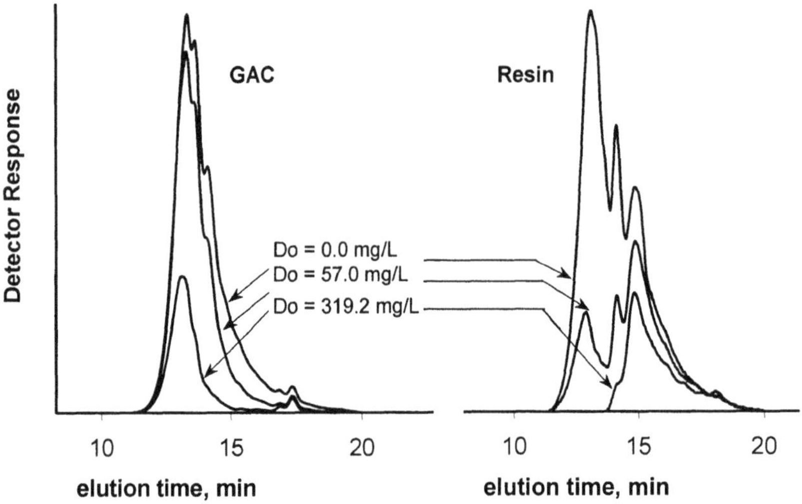

Figure 2. Size exclusion chromatograms of TMK DOM remaining in solution after sorption by TOG GAC (C_0 = 4.7 mg/L), left, and M43 resin (C_0 = 3.94 mg/L), right.

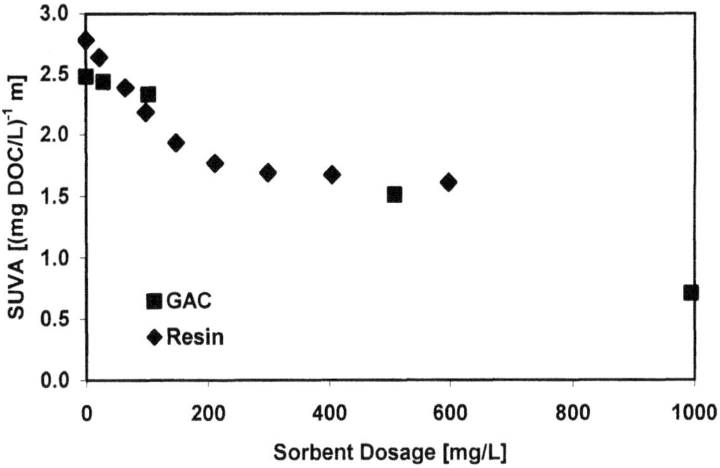

Figure 3. Effect of sorbent dose on $SUVA_{254}$ of TMK DOM components remaining in solution after sorption by M43 resin (C_o = 4.2 mg/L) and TOG GAC (C_o = 4.7 mg/L).

Bromide removal by M43 Resin

Bromide ion removal from TMK-Br water by the M43 resin was <5% below resin dosages of about 65 mg/L, probably due to competition from DOM and sulfate. As a result, the Br^-/DOC ratio increased sharply; although the same trend was observed for the GAC, it occurred to a lesser degree, because uptake of DOM by the GAC was lower. However, as shown in Figure 4, bromide removal by the resin was about 79% at a dosage of 600 mg/L, even in the presence of DOM. Therefore, as the resin dosage and sorption sites increased, and competition from DOM was reduced, Br^- removal increased, causing the Br^-/DOC ratio to go through a maximum and then decrease. In contrast, the Br^-/DOC ratio increased monotonically after GAC adsorption.

The Br^-/DOC ratio after resin sorption was modeled with relatively simple functions for bromide concentration and DOC uptake, shown as the solid line in Figure 4. The bromide data was fitted with a second order polynomial,

$$Br^- \text{ [mg/L]} = -8.44\times10^{-7}\times D_o^2 - 1.01\times10^{-3}\times D_o + 1.15$$

where D_o is the resin dose in mg/L. The DOC uptake was modeled with a Freundlich isotherm incorporating a non-adsorbing fraction, $q_e = K_F(C_e - C_{non})^n$ where K_F = 30.40, C_{non} = 0.67 mg/L, and n = 0.37. Note that these parameters are likely unique to the initial DOC concentration used in this experiment.

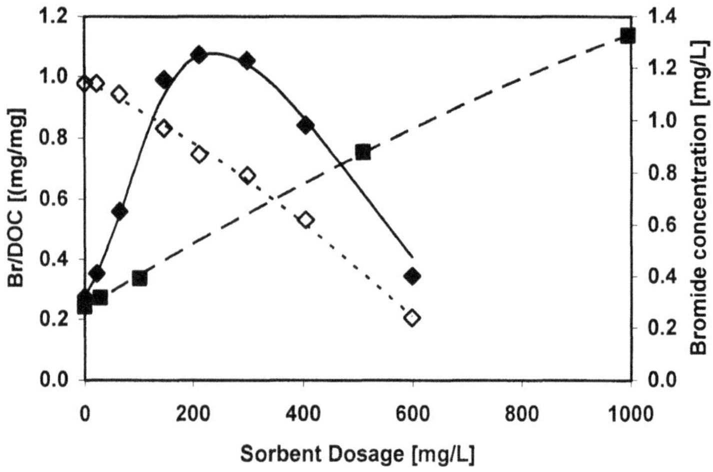

Figure 4. Effect of sorbent dose on bromide (open symbols) and Br/DOC ratio (closed symbols) after sorption by M43 resin (diamonds, DOC_0 *= 4.2 mg/L) and TOG GAC (squares,* DOC_0 *= 4.7 mg/L). Solid line is the model described in the text.*

DBP Formation

As shown in Figure 5, resin dosages of <100 mg/L were sufficient to reduce THM4 below 80 μg/L whereas a GAC dosage of >500 mg/L was required. (HAA5 was not formed to the same extent as THM4, and the dosages of both sorbents required to reduce concentrations to below 60 mg/L were much lower). The Br^-/DOC ratio increases initially for both sorbents, and although anion exchange removes bromide from solution while reducing DOC, the Br^-/DOC ratio after resin sorption was higher than after GAC sorption over much of the sorbent dosage range studied (Figure 4), due to preferential removal of DOC. Yet, THM4 formation after resin sorption decreases, whereas after GAC sorption it initially increases slightly before it decreases. For both sorbents, HAA6 formation decreases monotonically with increasing dosage, although the rate of decrease is greater for the resin.

At high resin dosages, the Br^-/DOC ratio decreases, whereas it continues to increase at high carbon doses. Even so, the greatly reduced DOC concentration limits the formation of total THMs and HAAs, and the behavior of the resin and the carbon tend to converge.

The effect of sorbent dosage on the reactivity (yield) of DOM components, expressed in terms of DBP formation normalized by equilibrium DOC concentration, is shown in Figure 6. The results showed that the DOM remaining in water after resin treatment had significantly higher THM and HAA yields

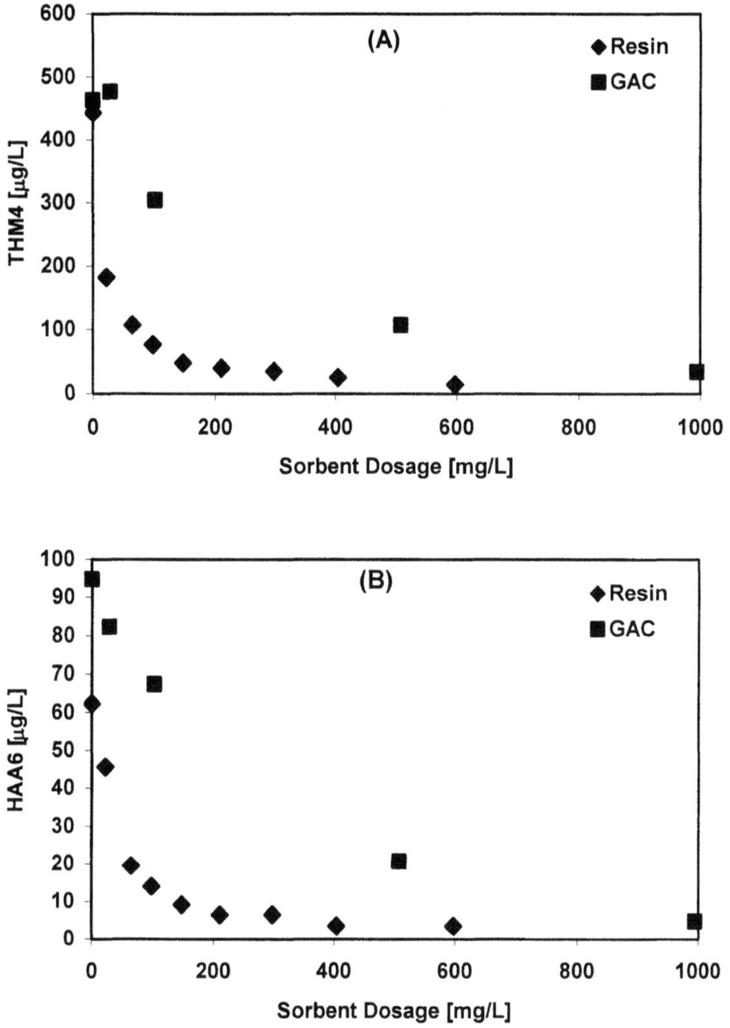

Figure 5. Effect of sorbent dose on THM4 (A) and HAA6 (B) formation after chlorination of DOM remaining in solution after sorption by M43 resin (C_o = 4.2 mg/L) and TOG GAC (C_o = 4.7 mg/L).

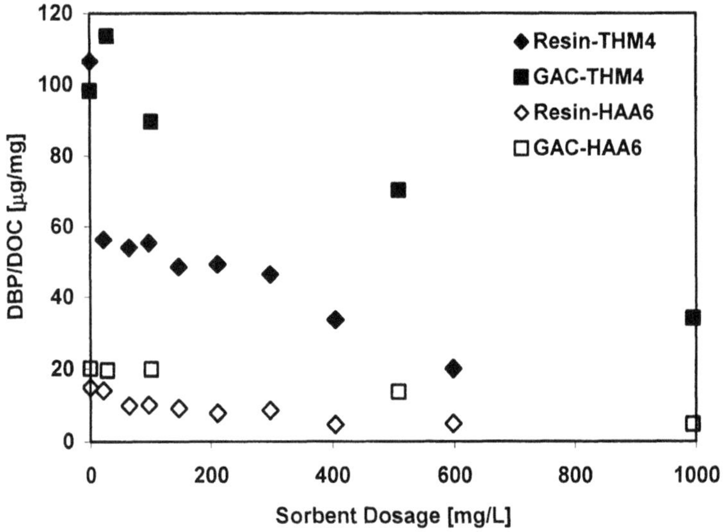

Figure 6. Effect of sorbent dose on DBP (THM4, filled symbols; HAA6, open symbols) reactivity (yield) after chlorination of DOM remaining in solution after sorption by M43 resin (C_o = 4.2 mg/L) and TOG GAC (C_o = 4.7 mg/L).

as compared to the DOM remaining after GAC adsorption at the same adsorbent doses. Since the Cl_2/DOC ratios at the same adsorbent doses were similar in these experiments, the results indicate that at the same adsorbent doses, ion exchange resin removed more DOM from water as shown in Figure 5, and also more reactive DOM species as compared to GAC.

DBP Speciation and Bromine Incorporation

In addition to total DBP concentrations, DBP speciation is of interest, especially in light of the higher toxicity of brominated DBP species, and the potential for future regulations on individual DBP species. In TMK-Br water before treatment, all four THM species are present in high concentrations, and the brominated species dominate, as expected given the high bromine concentration: TBM > DBCM > BDCM > TCM (data not shown). This trend does not change upon treatment by either the resin or the GAC for low sorbent dosages (100 to 150 mg/L). The TBM species remains dominant for the entire carbon dosage range, comprising the majority of the HAA4, whereas TBM becomes the minority species as the resin dosage is increased above about 200 mg/L. TBM becomes non-detectable at resin dosages between 300 and 400 mg/L, and the DBCM becomes non-detectable at resin dosages between 400 and

600 mg/L. In contrast, all brominated species are detectable after GAC treatment, even up to doses of 1000 mg/L. This rather striking difference in behavior is likely due to the removal of bromide by the resin, and lack of bromide removal by the GAC. Five of the six HAA species measured (HAA6 less MCAA) are present in TMK-Br water fractions, and the concentrations before resin treatment are in the order of: DBAA > BCAA > DCAA > MBAA > TCAA (data not shown). These trends hold up to a resin dosage of about 200 mg/L, and a carbon dosage of about 500 mg/L. At higher resin dosages, the MBAA species becomes dominant, whereas at high carbon dosages the DBAA species remains dominant, likely for the reasons discussed above.

The degree of bromine substitution in THM species can be evaluated by the bromine incorporation factor (BIF), defined as:

$$BIF = \frac{\sum_{i=1}^{3} i \times CHCl_{3-i}Br_i}{THM4}$$

where the concentrations of THM4 and species are in μM. BIF values range between 0 and 3, corresponding to exclusive formation of chloroform and bromoform respectively. BIF values of chlorinated TMK-Br DOM solutions after sorption by the resin and the carbon are shown in Figure 7. BIF values for both sorbents increase with resin dosage initially, due to preferential sorption of DOC. However, BIF values after resin sorption reach a maximum of 2.1 at a dosage of 22 mg/L, and then decrease sharply at dosages above about 65 mg/L. In contrast, after carbon adsorption BIF values of 2.4 are maintained over a large range in dosage, from about 100 to 500 mg/L. These trends in bromine incorporation in THM-Br water can be explained only in part by the changes in Br^-/DOC ratio with resin dosage, as shown in Figure 4, because the peak in BIF occurs at a lower dosage than the peak in Br^-/DOC.

Conclusions

The uptake of TMK DOM by the weak base ion exchange resin was significantly greater than by the GAC. GAC preferentially removed DOM components with lower MW, whereas components having intermediate and higher molecular weights were removed preferentially by the M43 resin. For both the resin and the GAC, the $SUVA_{254}$ values decreased with increasing sorbent dose, indicating a preferential removal of high $SUVA_{254}$ components from solution. However, the pattern of SUVA removal was different. The rate of decrease in SUVA with adsorbent dose was greater for the M43 resin, up to dosages of about 200 mg/L. At higher dosages, SUVA removal by the resin was minimal, whereas removal by the GAC continued. An advantage of the ion

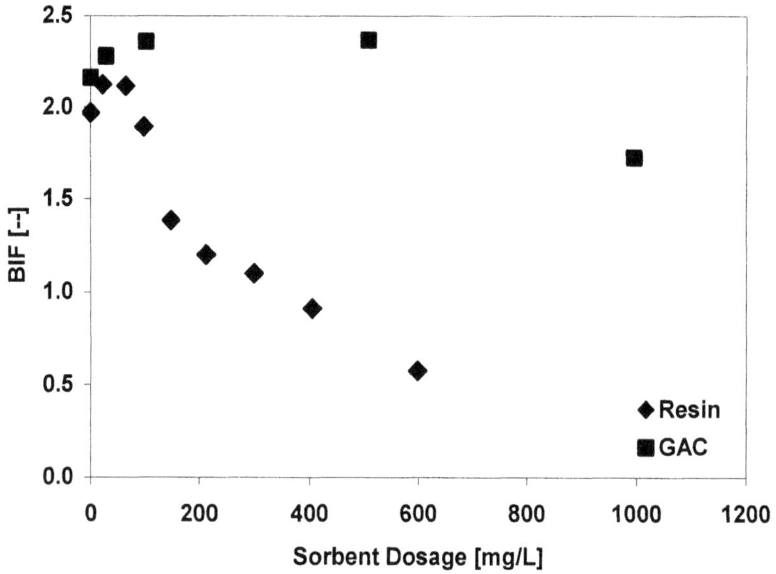

Figure 7. Effect of sorbent dose on BIF after chlorination of DOM remaining in solution after sorption by M43 resin (C_o = 4.2 mg/L) and TOG GAC (C_o = 4.7 mg/L).

exchange resin was its ability to remove bromide; however, DOC was removed preferentially, so at low doses the Br^-/DOC ratio increased for both the resin and GAC. As the sorbent dosage increased, the Br^-/DOC ratio after ion exchange went through a maximum and then decreased, whereas it increased monotonically for the GAC. THM4 and HAA6 concentrations were reduced by both sorbents, although the resin was capable of meeting drinking water regulations at much lower doses – resin dosages of <100 mg/L were sufficient to reduce THM4 below 80 μg/L whereas a GAC dosage of >500 mg/L was required. The effect of ion exchange and carbon adsorption on THM4 reactivity (yield) was markedly different. Patterns in DBP speciation were similar, although for the GAC, TBM comprised the majority of THM4 and was dominant throughout the dosage range, likely as a result of poor bromide removal. BIF values for both sorbents increase with resin dosage initially; however, BIF values after resin sorption reached a maximum at a dosage of 22 mg/L, and then decreased sharply above dosages of about 65 mg/L, whereas after carbon adsorption high values were maintained over a large range in dosage. Although some of the observed effects can be attributed to different patterns in Br^-/DOC and Br^-/Cl_2 ratios, it seems clear that ion exchange removes reactive DOM species that GAC does not.

Acknowledgements

This work was supported in part by a Research Grant (R-828-045-01-0) from the United States Environmental Protection Agency's Science to Achieve Results (STAR) program (Office of Research and Development, National Center for Environmental Research and Quality); however, it has not been subjected to the Agency's peer and policy review and therefore does not necessarily reflect the views of the Agency and no official endorsement should be inferred. The authors thank the administrative and technical staffs of the Troy, NY, drinking water treatment plant.

References

1. Rook, J. J. *Environmental Science & Technology* **1977**, *11*, 478-482.
2. Krasner, S. W.; Sclimenti, M. J.; Chinn, R.; Chowdhury, Z. K.; Owen, D. M. In *Disinfection By Products in Water Treatment: The Chemistry of their Formation and Control.* Minear, R.A and Amy, G.L., Eds. CRC Press Inc. 1996; p 59.
3. USEPA. National Primary Drinking Water Regulations: Stage 2 Disinfectants and Disinfection Byproducts Rule; Final Rule, Federal Register. Part II, 40 CFR Part 9, 141 and 142, 71: 2: 388, January 4, 2006.
4. Amy, G. L.; Siddiqui, M.; Zhai, W.; DeBroux, J.; Odem, W. *National Survey of Bromide Ion in Drinking Water Sources and Impacts on Disinfection by-product Formation.* AwwaRF, Denver, CO, 1995.
5. Westerhoff, P.; Chao, P.; Mash, H. *Water Research* **2004**, *38*, 1502-1513.
6. Summers, R. S.; Benz, M. A.; Shukairy, H. M.; Cumming, L. *Journal American Water Works Association* **1993**, *85*(1), 88-95.
7. Cowman, G. A.; Singer, P. C. *Environmental Science & Technology* **1996**, *30*, 16-24.
8. Kitis, M.; Karanfil, T.; Kilduff, J. E.; Wigton, A. *Water Science & Technology* **2001**, *43*(2), 9-17.
9. Liang, L.; Singer, P. C. *Environmental Science & Technology* **2003**, *37*, 2920-2928.
10. Croue, J-P.; Violleau, D.; Bodaire, C.; Legube, B. *Water Science & Technology* **1999**, *40*, 207-214.
11. Bolto, B.; Dixon, D.; Eldridge, R.; King, S. *Water Research* **2002**, *36*, 5066-5073.
12. Kitis, M.; Karanfil, T.; Wigton, A.; Kilduff, J. E. *Water Research* **2002**, *36*, 3834-3848.
13. Tan, Y.; Kilduff, J. E.; Kitis, M.; Karanfil, T. *Desalination* **2005**, *176*, 189-200.

14. Amy, G. L.; Tan, L.; Davis, M. K. *Water Research* **1991**, *25*, 191-202.
15. Bolto, B.; Dixon, D.; Eldridge, R.; King, S.; Linge, K. *Water Research* **2002**, *36*, 5057-5065.
16. Drikas, M.; Chow, C. W.; Cook, D. *AQUA* **2003**, *52*, 475-487.
17. Fearing, D. A.; Bank, J.; Guyetand, S.; Eroles, C. M.; Jefferson, B.; Wilson, D.; Hillis, P.; Campbell, A. T.; Parsons, S. A. *Water Research* **2004**, *38*, 2551-2558.
18. Boyer, T. H.; Singer, P. C. *Water Research* **2005**, *39*, 1265-1276.
19. Fu, P. L.-K.; Symons, J. M. *Journal American Water Works Association* **1990**, *82*(10), 70-77.
20. Afcharian, A.; Levi, Y.; Kiene, L.; Scribe, P. *Water Research* **1997**, 31, 2989-2996.
21. Heijman, S. G. J.; van Paassen, A. M.; van der Meer, W. G. J.; Hopman, R. *Water Science & Technology* **1999**, *40*, 183-190.
22. Chen, P. H. *Environment International* **1999**, *25*, 655-662.
23. Kitis M.; Kilduff, J.E.; Karanfil. T. *Water Research* **2001**, *35*, 2225.
24. Kilduff, J. E.; Mattaraj, S.; Wigton, A.; Kitis, M.; Karanfil, T. *Water Research* **2004**, *38*, 1026-1036
25. Kilduff, J. E.; Karanfil, T.; Chin, Y.; Weber, W. J. *Environmental Science & Technology* **1996**, *30*, 1336-1343.
26. Summers, R. S., Hooper, S. M., Shukairy, H. M., Solarik, G. and Owen, D. *Journal American Water Works Association* **1996**, *88*(6), 80-93.

Chapter 18

Fate and Transport of Wastewater-Derived Disinfection By-Products in Surface Waters

Baiyang Chen[1], Paul Westerhoff[1], and Stuart W. Krasner[2]

[1]Department of Civil and Environmental Engineering, Arizona State University, Tempe, AZ 82581
[2]Water Quality Standards Branch, Metropolitan Water District of Southern California, 700 Moreno Avenue, La Verne, CA 91750–3303

Disinfection by-products (DBPs) discharged into surface waters from upstream wastewater treatment plants (WWTPs) may result in occurrence of DBPs in downstream drinking water treatment plants (DWTPs). This article evaluates the effects of five biogeochemical mechanisms (biodegradation, photolysis, hydrolysis, volatilization, and adsorption) on the fate of several types of DBPs in surface waters that receive disinfected wastewater discharges. Pseudo-first order rate constants of the five mechanisms were obtained from literature reviews, laboratory experiments and/or quantitative structure-activity relationships. The dominant removal mechanism for each DBP class is: 1) volatilization for trihalomethanes (THMs), 2) biodegradation for dihalogenated haloacetic acids (HAAs), 3) hydrolysis for haloacetonitriles (HANs), and 4) photolysis for nitrosoamines (NAs), while adsorption of DBPs onto suspended solids is not important for all classes. The receiving waterbody geometry (width, depth), flowrate, and meteorological conditions are also important factors affecting the fate of DBPs.

Many surface waters that serve as potable water supplies may seasonally contain significant levels of upstream WWTP discharges. These wastewater effluents—which contain effluent organic matter (EfOM) that consists of bulk organic matter, preformed DBPs, DBP precursors, and emerging contaminants such as endocrine disrupting chemicals and pharmaceuticals (*1-3*)—are discharged to lakes, rivers, and streams (i.e., receiving waters). THMs and HAAs have been detected in wastewater effluents with median concentrations of 57 and 70 µg/L, respectively, when free chlorine residual were present; while median and 75th percentile concentrations of N-nitrosodimethylamine (NDMA) were 11 and 24 ng/L, respectively, when chloramines were present (*4-5*). However, studies are lacking regarding the fate and transport of wastewater-derived DBPs in rivers and lakes. Although previous studies have examined drinking water treatment processes for the removal of preformed DBPs, studies on the significance of natural biogeochemical processes (biodegradation, photolysis, hydrolysis, volatilization, and adsorption) on the fate of DBPs in surface waters are still limited. This article hereby attempts to quantify the fate and transport of DBPs by these biogeochemical processes in several representative waters characterized as different geometries and flowrates, so as to investigate the ultimate impact of wastewater-derived DBPs on downstream drinking water supplies.

Materials and Methods

Modeling Framework & Estimation of DBP Loss Rate Parameters

A fate-and-transport model (Equation 1) was adapted from a dissolved oxygen model (Streeter-Phelps) using a commercial software program (AQUASIM). In essence, the simulation mixed the water from the WWTP effluent and the river, where losses of DBPs may occur.

$$V \cdot \frac{\partial C}{\partial t} = C_{in,wwtp} \cdot Q_{in,wwtp} + C_{in,River} \cdot Q_{in,River} - V \frac{\partial}{\partial x}\left(U_x \cdot C\right) + r_T \cdot V \qquad (1)$$

where C is the DBP concentration (µg/L or ng/L); t is time (hr); Q is flow rate (m^3/hr); U_x is advection velocity (m/hr); x is longitudinal distance (m); and V is volume of a given river reach (m^3). r_T is net reaction term (mg/L-hr), which includes the sum of kinetics of biodegradation (r_B), photolysis (r_P), hydrolysis (r_H), volatilization (r_V), and adsorption (r_A).

Stream geometry, flowrate, and meteorological conditions can affect the five DBP loss mechanisms. Therefore, three representative scenarios were modeled (Table I). A common set of parameters are needed for modeling these loss mechanisms and these are summarized in Table II. The following values were used with the models summarized in Table II to simulate the fate of DBPs based upon the three scenarios identified in Table I (*5*): temperature (20°C), dynamic viscosity of water (0.014 g/cm-s), SS concentration (1,000 mg/L), diameter of SS (10 μm), fraction of organic carbon in SS (f_{OC}) (0.01%), pH 7.0, and wind velocity (U_w) (10,000 m/hr).

Table I. Three Representative Watershed Scenarios

Key Parameters	*Scenario I*	*Scenario II*	*Scenario III*
Water Body Characteristics	Low-flowing velocity/ deep river	Intermediate	High-flowing velocity/ shallow river
Flow Velocity (U) (m/hr)	442	830	1227
Depth (D) (m)	4.5	0.24	0.13
Flowrate (Q) (m^3/hr)	10,000	1,200	800
Width (W) (m)	5	5	5

Table II. Equations of the DBP Loss Processes

Mechanism	**Generalized modeling reaction**
Volatilization	$r_V = \frac{k_l}{H_1} \cdot (\frac{[C_g]}{H_c} - [C])$
Photolysis	$r_P = -K_P \cdot [C]$
Hydrolysis	$r_H = -K_H \cdot [C] = -[C] \cdot (K_a \cdot [H^+] + K_n + K_b \cdot [OH^-])$
Biodegradation	$r_B = -K_B \cdot [C]$
Sorption to suspended solids [SS]	$r_A = -\frac{v_s}{H} \cdot \frac{\Pi[SS]}{1+\Pi[SS]} \cdot [C]$

Biodegradation

A combination of modeling and experimentation were employed to obtain estimates for the rate of DBP loss. A U.S. Environmental Protection Agency (USEPA) software program (Biowin version 1.4, embedded in EPAWIN 3.2) was used to estimate the relative biodegradability of DBPs. The software program took into account the effects of functional group, molecular weight, and chain structure, and describes the biodegradability as a bioindex. A bioindex value >0.5 denotes high biodegradability with a half-life from hours to days, whereas bioindex <0.5 indicates biodegradability with a half-life from days to weeks. Although other factors, such as microbe types, oxygen presence, and nutrients availability, can also control biodegradability, the bioindices may represent the biodegradabilities of chemicals in a more universal condition and not in laboratory cultures.

To obtain biodegradation rate constants, experiments were conducted using acclimated biologically active sand (BAS) reactors (200 mL of sand, 500-mL water sample in a 1-L amber glass bottle) following a protocol used for determining biodegradable organic carbon (BDOC) (*6*). THMs, HAAs, and HANs were pre-formed by adding sufficient chlorine to EfOMs from four different WWTPs and reacting for 24 hr, while NAs were formed by applying chloramines for 72 hr. What followed was a dechlorination process in order to consume the residual disinfectants that may continue to form the DBPs. The pre-formed DBPs were then poured into the bioreactors (and control reactors with non-biologically active sand) to biodegrade over a 5-day period.

Photolysis

Photolysis rate constants for DBPs were obtained under natural sunlight conditions. Because such photolysis rate constants from the laboratory tests (K_0) represent only the near-surface river conditions, the rate constants in real water were adjusted using an empirical equation (Equation 2) to address the transmittance disparity caused by water depth:

$$K_p = K_0 \cdot e^{-k_\lambda \cdot D} \tag{2}$$

where K_D denotes the rate constant of a DBP at depth D (unit is hr^{-1}), K_0 denotes the rate constant at depth zero (hr^{-1}); k_λ is a constant with an average value of 0.1 cm^{-1}, and D is the average depth of the river (unit is cm) (*7*).

Values for K_0 were obtained from natural sunlight photolysis tests (conducted in May and August 2005). Sunlight photolysis of DBPs (spiked into organic-free water) was evaluated in 9-mL quartz test tubes (outside diameter = 13 mm). The small volume was selected to maximize solar irradiance, and this test probably represented a near-surface photolysis potential of chemicals. Sample vials were capped to prevent volatilization of DBPs or evaporation. The quartz tubes were placed on an inclined platform covered with aluminum foil. The inclined platform was placed on the laboratory roof (Tempe, Ariz.). Light intensity was monitored by a photometer (IL 1700, International Light Technologies, Inc., Peabody, Mass.) and verified by a database from the Arizona Meteorological Network website (http://ag.arizona.edu/azmet). The DBP concentrations were measured over time.

Hydrolysis

Hydrolysis kinetics of DBPs were obtained mainly from a literature review and two USEPA-released software, EPAWIN (version 3.2) and SCDM (version 1.0). Quantitative structural activity relationship (QSAR) models (*8-10*) were further developed to predict pseudo-first order rate constants.

The basic QSAR relationships shown in Equations 3 and 4 relate the rate constants for the hydrolysis reactions with electronic and steric effect constants (σ and E_s, respectively) and three unknown parameters (δ, ρ, c'). In this study, 0.50, 0.47, 0.42, and 0.38 served as the σ values and -0.46, -0.96, -1.16, and -1.40 served as E_s values for fluoride, chloride, bromide, and iodide, respectively (*10*). Effects of multiple halides were assumed to be the sum of single halides, based upon prior work (*11*).

$$Log(K_t) - Log(K_{ref}) = \delta \cdot E_s + \rho \cdot \sigma + c \tag{3}$$

$$Log(K_t) = \delta \cdot E_s + \rho \cdot \sigma + c' \tag{4}$$

where K_t is the rate constant of the target chemical of the group for any reaction; K_{ref} is the rate constant of the reference chemical of the group; δ is a sensitivity parameter of steric effects of the substituent; ρ is a sensitivity parameter of the electronic effects of the substituent; c and c' are constants.

Volatilization

For volatilization there are several controlling factors, including chemical properties (e.g., acid dissociation constant [pK_a], Henry's law constant [H_e], molecular weight [MW]), water conditions (e.g., flow velocity, water depth), and environmental characteristics (e.g., wind velocity). Henry's constants were obtained from the literature and reference materials (EPAWIN 3.2).

Adsorption

In rivers, DBP removal via adsorption processes was assumed to occur in two steps: 1) adsorption of DBPs onto SS, followed by 2) sedimentation of solids with sorbed DBPs out of the water column. The adsorption capability is controlled by SS concentrations in the river, f_{OC}, the DBP partition coefficient (K_{ow}), and pK_a. Settling velocity is controlled by suspended sediment particle size and the river depth.

DBP Analysis for Samples from Biodegradation and Photolysis Experiments

Four regulated THMs (THM4), four HANs (i.e., dichloro-, bromochloro-, dibromo-, and trichloroacetonitrile) (HAN4), two haloketones (i.e., dichloro- and trichloropropanone), and chloropicrin were measured using a salted liquid/liquid extraction and gas chromatograph (GC)/electron-capture detection (ECD) method (*12*). Other halonitromethanes (HNMs), as well as haloacetaldehydes, were analyzed using either liquid/liquid extraction–GC/ECD or solid-phase extraction–GC/mass spectrometry (MS) method (*13*). Nine HAAs (HAA9) were measured using an acidic and salted liquid/liquid extraction, plus derivatization with acidic methanol, followed by GC/ECD analysis (*14*). Eight NAs were pre-concentrated using a micro liquid-liquid extraction method (*15*) followed by analysis using chemical ionization GC/MS method (*16-17*).

Results and Discussion

Estimation of Rate Constants for DBP Loss Processes

Biodegradation Experiments and Rate Constants

THMs exhibited negligible biodegradation under aerobic conditions and were therefore assumed to have a pseudo first-order rate constant (K_B) of $< 0.02\ d^{-1}$. Figure 1 shows the results of HAA biodegradation in the bioreactors and control samples. Dihalogenated HAAs (DXAAs) (i.e., dichloro-, bromochloro-, and dibromoacetic acid [DCAA, BCAA, DBAA]) showed more than a 50% loss, whereas trihalogenated HAAs (TXAAs) (i.e., trichloro-, bromodichloro-, chlorodibromoacetic acid [TCAA, BDCAA, CDBAA]) were relatively stable. This finding is consistent with previous research conducted in a simulative drinking water distribution system (*18*). In addition, trichloroaldehyde, dichloroacetonitrile (DCAN), and 1, 1-dichloropropanone were readily biodegraded. Table III summarizes K_B values determined from first-order analysis of the experimental data.

The trend of bioindices for the DBPs corresponded with the trends observed in K_B for each class of DBP. For example, EPIWIN bioindices of DXAAs were 0.54, 0.58, and 0.62 for DCAA, BCAA, and DBAA respectively, which are significantly higher than the bioindices for TXAAs (0.09 to 0.22) or for the THMs (0.39 to 0.49). Biodegradation potential increased with decreasing levels of halogenation or more bromine substitution. Therefore, EPWIN bioindices were used to estimate the biodegradation for DBPs not explicitly studied in the bioreactors. For example, the bioindex of NDMA is 0.19, which is even smaller than chloroform (0.39). It was assumed that aerobic biodegradation of NDMA is very slow. This assumption appears consistent with the literature for aerobic

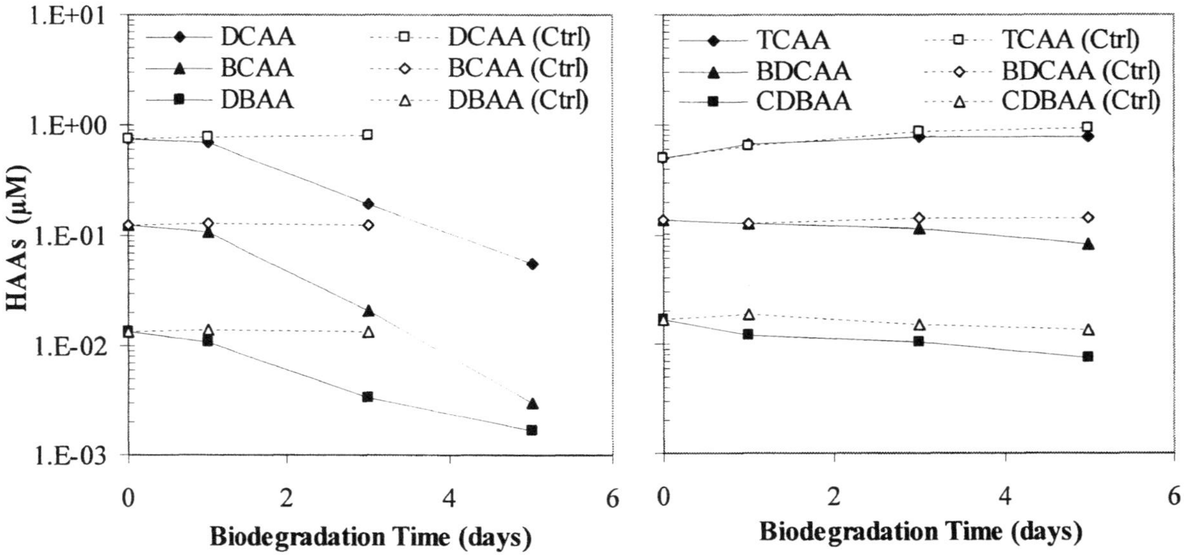

Figure 1. Biodegradation of preformed DXAAs and TXAAs in wastewater EfOM (Ctrl = control sample; results for DXAA control sample for day 5 not available) (Reproduced with permission from Reference 5. Copyright 2008 Awwa Research Foundation.)

Table III. Key parameters used for DBP fate-and-transport modeling

DBP	MW g/mole	H_e atm-m^3/mole	Log K_{ow} unitless	K_B 1/day	K_P^* 1/hr	K_H 1/hr
Chloroform	119	3.67×10^{-3}	1.97	<0.02	1.8×10^{-4}	5.8×10^{-5}
DCAA	129	3.52×10^{-7}	0.92	0.55	6.0×10^{-3}	6.6×10^{-12}
NDMA	74	1.82×10^{-6}	-0.57	<0.02	4.81	$<5.3 \times 10^{-3}$
DCAN	127	3.79×10^{-6}	0.29	0.26	<0.013	2.3×10^{-2}

NOTE: K_P^* represents the near-surface sunlight photolysis rate.
SOURCE: Reproduced with permission from Reference 5. Copyright 2008 Awwa Research Foundation.

NDMA biodegradation. In one study, the biodegradation half-life was 16 days under anaerobic conditions and was almost negligible under aerobic conditions (*19*). Monohalogenated HAAs (MXAAs) have higher bioindex values (0.66 to 0.71) than DBAA, whose bioindex was 0.62 and biodegradation kinetics was 0.56 day^{-1}. So MXAAs are more biodegradable than DBAA and probably have biodegradation kinetics greater than 0.56 day^{-1}. Trichloronitromethane (TCNM) has a lower bioindex (0.15) than chloroform (0.36), which reported less than a 10% removal within 5 days (or equivalent to a rate constant <0.02 day^{-1}), so the biodegradation kinetics of TCNM are expected to be lower than 0.02 day^{-1}.

Photolysis Experiments and Rate Constants

Figure 2 demonstrates the removal trends for nitrosamines (NAs) during sunlight photolysis. The half-lives of the NAs were less than 60 min under the conditions of this study (4.8 hr^{-1}). In contrast, photolysis of chloroform was slow under solar conditions, although noticeable losses of the three brominated THMs were observed. Neither MXAAs nor DXAAs were photolyzable under the solar condition of this study, whereas TXAAs exhibited a small loss (<20%). Most haloketones and haloacetaldehydes were stable too, except for brominated trihaloacetaldehydes. Except for DCAN, brominated dihalogenated HANs were very susceptible to sunlight photolysis, with pseudo first-order rate constant of 0.2 to 0.3 hr^{-1}. Table III summarizes K_P values for four specific DBPs later used in the fate-and-transport modeling.

The rates of sunlight photolysis increased for more halogen substitution in DBPs, bromine-substituted DBPs degraded faster than chlorinated analogs, and nitrogenous DBPs were more unstable than carbonaceous DBPs. The slight structural difference of NAs appeared insignificant in controlling the photolysis kinetics under the conditions of these tests. The photolysis kinetics of NAs with

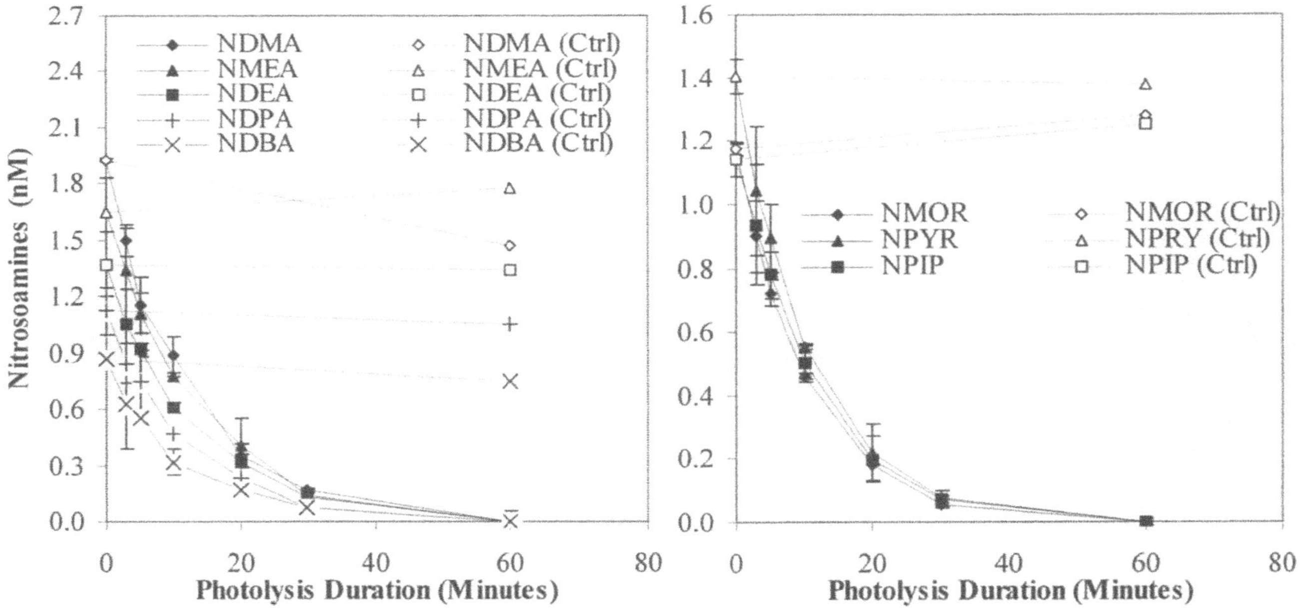

Figure 2. Natural sunlight photolysis of NAs (August 2005) (N-nitrosomethylethylamine [NMEA], N-nitrosodiethylamine [NDEA], N-nitrosomorpholine [NMOR], N-nitroso- pyrrolidine [NPYR], N-nitrosodipropylamine [NDPA], N-nitrosopiperidine [NPIP], N-nitrosodibutylamine [NDBA]). *(Reproduced with permission from reference 5. Copyright 2008 Awwa Research Foundation.)*

cyclic C-C chains (i.e., NMOR, NPYR, and NPIP) were almost equal to those species containing saturated straight C-C chains (i.e., NDMA, NMEA, NDEA, NDPA, and NDBA). Another laboratory study using a lamp source that simulated natural sunlight, where temperature was better controlled, was conducted for NAs (*20*). In that study, the photodegradability of the straight-chain NAs typically decreased with increasing number of carbon atoms, probably due to a steric effect. In addition, cyclic NAs were faster photolyzed, with rates almost twice that of NDMA (*20*).

Hydrolysis Rate Constants

Rates of hydrolysis of DBPs increase with higher temperature, higher pH, greater halogenation (except for THMs), and more bromine substitution (except for HANs). Figure 3 shows a comparison between published hydrolysis rate constants (*21-23*). A QSAR model was then calibrated to fit the data. Steric and electronic effect sensitivity coefficients were found to be $\delta = 0.53$ and $\rho = 3.62$, respectively, with an R^2 value of 0.94. The QSAR results, in turn, predicted the rate constants of two other species for which there were no literature values, i.e., monochloroacetonitrile (MCAN) and monobromoacetonitrile (MBAN), as 3.3×10^{-4} and 2.2×10^{-4} hour^{-1}, respectively. Likewise, K_H values were estimated for other DBPs (e.g., Table III).

Volatilization Rate Constants

Henry's constants (H_c) were obtained for each DBP class: H_c for THMs ranged from 4.35×10^{-4} to 7.83×10^{-3} atm-m^3/mole; H_c for HAAs ranged from 8.4×10^{-10} to 3.5×10^{-7} atm-m^3/mole; H_c for HANs ranged from 4.68×10^{-8} to 1.34×10^{-6} atm-m^3/mole; and H_c for NAs ranged from 2.45×10^{-8} to 1.32×10^{-5} atm-m^3/mole. Reported Henry's constants are usually for neutral species. Ionic species have significantly lower H_c values, which is controlled by their pK_a values and pH of the water. This is especially important for the HAAs. THMs had the highest H_c values and are likely to volatilize more than other DBP classes.

Figure 4 demonstrates the effects of chemical property (in terms of H_c), wind velocity (U_w) on volatilization kinetics of DBPs under two waterbody scenarios. Wind velocity affects the gas film exchange coefficient (K_g) and subsequently the k_l term expressed in Table II. Increasing H_c values enhanced the rate of volatilization within certain limits at a fixed U_w. Increasing H_c did not necessarily lead to high volatilization kinetics at a U_w of 10^5 m/hr under scenario I. Under scenario I, increasing U_w can facilitate volatilization significantly for low H_c DBPs, but not for more volatile DBPs with $H_c > 10^{-3}$ atm-m^3/mole. Volatilization was significantly lower under scenario (I) than with scenario (III) (Figure 4).

Significance of Adsorption on DBP Fate

K_{ow}, SS concentration, and f_{oc} in SS can be used to estimate the rate of sorbed DBP removal from the water column. Log K_{ow} values ranged from -0.57 to 2.63 (*5*). Figure 5 shows the adsorption capacity of DBPs on SS at a simulative condition with 10,000 mg/L of SS at most and f_{OC} of 0.01. NDBA, which has the highest log K_{ow} value (2.63), was found to have an adsorption capacity smaller than 2.6% based upon the equation in Table II. Because other DBPs have lower K_{ow} values than NDBA, adsorption onto depositing SS was unimportant for DBP removal in surface waters. Furthermore, the minor removal potential of DBPs by this mechanism made further investigation of their removal kinetics unimportant.

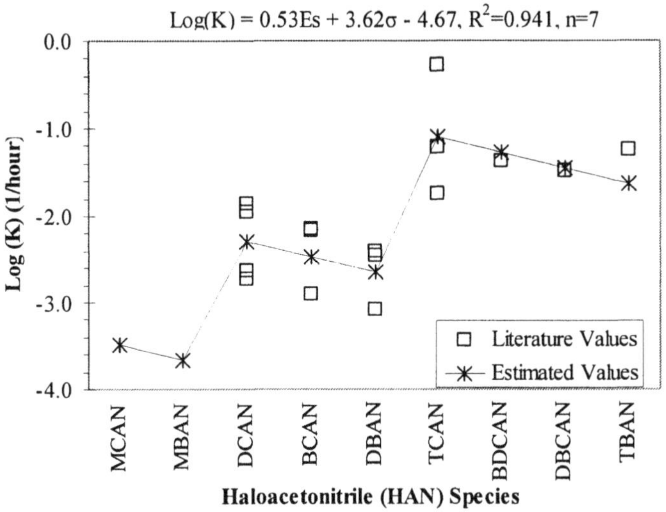

Figure 3. Comparison of literature results (□) and QSAR prediction results (*) for hydrolysis rate constants for HANs [AN = acetonitrile, B = bromo, C = chloro, D = di, M = mono, and T = tri).

Fate–and-Transport Modeling in Receiving Waters

Four DBP compounds (chloroform, DCAA, DCAN, and NDMA) were selected to evaluate the range of DBP loss in surface waters receiving wastewater effluent discharges containing DBPs. Relationships and parameters from Tables I through II and Equation 1 (and References 24 and 25) were used for these simulations. In all cases, DBP fate over a 100-km reach was simulated.

For chloroform, a high flowing velocity and shallow stream (scenario III) resulted in the most rapid chloroform loss (K_T = 1.38 hr^{-1} and >99% removal),

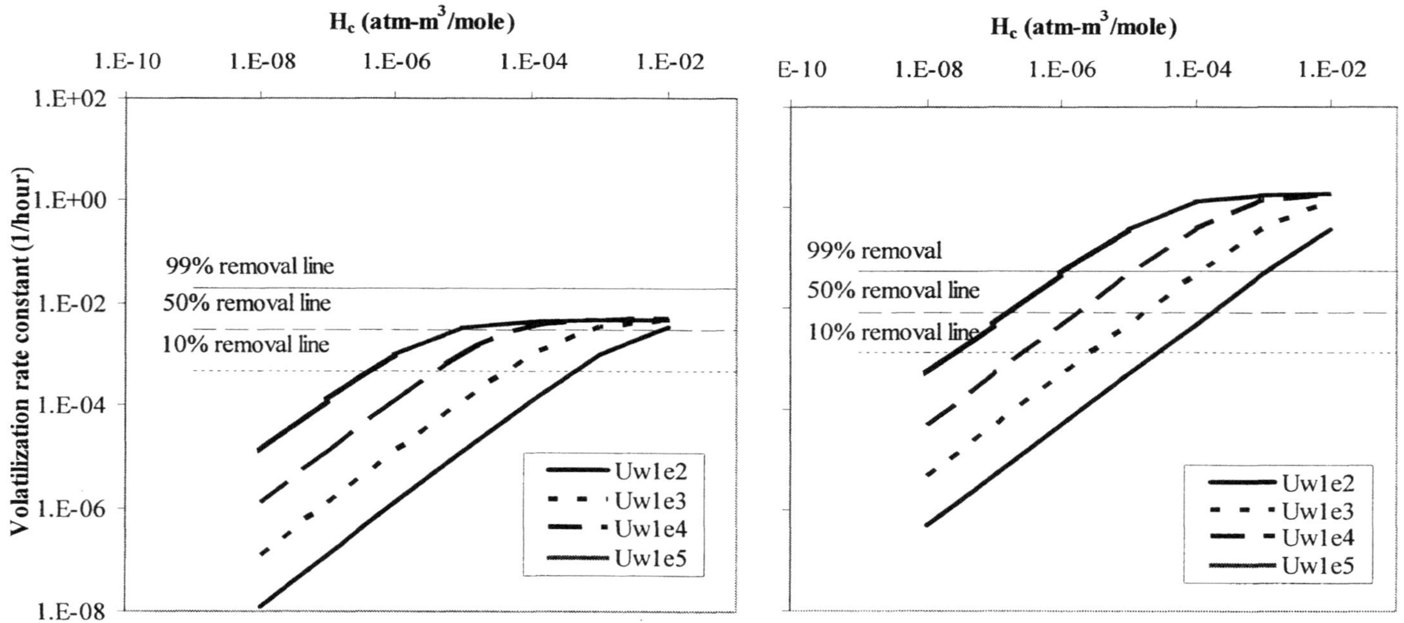

Figure 4. Influence of H_c, U_w, and river streamflow on DBP volatilization for representative rivers for Scenario III (i.e., river with high flow velocity and low depth; left) and for Scenario I (low flow velocity and high depth; right)
NOTE: U_w evaluated from 1×10^2 to 1×10^5 m/hr (Uw1e2 - Uw1e5). (Reproduced with permission from reference 5. Copyright 2008 Awwa Research Foundation.)

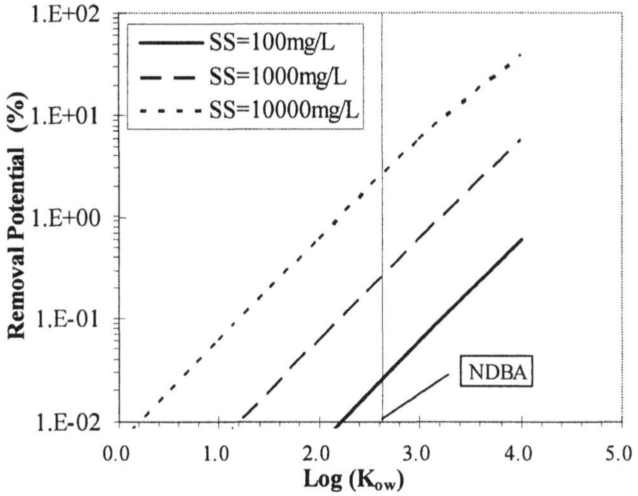

Figure 5. Influence of K_{ow} and SS concentration on DBP removal. (Reproduced with permission from reference 5. Copyright 2008 Awwa Research Foundation.)

whereas scenario I had a lower K_T (4.3 x 10^{-3} hr^{-1}) and less than 80% removal in the same travel distance. Regardless of hydrogeological conditions, volatilization accounted for >99% of the chloroform loss and other loss mechanisms were negligible.

Figure 6 illustrates the trends for DCAA removal in different river scenarios and via different loss mechanisms. Scenario I, with a slow flow velocity and high water depth, resulted in the least DCAA remaining, with a K_T value of 2.23 x 10^{-2} hr^{-1}. HAAs exist in the deprotonated form (i.e., pK_a values <3.0) in rivers with near-neutral pH levels. Consequently, volatilization or adsorption will have little effect on HAA removal. Biodegradation was the dominant loss mechanism for certain mono- and di- HAAs. K_B for DCAA was almost two orders of magnitude higher than the other loss mechanisms despite the river scenario. Because biodegradation was the dominant mechanisms for DCAA, its removal over a fixed distance was a function of flowrate (i.e., residence time in the river).

DCAN was predicted to undergo biodegradation (K_B = 1.1 x 10^{-2} hr^{-1}), but sunlight photolysis was negligible (K_P <1.76 x 10^{-2} hr^{-1}). Hydrolysis was the most critical mechanism for DCAN removal (K_B= 2.3 x 10^{-2} hr^{-1}). Volatilization can contribute to DCAN removal for up to 30% under scenario III (K_V = 1.46 x 10^{-2} hr^{-1}), which was even greater than biodegradation.

NDMA has a high potential for loss from photolysis (>99%) in scenario III (average K_P = 2.54 hr^{-1}), but this mechanism became less significant when the depth of the water increased (e.g., Scenario I or II). For example, ~1.5 order of magnitude increase in the water depth (from 0.13 to 4.5 m) would lead to 9.5

orders of magnitude decrease of K_P (2.54 to 7.3 x 10^{-10} hr^{-1}). In scenario I, a small NDMA reduction (8%) was predicted over 100 km, and the dominant mechanism was volatilization (K_V = 2.18 x 10^{-4} hr^{-1}). This finding thus suggests that NDMA might persist at conditions where photolysis was absent, such as deep lakes and reservoirs or groundwaters, unless anaerobic biodegradation occurred in the groundwater (*19*).

Figure 7 summarizes the half-lives ($t_{1/2}$ = ln 2 / K_T) for the four DBPs under the three river scenarios over a 100-km river reach. Half-life values, the time it takes for the DBP concentration to decrease by 50%, ranged from 5 hr to over 2,000 hr for the four DBPs and the three river scenarios. The travel time for scenario I, over 100 km (226 hr), was significantly longer than Scenario II (120 hr) or III (80 hr). When the half-life is lower than the travel time, greater than a 50% loss in DBP concentration is expected. This is always the case for chloroform and for NDMA under scenarios II and III, but not for NDMA in the deeper, slower moving condition, simulated by scenario I. DCAA and DCAN will experience >50% removal under scenario I and II due to the longer travel time.

This overall approach was subsequently extended to all THM, HAA, HAN, and NA species. Similar to chloroform, volatilization was the most important mechanism for the other three THMs. Biodegradation was the dominant loss mechanism for MXAAs and the other DXAAs, whereas photolysis and hydrolysis were more important in terms of TXAA losses.

In this study, limited field tests were conducted on the fate and transport of DBPs in receiving waters (*5*). The results of this study, nonetheless, may promote future studies to investigate systematically the distributions of DBPs in natural waters, determining in what conditions DBPs are expected to occur significantly relative to regulatory limits. In addition, the modeling methods applied in this paper may be applicable to other organic pollutants.

Conclusions

This study evaluated the impact of biodegradation, photolysis, hydrolysis, volatilization, and adsorption on the fate of several types of DBPs in representative watershed scenarios. The dominant removal mechanism for THMs was volatilization, for MXAAs and DXAAs it was biodegradation, for NAs it was photolysis, and for HANs it was hydrolysis, whereas the adsorption capacity of DBPs on SS was negligible for all of the DBP species studied. Receiving waters with high flow velocity and low water depth (i.e., scenario III) favor the removal of THMs and NAs; whereas watersheds with low flow velocity and high depth (i.e., scenario I) favor the removal of the MXAAs, DXAAs, and HANs. High losses of DBPs (>50%) may occur in surface waters under most scenarios of interest, depending upon numerous controlling factors, such as initial DBP amount in WWTP effluent, waterbody dilution effect,

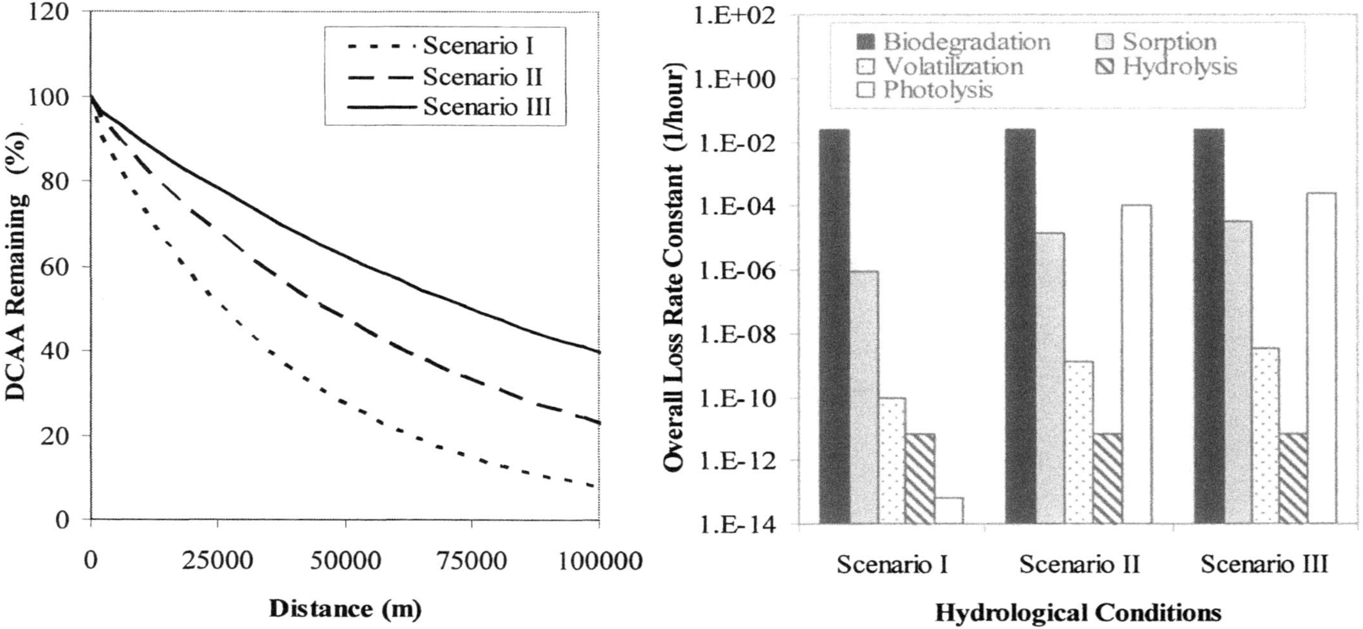

Figure 6. Fate of DCAA in three hydrological scenarios and the relative significances of natural mechanisms for DCAA removal. (Reproduced with permission from reference 5. Copyright 2008 Awwa Research Foundation.)

hydrogeological conditions, river profile, wind velocity, etc. Overall, the fate-and-transport model was successfully used to integrate these factors and, thus, estimate the relative losses of DBPs in receiving waters. In the future, field studies may be needed to verify the results of this study.

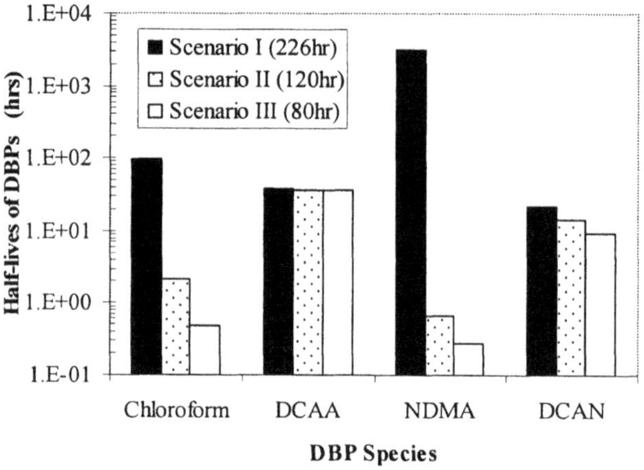

Figure 7. DBP half-lives under different river conditions (numbers in parentheses denote the retention time within the selected river scenarios over a 100-km reach). (Reproduced with permission from reference 5. Copyright 2008 Awwa Research Foundation.)

Acknowledgements

The authors gratefully acknowledge that the Awwa Research Foundation is the joint owner of the technical information upon which this publication is based. The author thanks the Foundation and the U.S. government, through the USEPA for its financial, technical, and administrative assistance in funding and managing the project through which this information was discovered. The comments and views detailed herein may not necessarily reflect the views of the Awwa Research Foundation, its officers, directors, affiliates or agents, or the views of the U.S. Federal Government. The project manager was Alice Fulmer.

References

1. Kolpin et al. Environ. Sci. Technol. **2002**, *36*, 1202-1211.
2. Heberer, T. Toxicology Letters **2002**, *131*, 5-17.
3. Ternes, T. A. Wat.Res **1998**, *32*, 3245-3260.

4. Chen, B., PhD dissertation, Arizona State University, **2007**.
5. Krasner, S. W.; et al. The Contribution of Wastewater to DBP Formation; Awwa Research Foundation: Denver, Colo., **2008** (in press).
6. Allgeier, S. C.; et al. In Proc. of 1996 Wat. Qual. Technol. Conf. (WQTC); Amer. Water Works Assoc. (AWWA): Denver, Colo., **1997**.
7. Boethling, R. S.; Mackay, D. Handbook of Property Estimation Methods for Chemicals in Environmental and Health Sciences; Lewis Publishers: Boca Raton, Fla., **2000**.
8. Hammett, L. P. J. Am. Chem. Soc. **1937**, *59*, 96.
9. Taft, R. W. J. Am. Chem. Soc. **1956,** *58*, 2436.
10. Hansch, C. Exploring QSAR: Fundamentals and Applications in Chemistry and Biology; American Chemical Society: Washington, DC, **1995**.
11. Zhang, X.; Minear, R. A. Wat. Res. **2002,** *36*, 3665-3673.
12. Munch, D. J., Hautman, D. P. In Methods for the Determination of Organic Compounds in Drinking Water; USEPA Office of Res. and Development, National Exposure Research Laboratory: Cincinnati, OH, **1995**.
13. Krasner, S. W.; et al. In Proc. of WQTC; AWWA: Denver, Colo., **2001**.
14. Munch, D. J.; Munch, J. W.; Pawlecki, A. M. In Methods for the Determination of Organic Compounds in Drinking Water; USEPA Office of Research and Development, National Exposure Research Laboratory: Cincinnati, Ohio, **1995**.
15. Cheng, R. C.; et al. In Proc. of 2004 WQTC; AWWA: Denver, Colo., **2004**.
16. Taguchi, V.; et al. Canadian J. Applied Spectroscopy **1994**, *39*, 87-93.
17. Yoo, L. J.; et al. In Proc. of 2000 WQTC; AWWA: Denver, Colo., **2000**.
18. Baribeau, H.; et al. J. AWWA **2005**, *97*, 69-81.
19. Gunnison, D.; et al. J. Hazardous Materials **2000,** *B73*, 179-197.
20. Dotson, A. In Proc. of AWWA Annual Conference; AWWA: Denver, Colo., **2007**.
21. Croué, J.-P.; Reckhow, D. A. Environ. Sci. Technol. **1989,** *23*, 1412-1419.
22. Glezer, V.; et al. Wat. Res. **1999,** *33*, 1938-1948.
23. Nikolaou, A. D.; et al. Chemosphere **2000,** *41*, 1149-1154.
24. Thomann, R. V.; Mueller, J. A. Principles of Surface Water Quality Modeling and Control; Harper & Row: New York, **1987**.
25. Lyman, W. J.; et al. Handbook of Chemical Property Estimation Methods; McGraw-Hill: New York, **1982**.

Chapter 19

Organic Nitrogen Occurrence and Characterization

Aaron D. Dotson[1], Paul Westerhoff[1], Baiyang Chen[1,2], and Wontae Lee[1,3]

[1]Department of Civil and Environmental Engineering, Arizona State University, Tempe, AZ 85287–5306
[2]Current address: DEI Professional Services, LLC, 6225 N 24th Street, Suite 200, Phoenix, AZ 85016
[3]Current address: HDR Engineering, Inc., 3200 East Camelback Road, Suite 350, Phoenix, AZ 85018

Dissolved organic nitrogen (DON) was measured in drinking waters and wastewaters during multiple sampling campaigns. DON analysis used a new dialysis based pretreatment method. On average, DON measured in 28 drinking water treatment plant influents was determined to be 0.16 mg-N/L. In wastewater effluents, DON concentrations varied between 0.51 – 1.1 mg-N/L. Additionally, organic nitrogen in dissolved organic matter isolates was characterized. A significant fraction (20 – 65 %) of the dissolved organic matter present in samples rich in organic nitrogen samples was of colloidal nature having an average carbon to nitrogen ratio of 7.7.

Organic nitrogen has gained a considerable amount of attention in recent years due to its potential reactivity with oxidants, such as chlorine and chloramines, to form nitrogenous disinfection by-products (DBPs). Some nitrogenous-DBPs have been found to be more cytotoxic and genotoxic than the currently regulated carbonaceous DBPs (*1*). Recent AwwaRF projects have determined occurrence and reactivity of DON in water, wastewater, and dissolved organic matter (DOM) isolates (*2-5*). This chapter discusses results from three AwwaRF projects focused on DON occurrence and characterization.

Analytical Techniques

Three analytical techniques described below provide the basis for quantifying organic nitrogen and its components. DON is calculated after dialysis pretreatment (*6*). Amino acids and sugars are measured by high performance liquid chromatography (HPLC) with or without hydrolysis using the Waters Accq·Tag chemistry package. Finally, nitrogen enriched dissolved organic matter (DOM) is isolated by the recently developed resin isolation scheme using macroporous styrene/divinylbenzene resins, XAD-1 and XAD-4, and followed MSC-1H a cation-exchange resin (*7*).

Measurement of Dissolved Organic Nitrogen

Currently, no direct analytical technique is available to measure DON directly in water. The most common technique, total Kjeldahl nitrogen (TKN), utilizes a classic chemical digestion with hot sulfuric acid. This method oxidizes DON to NH_4^+ followed by subsequent analysis of ammonium, determining the sum of ammonia and reduced organic nitrogen species. Although valid in many instances, the TKN measurement has three key drawbacks: (1) incomplete digestion of some organic nitrogen compounds, (2) lack of precision with a method detection limit (MDL) of 0.1 – 0.2 mg-N/L, and (3) inability to accurately report DON in waters containing high ammonia concentrations. To overcome these drawbacks, DON can be calculated by subtracting dissolved inorganic nitrogen from total dissolved nitrogen (TDN) when dissolved inorganic nitrogen (DIN) species are minimized through the use of dialysis (*6*):

$$DON = TDN - DIN \quad (1)$$

$$DIN = NO_3^- + NO_2^- + NH_3 \quad (2)$$

The method developed by Lee and Westerhoff (*6*) illustrates that when DIN/TDN ratios are less than 0.6, direct measurement of all nitrogen species can be used to accurately report DON based on equations 1 and 2. TDN (MDL = 0.009 – 0.019 mg-N/L) is measured simultaneously with dissolved organic carbon (DOC) by a Shimadzu TOC-V_{CSH} equipped with a TNM-1 total nitrogen (TN) unit. Nitrate (MDL = 0.005 mg-N/L) and nitrite (MDL = 0.005 mg-N/L) are measured on a Dionex DX-120 ion chromatography system. Ammonia (MDL = 0.005 mg-N/L) is measured on a TRAACS 800 autoanalyzer by the automated phenate method (SM 4500-NH_3 G). In some instances, ammonia was measured colorimetrically by a TNTplus ultra low range ammonia test kit by Hach.

However, propagation of error attributed to measuring each parameter in equations 1 and 2 separately lead to high variance in the calculated DON value when DIN/TDN ratios exceed 0.6. To reduce DIN and thus reduce variance, samples are dialyzed against distilled water (acceptor solution) using a 100 dalton (Da) molecular weight cut-off (MWCO) cellulose ester dialysis tube (Spectrum Laboratories). Dialysis is performed by submerging the dialysis tubes in rectangular high density polyethylene tank with distilled water flowing through, Figure 1. The samples typically undergo dialysis for 24 to 48 hours in order to remove DIN based on estimated nitrate and ammonia first-order rate constants of 4.0 x 10^{-2} and 3.5 x 10^{-2} per hour, respectively. Due to differences in conductivity between the sample and the acceptor solution conductivity, correction may be required due to changes in dialysis tube retentate.

To identify the molecular weight size distribution of DON, stirred cell membrane units (Amicon 8400, Millipore Corp., MA) were loaded with ultrafiltration membranes of two pore sizes. The samples were processed through two parallel systems, one equipped with a YM1 membrane (1,000 Da MWCO) and the other with a YM2 membrane (10,000 Da MWCO). Both the retentate and the permeate samples are analyzed for DON and DOC. A more detailed description of this process is found elsewhere (*8*).

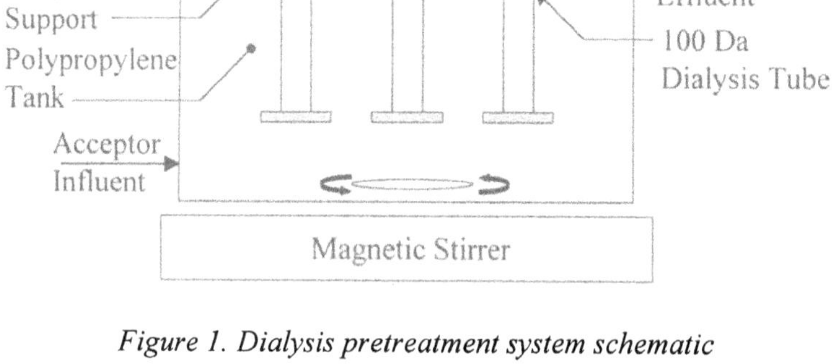

Figure 1. Dialysis pretreatment system schematic

Measurement of Amino Acids and Amino Sugars

Amino acids and sugars (AAS) are measured by high performance liquid chromatography (HPLC) followed by fluorescence detection. Amino acids and sugars are present in two forms, free amino acids and sugars (FAAS) and combined amino acids and sugars (CAAS). The majority of amino acids and sugars are known to be present as CAAS in water samples, which are combined into proteins and polypeptides (*9*). In order to determine the CAAS, FAAS are subtracted from total amino acids and sugars (TAAS). To liberate the CAAS, acid hydrolysis is used to break the peptide bonds that results in a hydrolysate containing TAAS. Various hydrolysis protocols (methanesulfonic acid, liquid phase hydrochloric acid, gas phase hydrochloric acid and phenol) exist that yield varying recoveries of CAAS and in some cases ammonia (*10-13*).

Due to the low concentrations of AAS in water, a pre-concentration is required. One liter of sample is evaporated at 60 °C under vacuum with a Yamato RE-200 rotary evaporator to less than 300 mL. The concentrated sample is freeze dried by a Labconco Freezone6 freeze drier. Up to ten milligrams of the lyophilized material is added to a hydrolysis tube. Up to eight hydrolysis tubes are loaded into a Pico·Tag reaction vial and undergo hydrolysis per the Pico·Tag Manual (*13*). Briefly, a strong acid (6N hydrochloric acid) vapor phase hydrolysis is performed under vacuum for 24 hours at 105 °C in a Pico·Tag workstation. The samples are dried under vacuum and diluted with borate buffer. The samples are derivatized by the addition of Accq·Fluor reagent per the Waters Accq·Tag method. Determination of FAAS is performed by omitting the hydrolysis step. The derivatized AAS in the sample are quantified by measuring the fluorescence response on a Waters 2475 Multi Wavelength Fluorescence detector at an excitation of 250 nm and an emission of 395 nm. Instrument calibration was performed by injecting 0.2 to 20 µL of 10 pmol/µL (cystine 5 pmol/µL) of diluted hydrolysate amino acid standard H (Waters catalog no. NCI0180), glucosamine (Sigma catalog no. G4875), and galactosamine (Sigma catalog no. 138495). A three eluent (diluted Waters catalog no. WAT052890, acetonitrile, and water) mobile phase gradient was utilized. A more detailed discussion of the AccQ·Tag method has been previously published (*11*).

Resin Isolation of Nitrogen Rich Natural Organic Matter

Previously, Leenheer et al. (*14*) developed a resin based process to isolate natural organic matter (NOM) or DOM considering organic carbon recovery. The modified method provides enrichment of organic nitrogen in select DOM fractions (*7*). This method will be briefly described here but the reader is directed to the references for a more detailed discussion.

This isolation procedure requires a large volume of water in order to fractionate measurable mass of each DOM isolate. A water sample, 60 – 100 L, is collected and filtered by a glass fiber filter with a nominal pore size of 1 micrometer (μm) and the pH is adjusted to 4. The sample is evaporated under vacuum to a salt slurry by a large rotary evaporator. The salt slurry is loaded into a 3500 Da MWCO regenerated cellulose dialysis tube (Spectra/Por-3, Spectrum Labs). The dialysis tube containing the salt slurry is dialysized against two solutions, initially against 0.1 M hydrochloric acid and followed by 0.2 M hydrofluoric acid. The contents of the dialysis tube, organic colloids, are freeze dried after dialysis against hydrofluoric acid.

The remaining DOM and salt permeate through the dialysis membrane during the hydrochloric acid dialysis step. To fractionate the DOM present in the dialysis permeate a series of three primary resin columns will be used. The first column contains a macroporous styrene/divinylbenzene XAD-1, the second is a similar XAD-4 resin, and the final column is a cation exchange resin, MSC-1H. The dialysis permeate collected from the hydrochloric acid dialysis phase is passed through the resins. The hydrophobic NOM will adsorb onto the XAD-1 resin, the amphiphilic NOM will absorb onto the XAD-4 resin, and the hydrophilic bases will be absorbed on the MSC-1H resin. The remaining hydrophilics, acids and neutrals, are not adsorbed on any of the columns.

The general compound classes adsorbed onto each of the resins can be further fractionated by passing differing eluents through each column. Table 1 identifies the eluent associated with each subfraction that can be desorbed from the resins. Briefly, the hydrophobic acids (HPO-A) and amphiphilic acids (AMP-A), are eluted from the XAD-1 and XAD-4 columns. The hydrophobic base and neutral (HPO-B/N) and amphiphilic base and neutral (AMP-N/B) are eluted from the XAD-1 and XAD-4 columns. Only the hydrophilic base (HPI-B) is adsorbed on the MSC-1H resin requiring only one elution step. During the elution process the hydrophobic and amphiphilic amino acids (HPO-AMP AA) can be isolated by attaching another MSC-1H column in series to the XAD-1 and XAD-4 resins while the acids, hydrophobic and amphiphilic, are being desorbed.

The hydrophilic acids plus neutrals (HPI-A+N) did not adsorb on the resins and remain in the effluent of the series of columns. In addition to the DOM, this

Table 1. NOM Resin Fractionation Eluents

Resin	*Fraction*	*Eluent*
XAD-1	HPO-A	1 M Sodium Hydroxide
XAD-1	HPO-N/B	75% Acetonitrile/25% Water
XAD-4	AMP-A	1 M Sodium Hydroxide
XAD-4	AMP-N/B	75% Acetonitrile/25% Water
MSC-1H	HPI-B	3M Aqueous Ammonia
None	Colloids	Dialysis Retentate

water contains concentrated inorganic salts and requires significant purification. Briefly, the purification requires a zenotrophic distillation with acetic acid, selective precipitations, processing with smaller ion exchange columns, and centrifugation to remove precipitates. This purification selectively removes inorganic salts such as bromide, nitrate, sulfate, and phosphate without removing organic matter. Due to the complexity of these purification steps, they will not be described here but they are described elsewhere (*14*).

Results

Organic nitrogen comprises upwards of fifty percent of total dissolved nitrogen (TDN) present in many surface waters, except those with elevated nitrate. However, it makes up a significantly smaller percentage of DOM (0.5% - 10% by weight). Wastewater treatment plant (WWTP) effluents can contain a broad range of DIN concentrations (<1 to >20 mg-N/L) with the speciation dependant upon the type of biological treatment. Consequently, DON in wastewater can range from 0.5 – 2 mg-N/L, and make up a significant fraction of the TDN in highly biologically treated wastewaters. The data discussed in this section was collected as part of three national sampling campaigns as well as the isolation of nitrogen enriched DOM from sources known to have notable concentrations organic nitrogen.

Drinking Water Treatment

The investigators participating in the AwwaRF project entitled Organic Nitrogen in Drinking Water and Reclaimed Wastewater[5], performed two sampling campaigns that sampled twenty eight water treatment plants (WTP) located in Arizona, California, Illinois, Indiana, Kentucky, Massachusetts, Michigan, Missouri, Nevada, Pennsylvania, Texas, Virginia, and Wisconsin. Another AwwaRF project: Occurrence and Formation of Nitrogenous Disinfection By-Products (*4*) also discussed, has currently only performed the first of two sampling campaigns collecting samples from eleven WTPs located in California, Oklahoma, New Jersey, Michigan, Colorado, and Pennslyvania. The former project focused on the occurrence of DON in a variety of water sources where the latter has selected WTPs that are suspected to be impacted by elevated DON concentrations due to the influence of wastewater or algal blooms.

Dissolved Organic Nitrogen

The average DIN/TDN ratio was greater than 0.6 mgN/mgN, thereby requiring dialysis pretreatment of the water samples to accurately determine DON concentration. The average raw water DON and DOC concentrations of the 28 WTPs was 0.19 mg-N/L and 3.44 mg-C/L, respectively. The average DON/DOC ratio was 20 mg DOC/mg DON for raw water. The finished water DON/DOC ratio was 18.1 mg DOC/mg DON with an average DOC concentration of 2.44 mg-C/L and average DON concentration of 0.15 mg-N/L. These WTPs removed roughly on average 29% of DOC and 20 % of DON. Table 2 shows the variability of this DON and DOC data for the 28 WTPs sampled in this study. Further study of the size fractionation of the DON and DOC on the raw and finished waters was also performed during the second sampling campaign in 2003. Table 3 shows that water treatment reduces the molecular weight distribution of DON and DOC.

Upon further examination of DON occurrence, a reasonable relationship between temperature and DON concentration appears to exist. Figure 2 presents temperature variation in WTP samples from a wide variety of geographical locations and reservoir depth in three Arizona reservoirs. Regardless the scenario, increased temperature tended to result in increased DON concentration. The trend described seems to correlate well with the hypothesis of increased algal growth present at increased temperatures (e.g. summer). Furthermore, this supports the hypothesis that algal growth can transform DIN to DON and act as a source of organic nitrogen due to the production of soluble microbial products, chlorophyll, proteins, and colloidal cell wall material.

Table 2. DOC and DON concentrations for raw and finished waters*

	DOC (mgC/L)		*DON (mgN/L)*		*% Removal*	
	Raw	*Finished*	*Raw*	*Finished*	*DOC*	*DON*
Average Spring	3.54	2.48	0.20	0.16	30	21
Average Fall	3.34	2.40	0.17	0.14	28	19
Study Average	3.44	2.44	0.19	0.15	29	20
Minimum	1.14	1.23	0.02	0.02	0	0
Maximum	7.76	3.95	0.43	0.47	57	59

*Excludes six samples with low DON that resulted in abnormally high DOC/DON ratios

Table 3. Molecular Weight Distribution of DON and DOC

Species	*DON*	*DOC*	*DON*	*DOC*	*DON*	*DOC*
MWCO (kDa)	*< 1*		*1 – 10*		*> 10*	
Raw	37 %	32 %	30 %	36 %	33 %	32 %
Finished	47 %	43 %	29 %	34 %	24 %	23 %

Amino Acids and Sugars

The 11-plant AwwaRF study studied the influence of nitrogen rich source waters on water treatment and DBP formation. Sampling of the WTPs and their

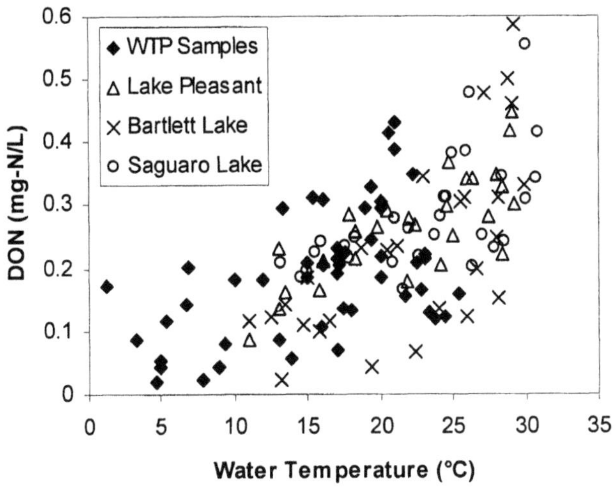

Figure 2. Temperature Effects on DON concentration. (Adapted with permission from reference 8. Copyright 2006 AWWA.)

upstream waters was performed when a significant algal bloom or low river flow, for wastewater influenced sources, was observed. The samples were collected upstream and throughout each of the WTPs process trains and analyzed for a multitude of parameters.

During the sampling campaign, at least one sample from each of the eleven participating WTPs was analyzed for both FAAS and TAAS. Samples were typically collected from the either the plant influent prior to chemical addition and/or the filter effluent before addition of disinfectant. An average raw TAAS concentration of 28 μg-N/L was measured, Table 4. Based on an average raw DON concentration of 0.24 mg-N/L, TAAS only made up approximately 12 percent of the measured DON and varied significantly (minimum 1%, maximum 42%). These results are of similar concentration to previously performed studies (*10*) upon conversion of concentration into similar units. The average filter effluent TAAS concentration account for only 6 percent of the average effluent DON concentration of 0.18 mg-N/L.

To illustrate typical removal of FAAS and TAAS by a conventional WTP, Figure 3 depicts removal of the top five TAAS and corresponding FAAS. A significant removal of TAAS is measured where removal of FAAS is observed to a lesser extent. Based on an average effluent DON concentration of 0.19 mg-N/L, TAAS only made up approximately 3 percent of the measured DON and showed moderate variation (minimum < 1%, maximum 7%). Of the 16 samples analyzed for TAAS, arginine and glycine were present in the highest concentrations in the most samples. Ranking the amino acids on concentration, arginine and glycine are ranked first and second, respectively. Table 5 provides

Table 4. Total Amino Acids and Sugars

	Total AA + AS (µg-N/L)		
WTP No.	*Influent*	*Settled*	*Effluent*
1	75		5
2	4		5
3		17	16
4	24		
5			[a]
6	17		5
7		10	
8	31		
9	34		14
10	14		
11			14
Average	28	14	10[b]

[a]Sample compromised; [b]Excluding WTP No. 5

a count of samples and the associated ranking (1-4) of the dominant amino acids and sugars. For example, arginine was present at the highest concentration in 9 samples, the second highest in 6 samples, and the third highest in only 1 sample. The first and second ranked amino acids are 1 of 3 amino acids (arginine, glycine, or alanine). The third and fourth ranked amino acids and sugars encompass a much broader group. For ranks lower than fourth, the group of amino acids broadens as measured concentration decreases.

Wastewater Effluent

Biological processes used in wastewater treatment significantly effect the concentrations of DIN and DON present. Table 6 describes typical concentrations of DON and DIN present in an array of biological wastewater treatment processes. Compared to concentrations of DON in drinking water, wastewater concentrations are much higher even after treatment (0.51 – 1.1 mg-N/L). Figure 4 illustrates that a facility using any sort of nitrification process discharges lower concentrations of DON on a central tendency basis. Additionally, a trend of increased variability in DON concentration is present in samples with a higher DIN/DON ratio, particularly in the effluents of WWTPs that perform no nitrification (NN) or good nitrification (GN).

In addition to DON removal, the importance of speciation of nitrogen is important. Wastewater samples were collected as part of an AwwaRF project identifying the impacts of wastewater on DBP formation (*3*). Shown on a ternary plot, Figure 5, the speciation of ~20 WWTPs sampled over three seasons of differing biological processes are presented. TDN was made up of a greater

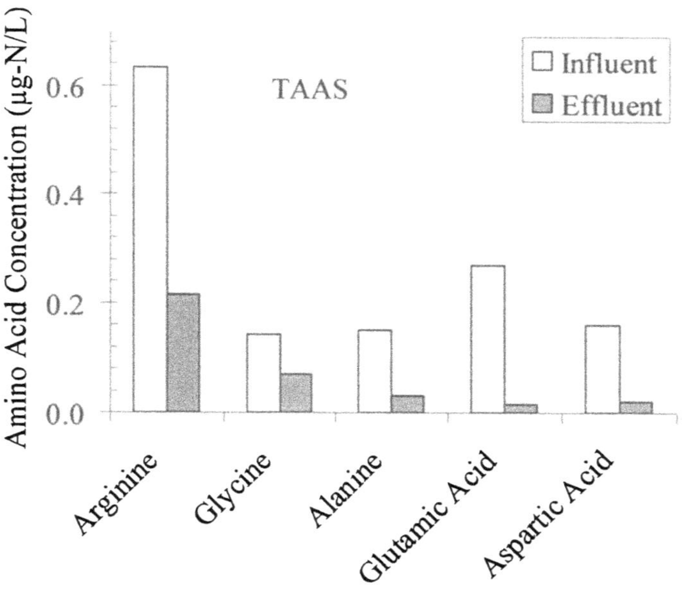

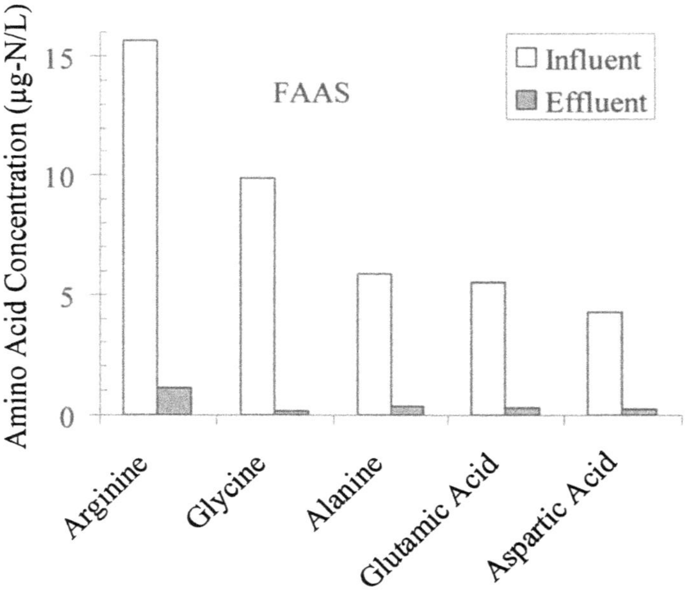

Figure 3. Removal of the 5 Most Prominent Amino Acids during Conventional Water Treatment (WTP No.1)

Table 5. Dominant TAAS and their Associated Occurrence by Number of Facilities

Amino Acid/Sugar	*Ranking** 1	2	3	4
Arginine	9	6	1	
Glycine	7	9		
Alanine		1	5	2
Serine			3	5
Threonine			3	5
Glutamic acid			2	3
Glucosamine			1	
Phenylalanine			1	
Proline				1

*16 samples from influent, settled, and effluent

Table 6. Biological Wastewater Treatment Criteria and Associated DON Concentration

Biological Treatment Type	NH_4^+	NO_2^-	NO_3^- *mg-N/L*	*DIN**	*DON*[†]
No Nitrification (NN)	> 10	< 1	< 2	> 10	1.1
Partial or Poor Nitrification (PN)	2 – 10	> 1	> 2	> 10	0.51
Good Nitrification (GN)	< 2	< 1	> 10	> 10	0.52
Partial Denitrification (PDN)	< 2	> 1	> 5	> 7	0.68
Good Denitrification (GDN)	< 2	< 1	< 5	< 7	0.70

*DIN per equation 2; [†]Median DON concentration

SOURCE: Adapted with permission from reference 3. Copyright 2008 AWWA.

percentage of DON in facilities practicing good denitrification (GDN) compared to other processes.

Dissolved Organic Matter

DOM was isolated from five nitrogen-rich waters. Of the five collected waters two were generated in the laboratory in a batch type reactor containing wastewater bacteria (activated sludge from a wastewater treatment plant; Bacteria) or blue-green algae (*Oscillatoria prolifera;* Algae), further description of these cultures can be found elsewhere (*7*). The remaining three waters were collected from Saguaro Lake which is a reservoir located in central Arizona(SAG), a wastewater plant in Colorado that provides partial nitrification

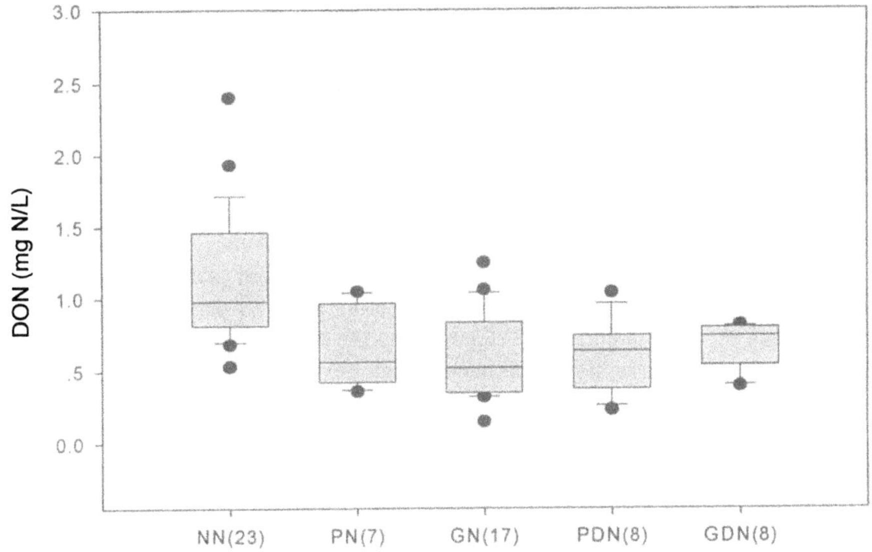

Figure 4. DON Occurrence based on Biological Treatment Type (3)

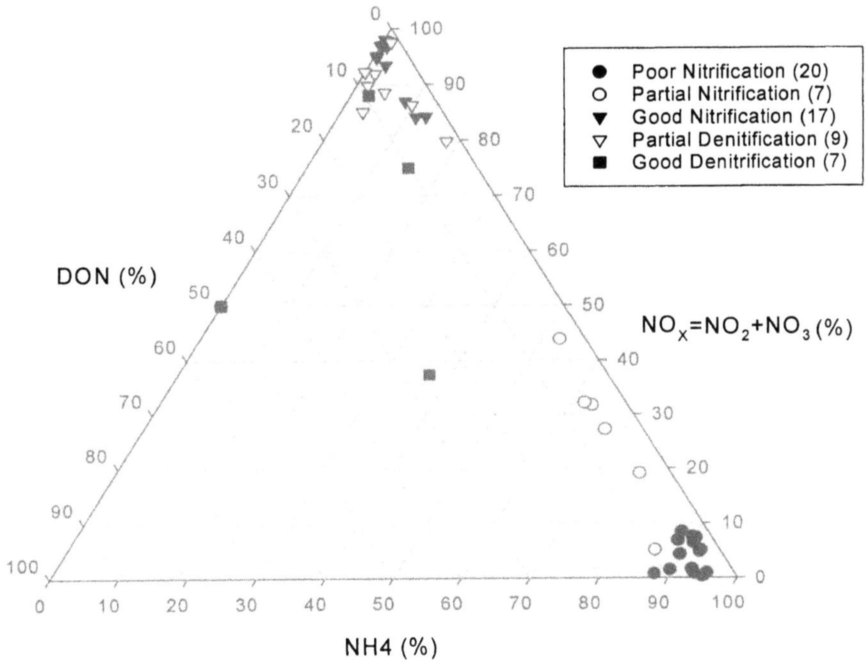

Figure 5. Distribution of Total Dissolved Nitrogen in WWTP Effluents

and utilizes trickling filters (PNTF), and a wastewater plant in southern Arizona that utilizes an aerated lagoon producing a poorly treated wastewater (PTW).

Table 7 summarizes the percent of DOM present as each fractionated isolate as well as the C:N ratio as calculated by elemental analysis. Due to the nature of the isolated mass (e.g. stickiness) or the small mass isolated, elemental analysis was not able to be performed on all the isolates. SAG is the only sample not dominated by DOM colloids and shows the greatest percentage of HPI-A+N which may be the result of biodegradation of DOM colloids (*14,15*). The large percentage of DOM colloids present in the remaining four samples support the environmental significance of dead cell material, such as peptidoglycan, from sources rich in bacteria and algae. Comparing the isolated masses of each of the fractions, the laboratory algae and bacteria samples are deficient in HPO-A and amphiphilics compared to the field collected samples. The PTW stands out compared to the other samples due to the high percentage of colloids.

Nitrogen enrichment follows HPI-B > Amino Acids > Colloids > AMP-B/N > HPO-B/N ≈ AMP-A ≈ HPI-A+N > HPO-A. Although, HPI-B fraction which is shown to have the lowest C:N ratio should be considered with great care due the elution process which may result in residual aqueous ammonia in the isolated sample. The amino acids fraction with a C:N ratio of around 6:1 is indicative of containing primarily proteins and peptides. Subfractionation is used to generate more nitrogen rich fractions. For example, the HPO-B/N fraction consists of

Table 7. Isolated DOM Percentage and C:N Ratios

Sample	*Colloids*	*HPO-B/N*	*HPO-A*	*AMP-B/N*	*AMP-A*	*Amino Acids*	*HPI-A+N*	*HPI-B*
Percent of DOM per Isolated Fraction								
Algae	60	22	0.4	1.8	0.5	0.2	12	2.9
Bacteria	40	26	1.3	8.2	1.2	3.1	9.9	11
SAG	20	30	21	4.6	7.2	2.2	14	1.1
PNTF	39	13	19	3.6	8.5	1.3	13	3.6
PTW	65	3.4	11	8.2	3.4	NI	4.6	4.4
Elemental C:N Ratio								
Algae	8.5	NA	NA	9.9	NA	NA	14	NA
Bacteria	7.1	5.5	NA	8.8	NA	6.7	NA	2.1
SAG	8.1	17	42	14	21	NA	16	NA
PNTF	9.2	13	20	9.5	8.8	6.0	NA	NA
PTW	5.8	25	28	9.0	12	NA	9.2	1.5

NI – Not Isolated; NA – Not Analyzed

lipid material and proteinaceous material that are depleted and enriched, respectively. Subfractionation of the SAG HPO-B/N sample resulted in subfractions are distinctly different with respect to C:N ratio of 68.6 for the lipid subfraction and 9.4 for the proteinaceous subfraction. Further analysis discussed elsewhere (*7*) of FTIR and ^{13}C-NMR sheds additional light on the structure of the carbon and nitrogen present. The key results are (*7*):

- Colloidal DOM, rich in Org-N, consists of peptidoglycan from algal and bacterial cell walls
- FTIR of samples not cultured in the lab show a higher content of carboxylic acids than cultured samples which are suspected be linked to the presence of degraded terrestrial plant material.
- The PNTF sample shows aromatic sulfonates, indicative of the degradation of wastewater surfactants, where SAG and lab cultures do not.

Conclusions and Research Needs

With the plethora of occurrence data presented here, further research should delve into the fundamentals with regard to characterization of organic nitrogen. Very little is known about many of the organic nitrogen rich DOM fractions such as the colloidal fraction. These fractions should be characterized through the use of NMR, FTIR, CHN, and other analytical tests. To further characterize the DOM isolates, the isolates should be subjected to relevant water and waste water treatment processes ranging from the most classical method (e.g. coagulation/sedimentation) to more innovative methods (e.g. advanced oxidation, membranes).

Acknowledgements

The authors gratefully acknowledge that the Awwa Research Foundation is the joint owner of the technical information upon which this publication is based. The author thanks the Foundation and the U.S. government, through the Environmental Protection Agency (USEPA) for its financial, technical, and administrative assistance in funding and managing the project through which this information was discovered. The comments and views detailed herein may not necessarily reflect the views of the Awwa Research Foundation, its officers, directors, affiliates or agents, or the views of the U.S. Federal Government.

References

1. Plewa, M. J.; Wagner, E. D.; Jazwierska, P.; Richardson, S. D.; Chen, P. H.; McKague, A. B., Halonitromethane drinking water disinfection byproducts:

Chemical characterization and mammalian cell cytotoxicity and genotoxicity. *Environmental Science & Technology* **2004**, *38*(1), 62-68.

2. Hwang, C.; Krasner, S.; Sclimenti, M.; Amy, G.; Dickenson, E.; Bruchet, A.; Prompsy, C.; Filippi, G.; Croue, J. P.; Violleau, D.; Leenheer, J., *Polar NOM: Characterization, DBPs, Treatment.* American Water Works Research Foundation: Denver, CO, 2001.
3. Krasner, S.W.; Westerhoff, P.; Chen, B.; Amy, G.; Nam, S.-N.; Chowdhury, Z.K.; Sinha, S.; Rittmann, B.E., *Contribution of Wastewater to DBP Formation.* American Water Works Association Research Foundation: Denver, CO, *in press*.
4. Mitch, W., et al., *Occurrence and Formation of Nitrogenous Disinfection By-Products*. American Water Works Association Research Foundation: Denver, CO, *In Progress*.
5. Westerhoff, P.; Lee, W.; Croué, J. P.; Gallard, H.; Amy, G., *Organic Nitrogen in Drinking Water and Reclaimed Wastewater*. American Water Works Research Foundation: Denver, CO, 2006.
6. Lee, W. T.; Westerhoff, P., Dissolved organic nitrogen measurement using dialysis pretreatment. *Environmental Science & Technology* **2005**, *39* (3), 879-884.
7. Leenheer, J. A.; Dotson, A.; Westerhoff, P., Dissolved organic nitrogen fractionation. *Annals of Environmental Science* **2007**, *1*, 45-56.
8. Lee, W.; Westerhoff, P.; Esparza-Soto, M., Occurrence and removal of dissolved organic nitrogen in US water treatment plants. *Journal American Water Works Association* **2006**, *98* (10), 102.
9. Thurman, E. M., *Organic Geochemistry of Natural Waters*. D. Reidel Publ. Co.: Dordrecht, The Netherlands, 1985; p 497.
10. Chinn, R.; Barrett, S. E., Occurence of Amino Acids in Two Drinking Water Sources. In *Natural Organic Matter and Disinfection By-Products*, Barrett, S. E.; Krasner, S. W.; Amy, G. L., Eds. American Chemical Society: Washington, DC, 2000; pp 96-108.
11. Cohen, S. A.; Deantonis, K. M., Applications of Amino-Acid Derivatization with 6-Aminoquinolyl-N-Hydroxysuccinimidyl Carbamate - Analysis of Feed Grains, Intravenous Solutions and Glycoproteins. *Journal of Chromatography A* **1994**, *661*(1-2), 25-34.
12. Martens, D. A.; Loeffelmann, K. L., Soil amino acid composition quantified by acid hydrolysis and anion chromatography-pulsed amperometry. *Journal of Agricultural and Food Chemistry* **2003**, *51*(22), 6521-6529.
13. Waters, In *PICO·TAG WORK STATION Operator's Manual* Waters Publications: Milford, MA, 1987 Rev. B; Vol. Rev. B, pp 3.2 - 3.4.
14. Leenheer, J. A.; Croué, J.-P.; Benjamin, M.; Korshin, G. V.; Hwang, C. J.; Bruchet, A.; Aiken, G. R., Comprehensive Isolations of Natural Organic Matter from Water for Spectral Characterizations and Reactivity Testing. In *ACS Symposium Series 761: Natural Organic Matter and Disinfection By-Products, Characterization and Control in Drinking Water*, Barrett, S. E.; Krasner, S. W.; Amy, G. L., Eds. American Chemical Society: Washington, DC, 2000; pp 68-83.
15. Leenheer, J. A., Progression from model structures to molecular structures of natural organic matter componenets. *Annals of Environmental Science* **2007**, *1*, 57-68.

Chapter 20

Analysis of *N*-nitrosamines Formed in Various Source Waters Treated with Eleven Disinfection Processes

Yuan Yuan Zhao[1‡], Jessica M. Boyd[1‡], Matthew Woodbeck[3], Robert Andrews[3], Steve Hrudey[2], and Xing-Fang Li[1,2]

[1]Division of Analytical and Environmental Toxicology, Department of Laboratory Medicine and Pathology, and [2]School of Public Health, University of Alberta, Edmonton, AB, T6G 2G3
[3]Department of Civil Engineering, University of Toronto, Toronto, ON, M5S 1A1
‡These authors contributed equally to this chapter

N-nitrosodimethylamine (NDMA) was the first nitrosamine identified as a disinfection by-product (DBP); other nitrosamines such as *N*-nitrosomorpholine (NMor) and *N*-nitrosopyrrolidine (NPyr) have recently been detected in drinking water and identified as DBPs. However there is little information on the formation and occurrence of non-volatile and thermally unstable nitrosamines because they are not detected by commonly used gas chromatography methods. Here we describe the development of a new solid phase extraction-high performance liquid chromatography-tandem mass spectrometry (SPE-HPLC-MS/MS) method for analysis of both volatile and non-volatile nitrosamines at low ng.L^{-1} concentrations. This method has successfully detected a non-volatile nitrosamine DBP (N-nitrosodiphenylamine) in a drinking water distribution system and was used to study nitrosamine formation when source waters of various qualities were disinfected by eleven different processes.

Disinfection of drinking water is one of the most successful public health measures ever enacted. Spread of many water borne diseases can be effectively eliminated by proper use of a disinfectant and maintenance of a residual in a distribution system. However, disinfectants can react with naturally occurring organic matter present in source water to form disinfection by-products (DBPs). Various disinfectants are used for the treatment of drinking water and potentially produce different types of DBPs. N-nitrosamines are a group of emerging DBPs. Most nitrosamines are classified as probable or possible human carcinogens by the International Agency for Research on Cancer (IARC) (1). *N*-nitrosodimethylamine (NDMA) was the first nitrosamine discovered in drinking water in Ontario, Canada (2) and California, U.S.A (3). As a result the Ontario Ministry of the Environment has established a drinking water quality standard of 9 $ng.L^{-1}$ for NDMA (4) while the California Department of Health Services has set a notification level of 10 $ng.L^{-1}$ for NDMA (5). The United States Environmental Protection Agency has recently proposed the inclusion of six nitrosamines (NDMA, *N*-nitrosopyrrolidine (NPyr), *N*-nitrosodiethylamine (NDEA), *N*-nitrosodibutylamine (NDBA), *N*-nitrosodi-*n*-propylamine (NDPA), and *N*-nitrosomethylethylamine (NMEA)) in the Unregulated Contaminant Monitoring Rule (6). This is the first federal monitoring proposal for nitrosamines in drinking water in either Canada or the United States.

Nitrosamine research has focused on volatile or semi-volatile nitrosamines such as NDMA. This is due to the use of gas chromatography (GC) as the nitrosamine analytical method of choice (2, 7, 8). However it is possible that non-volatile or thermally unstable nitrosamines are also formed during water disinfection. These nitrosamines may also be of toxicological relevance, but they cannot be detected by the common GC methods. Here we present a high performance liquid chromatography–tandem mass spectrometry (HPLC-MS/MS) method that can detect both thermally stable and unstable nitrosamines at low $ng.L^{-1}$ concentrations when combined with solid phase extraction (SPE) preconcentration step. This new method has been used to identify a new thermally unstable nitrosamine in a drinking water distribution system.

Another gap in current nitrosamine knowledge is lack of occurrence and formation data for different nitrosamines. Although a number of studies have investigated different precursors and possible nitrosamine formation pathways (9-20), they have focused mainly on NDMA. Little is known about how other nitrosamines form during water treatment. Understanding of formation of other nitrosamines is further complicated by the wide range of source water qualities and disinfection processes that are used by water treatment plants across North America. Each water treatment plant potentially produces its own unique suite of DBPs making it more difficult to elucidate which factors are important for nitrosamine formation. However it is likely that the underlying causes of nitrosamine formation are the same. Therefore it is important to assess how different source water qualities and treatment parameters affect nitrosamine formation. These results will be useful for developing strategies to reduce nitrosamine formation in treated drinking water. Here we describe a systematic investigation of the occurrence and formation of nitrosamines in source water of varied quality treated with eleven different disinfection processes.

HPLC-MS/MS analysis of nitrosamines

Before they were identified as DBPs, nitrosamines were discovered to be present at high concentrations in a number of commercial and food products such as beer, smoked meats and rubber (21-23). As a result, a number of different separation and detection techniques have been developed for nitrosamine analysis from different sample matrices including GC combined with thermal energy analysis (GC-TEA) (24) or /mass spectrometry (GC/MS) (25) and HPLC-MS/MS (26). GC/MS has become a common technique for nitrosamine analysis in drinking water (2, 7, 8). GC analysis cannot detect non-volatile and thermally unstable nitrosamines that may be present in water samples. However, non-volatile and thermally unstable nitrosamines may form during water treatment and may be toxicologically relevant.

An HPLC-MS/MS method has recently been developed for analysis of both thermally stable and unstable nitrosamines in drinking water (27). This method combines SPE enrichment and HPLC separation with tandem mass spectrometry using multiple reaction monitoring (MRM). Table I lists the nine target nitrosamines analyzed by this technique, including one thermally unstable nitrosamine (NDPhA). The SPE procedure has been previously described in detail (8). Briefly, the SPE cartridges contain Ambersorb 572 and LiChrolut EN in order to retain nitrosamines from a 500 mL water sample. The nitrosamines are eluted with dichloromethane and concentrated further, producing approximately 2500X preconcentration. Two standards are used for analysis. NDMA-d6 is used to determine extraction recovery while NDPA-d14 is used for sample quantification. The new HPLC-MS/MS method was developed using a capillary HPLC (Agilent) coupled with an API 4000 QTrap (Applied Biosystems/MDX Sciex) using MRM for detection. MRM transition ions and method detection limits (MDL) for the HPLC-MS/MS method are listed in Table I. The MDLs are in the low $ng.L^{-1}$ concentration range, making this method sensitive enough to detect nitrosamines at the concentrations that they are found in drinking water (28, 29). Recovery experiments showed average recovery was above 70 % at 10 and 40 $ng.L^{-1}$ concentrations (data not shown).

The combination of MRM and HPLC separation produced clean spectra from which clearly distinguished nitrosamine peaks from interference. Figure 1 shows a comparison of the chromatograms obtained from monitoring the transition ion pairs 131/89 (NDPA) or 199/169 (NDPhA) from a spiked pure water sample or from an authentic extracted water sample. For MRM ion pair

Table I. MRM ion pairs and method detection limits of the nine nitrosamines analyzed in this study

Nitrosamine	*Abbrev.*	*MRM ion pair*	*MDL[a] ($ng.L^{-1}$)*
N-nitrosodimethylamine	NDMA	75/43	3.1
N-nitrosomethylethylamine	NMEA	89/61	2.4
N-nitrosopyrrolidine	NPyr	101/55	2.1
N-nitrosodiethylamine	NDEA	103/75	10.6
N-nitrosopiperidine	NPip	115/69	0.9
N-nitrosomorpholine	NMor	117/87	0.2
N-nitrosodi-n-propylamine	NDPA	131/89	0.2
N-nitrosodibutylamine	NDBA	159/103	3.1
N-nitrosodiphenylamine	NDPhA	199/169	0.1
***N*-nitrosodipropylamine-d14[b]**	**NDPA-d14**	**145/97**	**N/A**
***N*-nitrosodimethylamine-d6[c]**	**NDMA-d6**	**81/46**	**N/A**

[a]Method detection limit: it is estimated as 3 times the standard deviation of 10 $ng.L^{-1}$ spiked pure water sample. Six samples were used to determine the MDL.
[b]The NDPA-d14 internal standard is spiked into extracts following SPE and used for sample quantification
[c]The NDMA-d6 internal standard is spiked into samples prior to extraction and used to determine extraction recoveries
SOURCE: Adapted from reference 27. Copyright 2006 American Chemical Society.

131/89 (Figure 1a), the peak observed in the authentic water sample was determined not to be NDPA based on the retention time. The peak in the authentic water sample had a retention time of 5.56 min versus 6.87 min for NDPA standard in the pure water sample. A similar situation was observed when ion pair 199/169 (NDPhA) was monitored (Figure 1b). The peak in the authentic water sample with retention time of 7.96 min was determined to be NDPhA as it corresponded with the retention time of the NDPhA standard peak in the pure water sample (7.94 min). The other peak observed in the authentic water sample was determined not to be NDPhA due to its retention time (5.83 min).

The developed HPLC method was used to analyze authentic water samples from a water distribution system previously identified to have elevated NDMA concentrations (8). This water system uses chloramination for water disinfection. Analysis identified NDPhA in the treated drinking water from this distribution system (Figure 2d). NDPhA was not detected in source water from this system (Figure 2b), identifying NDPhA as a new DBP. To our knowledge, this is the first thermally unstable nitrosamine identified as a DBP (27). Similar results were obtained for NDMA, NPyr and NPip indicating that they also resulted from

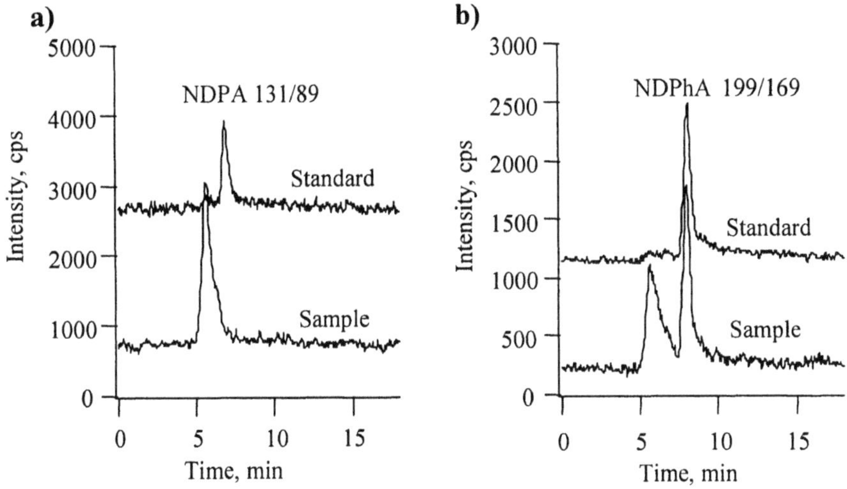

Figure 1. Comparison of extracts from an authentic drinking water sample (sample) and from pure water spiked with nitrosamine standards (standard). a) MRM ion pair 131/89 (NDPA), b) MRM ion pair 199/169) (NDPhA). The concentration of the spiked standards was 5 ng.L^{-1}. (Reproduced from reference 27. Copyright 2007 American Chemical Society.)

the disinfection process. This is also the first time that NPip has been reported as a DBP (27).

We further investigated the timeframe of nitrosamine formation within this distribution system. NDPhA was not detected at the water treatment plant (Figure 2c), but was detected in the distribution system (Figure 2d). Nitrosamine concentrations were determined in drinking water collected from two more points in the distribution system. The results are shown in Figure 3. Location 1 is the water treatment plant and locations 2-4 are increasing distance from the water treatment plant. Locations 2 is within the main community served by the water treatment plant (approximately 2 km from the treatment plant). Location 3 is on the outskirts of the community (approximately 4 km from the treatment plant) and location 4 is located approximately 20 km from the treatment plant.

NDMA, NPip and NPyr were all detected in water from the water treatment plant and their concentrations increased with increasing residence time (Figure 3a). NDMA and NPip concentrations were over 100 ng.L^{-1} at Location 4. NDPhA concentrations increased between the treatment plant and location 3 but decreased between locations 3 and 4 (Figure 3b). This suggests that NDPhA may not as stable in the distribution system as the other nitrosamines.

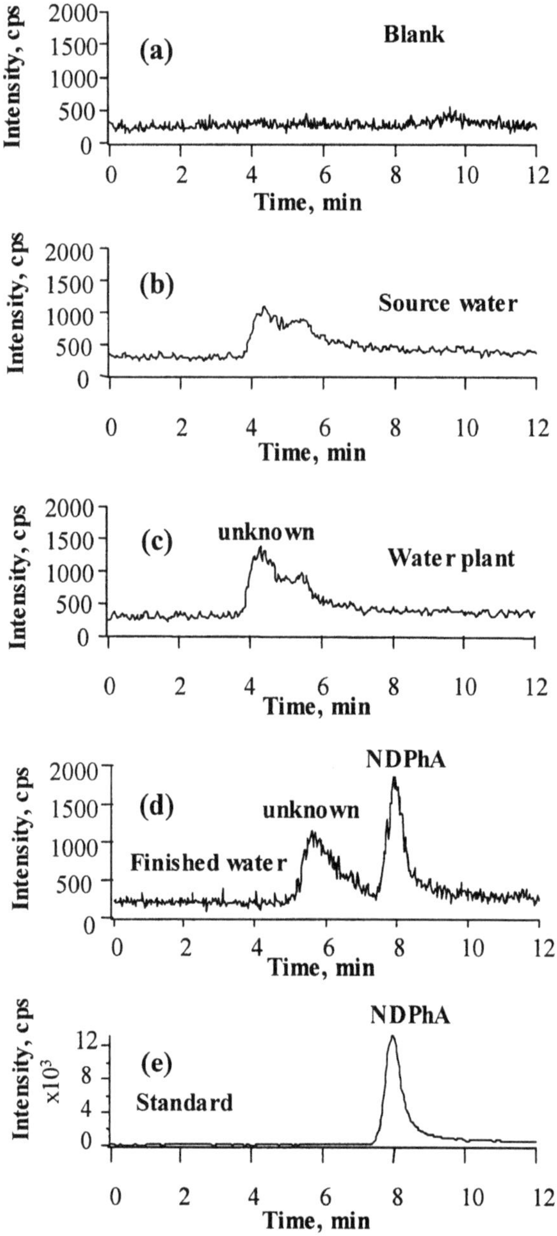

Figure 2. Identification of NDPhA as a disinfection by-product in a distribution system (a) pure water blank, (b) source water, (c) treated water sample collected from the water treatment plant, (d) treated water collected from a location within the distribution system, (e) NDPhA standard. (Reproduced from reference 27. Copyright 2006 American Chemical Society.)

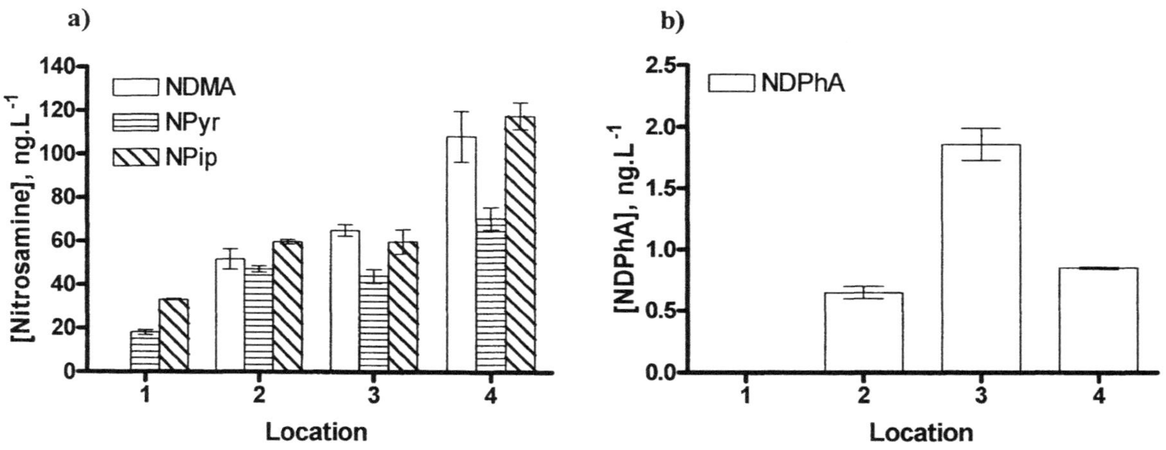

Figure 3. Change in nitrosamine concentration with increasing residence time within a distribution system for a) NDMA, NPyr, NPip and b) NDPhA. Location 1 is the water treatment plant, Locations 2-4 are increasing distance from the treatment plant. Nitrosamine concentrations are an average of triplicate samples. Error bars indicate standard deviation. (Adapted from reference 27. Copyright 2006 American Chemical Society.)

Nitrosamine formation during eleven water disinfection treatments

DBP formation is a complex process that is influenced by many variables including pH, temperature, turbidity and the concentration of natural organic matter (NOM) in source water, as well as the type and concentration of the disinfectants used and disinfectant contact time. This study is to investigate the formation of nine nitrosamines when source waters of various qualities are treated with eleven different disinfection processes. Our rationale is that each water source has its own mix of carbon and nitrogen containing precursors or available nitrosamine precursors, which can result in the formation of different nitrosamines at various concentrations during water disinfection. In addition, a number of treatment plants are switching from chlorination to alternative disinfection methods such as chloramination in order to lower regulated trihalomethane (THM) concentrations. This switch may result in a reduction in THMs, but could potentially cause the formation of other, non regulated DBPs such as nitrosamines (30). To our knowledge no previous studies have investigated the formation of nine nitrosamines when source water of differing quality was treated with eleven disinfection methods in parallel. In this study, raw water was collected from seven different locations in the U.S. and Canada and treated with eleven disinfection treatments. The raw water quality parameters are summarized in Table II. Turbidity, UV absorbance at 254 nm, colour, pH, and total organic carbon (TOC) were measured.

Table II. Summary of raw water quality for the 7 source waters investigated in this study

Location	*Source*	*TOC[a] (mg/L)*	*UV A254 (cm^{-1})*	*Turbidity (NTU[b])*	*pH*	*Color (TCU[c])*
1	River	2.0	0.039	1.8	8.3	14
2	Lake	5.7	0.329	6.1	7.3	118
3	River	5.7	0.275	25	8.3	210
4	River	5.9	0.127	0.75	8.0	18
5	River	7.3	0.223	0.84	7.7	35
6	Lake	15.9	0.464	30	8.8	230
7	River	23.9	0.933	4.68	7.4	235

[a]Total Organic Carbon
[b]NTU: Nephelometric turbidity units
[c]TCU: True colour units
SOURCE: Reproduced from reference 31

The eleven disinfection methods used in this study are shown in Scheme 1 and these include chemical (Cl_2, NH_2Cl, ClO_2, O_3) as well as alternative techniques such as ultra violet (UV) and advanced oxidation processes (AOP). Combinations of chlorine with O_3, UV, or AOP were also investigated. The concentrations of Cl_2, NH_2Cl, ClO_2 applied to each water sample depended on obtaining a disinfectant residual of 0.5, 2.0 and 0.5 mg/L, respectively, after 24 h contact time. The actual doses applied to the samples are listed in Table III. Low pressure (LPUV) and medium pressure (MPUV) UV radiation were both used in this study. The AOP treatment consisted of a combination of MPUV and H_2O_2. The dosages of O_3, UV and AOP disinfectants were the same for all source waters (Table IV). The chlorine dose used in the combination treatments including O_3/Cl_2, LPUV/Cl_2, MPUV/Cl_2 and AOP/Cl_2 experiments was the same as that used in the Cl_2 treatment for each location (see Cl_2, Table III).

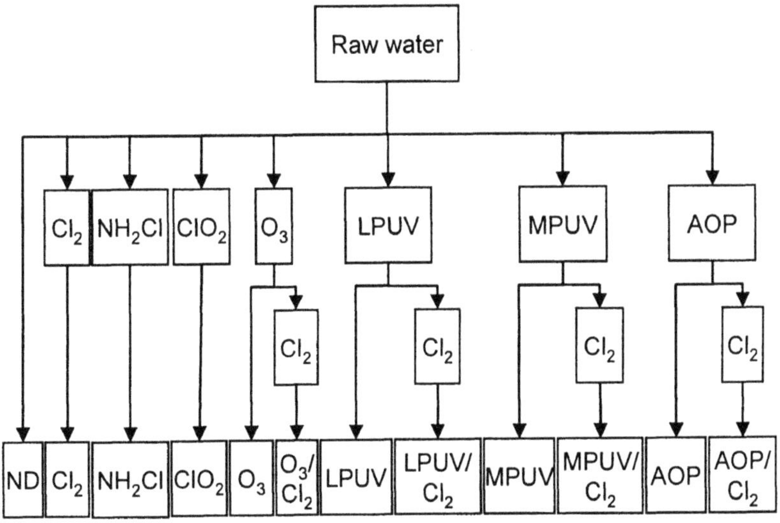

Scheme 1. The eleven different disinfection processes used in this study. ND = Not disinfected; (Reproduced from reference 31.)

Disinfected water samples were analyzed for the concentrations of the nine target nitrosamines (Table I) using the SPE-HPLC-MS/MS method. The results from locations 3, 6 and 7 are shown in Figure 4. Nitrosamines were detected in treated water samples from locations 2, 3, 4, 5, 6 and 7. Five nitrosamines including NMor, NDEA, NMEA and NDPhA along with NDMA were detected in some of the treated water samples. NDMA was the most frequently detected (6 of 7 locations) and its concentration was the highest of all the nitrosamines detected. In locations 2, 4 and 5 only NDMA was detected in the samples

Table III. Doses of Cl_2, NH_2Cl and ClO_2 applied to water samples

Location	*Disinfectant Dose**		
	Cl_2 ($mg.L^{-1}$)	*NH_2Cl ($mg.L^{-1}$)*	*ClO_2 ($mg.L^{-1}$)*
1	2	3.5	2.5
2	12	7.5	8.5
3	9.5	3	8
4	6	4.5	7
5	8	3	6.5
6	25	6	11
7	33	10	20

NOTE: Doses based on obtaining a disinfectant residual of $0.5 mg.L^{-1}$ for Cl_2, $2.0 mg.L^{-1}$ for NH_2Cl and $0.5 mg.L^{-1}$ for ClO_2 after 24 h contact time. SOURCE: Table adapted from reference 31

following disinfection treatment (data not shown). NDMA was also detected in several source waters [non disinfected (ND) samples] from locations 2, 3, 4, 6, and 7, indicating that its presence is not completely due to disinfection. NMor was detected in the treated samples when source water from location 6 was treated with Cl_2 or O_3 or O_3/Cl_2 (Figure 4b). O_3 and O_3/Cl_2 treatment produced more NMor than Cl_2 treatment. NDPhA was detected in the treated samples when source water from location 3 was treated with Cl_2 or NH_2Cl, and when source water from location 7 was treated with O_3 or MPUV/Cl_2 (Figure 4a and c). This confirms our previous finding that NDPhA can be produced from water disinfection. NMEA and NDEA were detected in some treated samples but were below the limit of quantification (NMEA – 8.1 $ng.L^{-1}$; NDEA – 32 $ng.L^{-1}$). NMEA was produced after source water from location 7 was treated with Cl_2, while NDEA was formed after the same source water was treated with AOP/Cl_2, or LPUV/Cl_2. These results suggest that other nitrosamines in addition to NDMA are produced depending on the quality of source water and disinfectants used. Formation of specific nitrosamines during different disinfection processes is summarized as follows.

Chloramination has been suspected to produce more NDMA than chlorination (17). Our results indicate that nitrosamine formation depends both on the disinfection method and source water quality. Both chlorination and chloramination produced a significant amount of NDMA compared to the controls in four locations (Figure 4 and data not shown). In location 6, chloramination produced the highest concentration of NDMA. Location 6 had the second highest values of TOC, UV A254, and colour among the seven locations, indicating that various organic components in the source water from location 6 may preferentially form nitrosamines with chloramine. However,

Table IV. Doses of O_3, UV and AOP applied to water samples

Disinfectant	*Dose*
O_3	10 mg·min/L
LPUV	100 mJ/cm^2
MPUV	1000 mJ/cm^2
AOP (MPUV/H_2O_2)	1000 mJ/cm^2, 10 mg/L H_2O_2

NOTE: The same chlorine dose was used for O_3, UV and AOP treatments combined with chlorine as for chlorine alone treatments as indicated in Table III for each water source

SOURCE: Table adapted from reference 31

chloramine disinfection did not produce as high level of NDMA as that with Cl_2 in location 7 (Figure 4c), although the source water from location 7 had higher values of TOC, UV A254 and color than those from location 6. These results suggest that the composition of organic precursors in the source waters from locations 6 and 7 may be different and nitrosamine formation is not simply based on TOC or UV A254. We also found that **ClO_2** treatment produced a significant amount of NDMA compared to the controls in location 3 (Figure 4a) and in location 5 (data not shown), which is consistent with previous study (19).

O_3 treatment produced significant concentrations of NDMA (compared to the control) only in location 2 (data not shown), while the combination of O_3 with Cl_2 treatment produced significant concentrations of NDMA in locations 2, 5 and 6. This indicates that the addition of Cl_2 after O_3 treatment was important in NDMA formation in locations 5 and 6. In location 6, O_3 and O_3/Cl_2 treatment also produced NMor (Figure 4b). This was the only time that significant amount of NMor was detected in this study. Our results here indicate that ozone treatment can produce NMor depending on the source water composition.

UV disinfection has previously been described as a potentially useful technique for reducing nitrosamine concentrations in drinking water (32-33). Our study found that in some source waters, UV light was able to reduce nitrosamine concentration. However this effect was negated when followed by chlorine disinfection (eg. MPUV vs. MPUV/Cl_2 in Location 7, Figure 4c), indicating that UV can destroy nitrosamines but not precursors of nitrosamines. Generally, UV light did not decrease nitrosamine concentration in source water with high turbidity and color.

AOP has been suggested to be effective for reduction of NDMA in water with low concentrations of natural organic matter (34). In the present study, AOP treatment consisted of a combination of MPUV and H_2O_2. AOP treatment produced a statistically significant concentration of NDMA in source water from locations 2 (data not shown) and 7 (Figure 4c). In addition, the combination of AOP and Cl_2 (AOP/Cl_2) produced a statistically significant concentration of

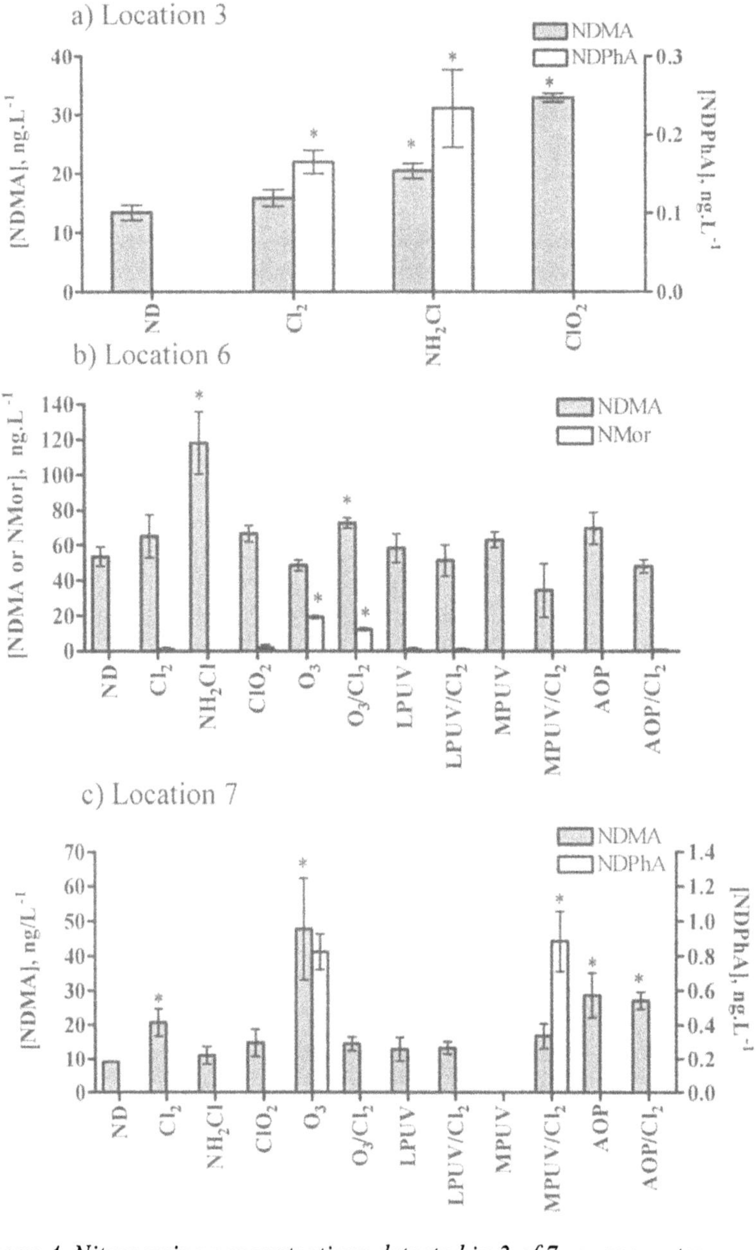

Figure 4. Nitrosamine concentrations detected in 3 of 7 source water samples treated with eleven disinfection methods. NDMA concentrations were calculated using both NDPA-d14 for quantification and NDMA-d6 for extraction recovery. Error bars indicate standard error. Asterisks (*) indicate the statistical significance compared to the control (ND) based on a two sided two sample t-test at 95 % confidence. (Adapted from reference 31.)

NDMA in locations 2, 5 and 7. In all three locations AOP produced more NDMA than was seen with MPUV alone, suggesting that H_2O_2 may be responsible for the NDMA formation.

In summary, an HPLC-MS/MS method was developed for analysis of thermally stable and unstable nitrosamines in drinking water at $ng.L^{-1}$ concentrations. This method was able to detect NDPhA, a thermally unstable nitrosamine, in the treated samples of source waters from two different locations, confirming the formation of NDPhA from water disinfection. NMor, NMEA, and NDEA were also detected in some treated water samples. This study demonstrated that source water quality, types of disinfectants, and disinfection processes strongly influenced nitrosamine formation in treated water. Chemical disinfectants such as Cl_2, NH_2Cl and ClO_2 are able to produce different nitrosamines depending on the source water quality. UV treatment may be effective for reducing/eliminating nitrosamines in water of low turbidity or color. Chlorination after UV may result in nitrosamine formation, limiting the effectiveness of UV for removing nitrosamines. H_2O_2 may contribute to nitrosamine formation, limiting the use of AOP for removing nitrosamines. Assessment of source water quality and disinfection process is important to eliminate/reduce nitrosamines in drinking water.

References

1. International Agency for Research on Cancer. IARC monographs on the evaluation of the carcinogenic risk of chemicals to humans: Some N-nitroso compounds. **1978**, *17*, 365.
2. Taguchi, V.; Jenkins, S. D. W.; Wang, D. T.; Palmentier, J. P. F. P.; Reiner, E. J. Determination of N-nitrosodimethylamine by isotope-dilution, high-resolution mass-spectrometry. *Can. J. Appl. Spectrosc.* **1994**, *39*, 87-93.
3. California Department of Health Services. A brief history of NDMA findings in drinking water supplies and related activities. http://www.dhs.ca.gov/ps/ddwem/chemicals/NDMA/history.htm (accessed October 15, 2005).
4. Government of Ontario. Safe Drinking Water Act. **2002**, *Ontario Regulation 169/03*, Schedule 2.
5. California Department of Health Services. California Drinking Water: Activities related to NDMA and other Nitrosamines. http://www.dhs.ca.gov/ps/ddwem/chemicals/ndma/NDMAindex.htm (accessed April 27, 2006).
6. US EPA. Unregulated contaminant monitoring regulation (UCMR) for public water systems revisions. *Fed. Regist.* August 22, 2005, *70*.
7. Munch, J. W.; Bassett, M. V. *Method 521: Determination of nitrosamines in drinking water by solid phase extraction and capillary column gas*

chromatography with large volume injection and chemical ionization tandem mass spectrometry (MS/MS); National Exposure Research Laboratory, USEPA, 2004.

8. Charrois, J. W. A.; Arend, M. W.; Froese, K. L.; Hrudey, S. E. Detecting N-nitrosamines in drinking water at nanogram per liter levels using ammonia positive chemical ionization. *Environ. Sci. Technol.* **2004**, *38*, 4835-4841.
9. Wilczak, A.; Assadi-Rad, A.; Lai, H. H.; Hoover, L. L. *et al.* Formation of NDMA in chloraminated water coagulated with DADMAC cationic polymer. *J. Am. Water Work Assoc.* **2003**, *95*, 94-106.
10. Mitch, W. A.; Gerecke, A. C.; Sedlak, D. L. A N-nitrosodimethylamine (NDMA) precursor analysis for chlorination of water and wastewater. *Water Res.* **2003**, *37*, 3733-3741.
11. Charrois, J. W. A.; Hrudey, S. E. Breakpoint chlorination and free-chlorine contact time: Implications for drinking water N-nitrosodimethylamine concentrations. *Water Res.* **2007**, *41*, 674-682.
12. Chen, Z.; Valentine, R. L. Modeling the formation of N-nitrosodimethylamine (NDMA) from the reaction of natural organic matter (NOM) with monochloramine. *Environ. Sci. Technol.* **2006**, *40*, 7290-7297.
13. Choi, J.; Duirk, S. E.; Valentine, R. L. Mechanistic studies of N-nitrosodimethylamine (NDMA) formation in chlorinated drinking water. *J. Environ. Monit.* **2002**, *4*, 249-252.
14. Choi, J.; Valentine, R. L. *Studies on the formation of N-nitrosodimethylamine (NDMA) in drinking water: A new chloramination disinfection by-product.* Proceedings of the AWWA Annual Conference, 2001.
15. Choi, J. H.; Valentine, R. L. N-nitrosodimethylamine formation by free-chlorine-enhanced nitrosation of dimethylamine. *Environ. Sci. Technol.* **2003**, *37*, 4871-4876.
16. Mitch, W. A.; Sedlak, D. L. Formation of N-nitrosodimethylamine (NDMA) from dimethylamine during chlorination. *Environ. Sci. Technol.* **2002**, *36*, 588-595.
17. Najm, I.; Trussell, R. R. NDMA formation in water and wastewater. *J. Am. Water Work Assoc.* **2001**, *93*, 92-99.
18. Schreiber, I. M.; Mitch, W. A. Nitrosamine formation pathway revisited: The importance of chloramine speciation and dissolved oxygen. *Environ. Sci. Technol.* **2006**, *40*, 6007-6014.
19. Andrzejewski, P.; Kasprzyk-Hordern, B.; Nawrocki, J. The hazard of N-nitrosodimethylamine (NDMA) formation during water disinfection with strong oxidants. *Desalination* **2005**, *176*, 37-45.
20. Lee, C.; Schmidt, C.; Yoon, J.; von Gunten, U. Oxidation of N-nitrosodimethylamine (NDMA) precursors with ozone and chlorine dioxide: Kinetics and effect on NDMA formation potential. *Environ. Sci. Technol.* **2007**, *41*, 2056-2063.

21. Lijinsky, W. N-Nitroso compounds in the diet. *Mutat. Res.* **1999**, *443*, 129-138.
22. Shank, R. C. Occurrence of N-nitroso compounds in the environment; In *Mycotoxins and N-nitroso compounds: Environmental risks;* Shank, R. C., Ed.; CRC Press, Inc. Boca Raton, FL, 1981; pp 155-183.
23. Osterdahl, B. G. N-nitrosamines and nitrosatable compounds in rubber nipples and pacifiers. *Food Chem. Toxicol.* **1983**, *21*, 755-757.
24. Billedeau, S. M.; Miller, B. J.; Thompson, H. C. N-nitrosamine analysis in beer using thermal-desorption injection coupled with GC-TEA. *J. Food Sci.* **1988**, *53*, 1696.
25. Yurchenko, S.; Molder, U. Volatile N-nitrosamines in various fish products. *Food Chem.* **2006**, *96*, 325-333.
26. Lee, H. L.; Wang, C. Y.; Lin, S.; Hsieh, D. P. H. Liquid chromatography/tandem mass spectrometric method for the simultaneous determination of tobacco-specific nitrosamine NNK and its five metabolites. *Talanta* **2007**, *73*, 76-80.
27. Zhao, Y. Y.; Boyd, J.; Hrudey, S. E.; Li, X. F. Characterization of new nitrosamines in drinking water using liquid chromatography tandem mass spectrometry. *Environ. Sci. Technol.* **2006**, *40*, 7636-7641.
28. Charrois, J. W. A.; Boyd, J. M.; Froese, K. L.; Hrudey, S. E. Occurrence of *N*-nitrosamines in Alberta public drinking-water distribution systems. *J. Environ. Eng. Sci* **2007**, *6*, 103-114.
29. Barrett, S.; Hwang, C.; Guo, Y.; Andrews, S. A.; Valentine, R. L. *Occurrence of NDMA in drinking water: a North American Survey, 2001-2002.* Proceedings of the AWWA Annual Conference, 2003.
30. Krasner, S. W.; Weinberg, H. S.; Richardson, S. D.; Pastor, S. J.; Chinn, R.; Sclimenti, M. J.; Onstad, G. D.; Thruston, A. D. Occurrence of a new generation of disinfection byproducts. *Environ. Sci. Technol.* **2006**, *40*, 7175-7185.
31. Zhao, Y. Y.; Boyd, J. M.; Woodbeck, M.; Andrews, R.; Qin, F.; Hrudey, S.; Li, X-F. Formation of N-nitrosamines during treatments of surface waters using eleven different disinfection methods. *Unpublished*.
32. Lee, C.; Choi, W.; Kim, Y. G.; Yoon, J. UV photolytic mechanism of N-nitrosodimethylamine in water: Dual pathways to methylamine versus dimethylamine. *Environ. Sci. Technol.* **2005**, *39*, 2101-2106.
33. Stefan, M. I.; Bolton, J. R. UV direct photolysis of N-nitrosodimethylamine (NDMA): Kinetic and product study. *Helv. Chim. Acta* **2002**, *85*, 1416-1426.
34. Landsman, N. A.; Swancutt, K. L.; Bradford, C. N.; Cox, C. R.; Kiddle, J. J.; Mezyk, S. P. Free radical chemistry of advanced oxidation process removal of nitrosamines in water. *Environ. Sci. Technol.* **2007**, *41*, 5818-5823.

Chapter 21

Formation of Nitrosamines in Effluent-Impacted Drinking Waters

Stuart W. Krasner, Michael J. Sclimenti, Chih Fen Tiffany Lee, and Jessica Schramm

Water Quality Standards Branch, Metropolitan Water District of Southern California, 700 Moreno Avenue, La Verne, CA 91750–3303

Treated wastewater discharges are a source of precursors for nitrosamines (e.g., N-nitrosodimethylamine [NDMA]). Drinking water treatment plants on effluent-impacted rivers or lakes may form NDMA if they chloraminate their water. The amount of treated wastewater in the drinking water supply can affect the level of NDMA that can form. However, prechlorination with a sufficient free-chlorine contact time can destroy or transform NDMA precursors, resulting in less NDMA upon chloramination.

Introduction

In recent years, greater portions of treated wastewater are contributing more toward the drinking water supplies through reclamation, recycling, and reuse (intentional and incidental) processes. Many rivers, lakes, and groundwaters that supply water to drinking water treatment plants (DWTPs) contain discharges from upstream wastewater treatment plants (WWTPs). It is not uncommon to have a significant portion of the source water for these DWTPs originally derived from the upstream wastewater contribution. Effluent organic matter (EfOM) from WWTPs is a source of disinfection by-products (DBPs), if chlorine disinfection is practiced, and DBP precursors (*1*).

Nitrosamines (e.g, N-nitrosodimethylamine [NDMA]) are found in treated wastewater and drinking water (*2*). NDMA is a by-product of the disinfection of some natural waters and wastewaters with combined chlorine (*3-4*). Although there is no U.S. regulation for any of the nitrosamines, the California Department of Health Services has notification levels (at 10 ng/L each) for the following species: NDMA, N-nitrosodiethylamine (NDEA), and N-nitrosodipropylamine (NDPA).

In addition to the presence of NDMA in chloraminated wastewaters, EfOM is also a source of NDMA precursors. On a central tendency basis, achieving some nitrification at WWTPs resulted in reducing the level of NDMA precursors (*1*). Nitrification reduced the level of hydrophobic and transphilic natural organic matter (NOM) to some extent, but really reduced the level of hydrophilic NOM considerably (*1*). These results suggest that the NDMA precursors in EfOM may have been primarily in the hydrophilic NOM fraction.

When a DWTP uses chloramines to control the formation of regulated DBPs (e.g., trihalomethanes [THMs]), there may be a trade-off issue with the formation of NDMA. Thus, the use of chloramines might be precluded in waters that are high in NDMA precursors—e.g., in source waters heavily impacted by EfOM—because of the formation of too high of a level of NDMA. Because only a percentage of NDMA precursors will be converted to NDMA during full-scale chloramination practices, it is unknown under what conditions too much NDMA (e.g., >10 ng/L) might be formed in effluent-impacted rivers. Therefore, research was conducted with a range of EfOM/river mixtures under drinking water chloramination conditions to access vulnerability to NDMA formation.

Experimental Methods

Eight nitrosamines were measured: NDMA, N-nitrosomethylethylamine (NMEA), NDEA, N-nitrosomorpholine (NMOR), N-nitrosopyrrolidine (NPYR), NDPA, N-nitrosopiperidine (NPIP), and N-nitrosodibutylamine (NDBA). Nitrosamine samples were concentrated using solid-phase extraction with Ambersorb (*5*) and were analyzed using chemical ionization gas chromatography (GC)/mass spectrometry (*6*). Selected samples were also analyzed for THMs. The THMs were measured using a salted liquid/liquid extraction and GC/electron-capture detection method (*7*).

Two EfOM samples were collected, one that was well nitrified and one that was poorly nitrified. Nitrification transforms ammonia and organic-nitrogen to nitrate. In addition, a river sample upstream of any WWTPs was collected for the "background" water quality. Mixtures of these samples were prepared to simulate a range of effluent-impacted waters (e.g., 25, 50, and 75 percent).

Each EfOM/river mixture was chloraminated under DWTP chloramination conditions. Typically, most DWTPs that use chloramines will first use free chlorine for primary disinfection. In the case in which there is a significant amount of ammonia in the plant influent, many DWTPs use breakpoint chlorination to achieve a free-chlorine residual. In this testing, the samples were prechlorinated to break out all of the ammonia and have a free-chlorine residual (e.g., ~2-5 mg/L) for the formation of chloramines during post-ammonia addition. In addition, the impact of free-chlorine contact time (e.g., 5, 20, and 60 min) was evaluated. Chloramines, with a chlorine/ammonia-nitrogen (Cl_2/NH_3-N) ratio of 3:1 or 4:1 (on a weight basis), were set up. In parallel, selected samples were chloraminated with no prechlorination step (where ammonia was added first). The final chloraminated samples were held for 2 or 3 days (at a pH ~8 and at 25°C).

Results

Table I shows the water quality (i.e., total organic carbon [TOC], ultraviolet absorbance [UVA], ammonia-nitrogen [NH_3-N]) of EfOM and river samples. For some experiments with the poorly nitrified EfOM, there was insufficient river water remaining. For those tests, surface water collected at the Weymouth DWTP influent (La Verne, Calif.) was used. That water was a mixture of California State Project water and Colorado River water. Subsequently, fresh samples of river water and EfOM were collected and used (Table II).

Table I. Water Quality of Samples Initially Used

Water Source	*TOC (mg/L)*	*UVA (cm^{-1})*	*NH_3-N (mg/L)*
River	3.0	0.057	0.03
Well nitrified EfOM	9.3	0.160	0.08
Poorly nitrified EfOM	12.2	0.181	~17

SOURCE: Reproduced with permission from reference 1. Copyright 2008 Awwa Research Foundation.

Table II. Water Quality of Samples Subsequently Used

Water Source	*TOC (mg/L)*	*UVA (cm^{-1})*	*NH_3-N (mg/L)*	*Bromide (mg/L)*
Weymouth DWTP influent	~3	--	~0	--
River	3.1	0.052	0.12	0.08
Well nitrified EfOM	9.8	0.146	0.04	0.20
Poorly nitrified EfOM	14.4	0.177	14	0.22

SOURCE: Reproduced with permission from reference 1. Copyright 2008 Awwa Research Foundation.

Impact of Free-Chlorine Contact Time

Figure 1 shows the impact of free-chlorine contact time (before ammonia was added to form chloramines) on NDMA formation during post-chloramination in a mixture of 75-percent river water and 25-percent well nitrified EfOM (water from Table I). The well nitrified EfOM had a background level of NDMA (i.e., 12.9 ng/L), which corresponded to 3.2 ng/L when the EfOM was diluted fourfold. As part of the post-chloramination, the chlorine dose ranged from 2 to 7 mg/L. In these tests, a Cl_2/N (weight) ratio of 4:1 was used and the samples were held for 2 days. The following observations can be made:

- Three tests with 1-hr prechlorination gave the same NDMA formation (7.0–8.6 ng/L). Of this, ~4–5 ng/L of NDMA was formed over the background amount.
- One test with 20-min prechlorination gave the same NDMA formation (7.5 ng/L) as the 1-hr prechlorination tests.
- Three tests with 5-min prechlorination gave highly variable, but high NDMA formation (14–86 ng/L).
- Three tests with pre-ammoniation gave highly variable and even higher NDMA formation (52–252 ng/L).

According to Schreiber and Mitch (*8*), the order of chlorine and ammonia addition to form chloramines can highly affect NDMA formation. According to their research, in which they conducted experiments with (1) dimethylamine in distilled water or (2) secondary municipal wastewater effluent, dichloramine forms much more NDMA than monochloramine. When chlorine is added first and ammonia is added second, one should get monochloramine and some NDMA formation (which is what happened in the 20-min and 1-hr prechlorination tests). When ammonia is added first and chlorine is added second, one can get a localized region in which dichloramine forms, and more NDMA is produced. Not only did the latter happen in the pre-ammoniation tests, it appeared to have happened in some of the 5-min prechlorination tests (perhaps due to inadequate mixing). This explains some of the variability in the latter tests.

According to Charrois and Hrudey (*9*) or Chen and Valentine (*10*), prechlorination can significantly reduce NDMA formation during subsequent chloramination. This may be due to the destruction or transformation of NDMA precursors by the free chlorine. This explains (in part) why NDMA formation was relatively low, compared to its pre-ammoniation, in the tests with a sufficient amount of free-chlorine contact time. However, with longer free-chlorine contact times, more THM formation can occur. For example, in 1–hr

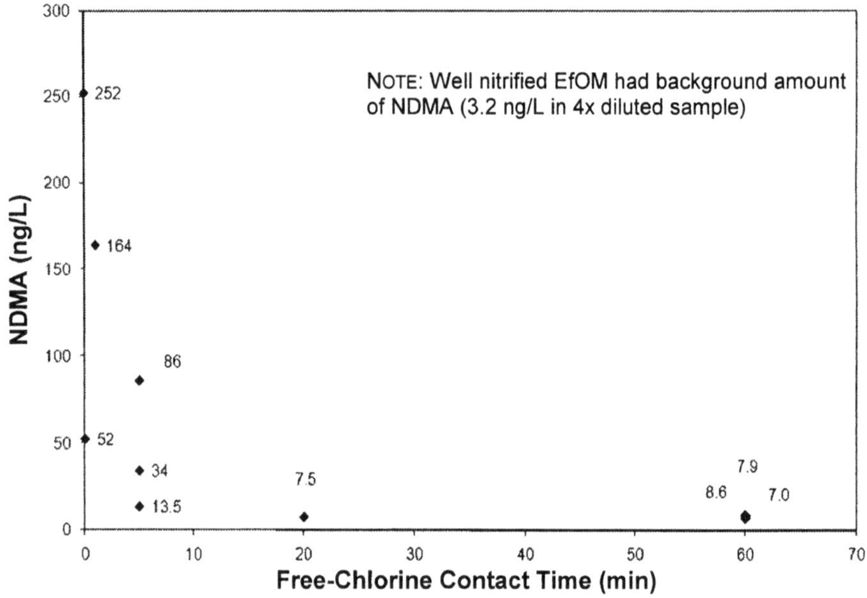

Figure 1. Impact of free-chlorine contact time (before ammonia was added to form chloramines) on NDMA formation during post-chloramination in mixture of 75-percent river water and 25-percent well nitrified EfOM. (Reproduced with permission from reference 1. Copyright 2008 Awwa Research Foundation.)

prechlorination/post-chloramination tests, THM formations for the 75/25 blend and a 50/50 mixture of river water and well nitrified EfOM were 78 and 99 μg/L, respectively. In order to reliably comply (e.g., 20-percent safety factor) with the U.S. THM maximumum contaminant level (MCL) of 80 μg/L, a DWTP would want to form ≤64 μg/L of THMs. Thus, a balance between halogenated DBP and NDMA formation may be reached by selecting an appropriate free-chlorine contact time. In addition, a certain free-chlorine contact time is normally required to meet disinfection requirements.

Standardization of the Nitrosamine Formation Test

A mixture of 75-percent river water and 25-percent well nitrified EfOM (water from Table I) was evaluated in two sets of tests, one with a 5-min and the other with a 20-min prechlorination time. Both sets were run with five replicates, and care was taken to ensure good mixing in each sample. The only other difference in this set of tests was the Cl_2/N (weight) ratio was set up at 3:1, which should favor monochloramine formation.

For the 20-min prechlorination tests, NDMA formation ranged from 13 to 18 ng/L, with a relative standard deviation of 16 percent. For the 5-min prechlorination tests, NDMA formation ranged from 51 to 58 ng/L for four of the tests and was 86 ng/L for the fifth test. The overall relative standard deviation was 23 percent, but was only 5 percent when the outlier was excluded. Although these results were not perfect, they were a significant improvement over the previous set of tests (Figure 1) in which the relative standard deviation for the three tests conducted with a 5-min free-chlorine contact time was 84 percent. Nonetheless, these limited data suggested that a short time for prechlorination may result in a significant amount of NDMA formation.

Breakpoint Chlorination of Poorly Nitrified EfOM

Figure 2 shows the chlorine demand during breakpoint chlorination of different blends of Weymouth DWTP influent and the poorly nitrified EfOM. The goal was to achieve breakpoint chlorination and to end up with several mg/L of free-chlorine residual after a 2- or 20-min contact time. For example, a mixture of 75-percent Weymouth DWTP influent and 25-percent poorly nitrified EfOM had an ammonia concentration of ~4.3 mg/L. Because it takes 7.6 mg/L of chlorine to break out every 1.0 mg/L of ammonia, it should have taken ~32 mg/L of chlorine to break out the ammonia in this mixture. A chlorine dose of 45 mg/L achieved a free-chlorine residual of 3.7 and 2.2 mg/L after 2 and 20 min, respectively. The chlorine demand (41–43 mg/L) was somewhat higher than the theoretical inorganic (ammonia) demand. In actual WWTP practice, a higher chlorine dose (e.g., 10 mg/L for each 1.0 mg/L of NH_3-N) is required (*11*), which matched the chlorine demand in this test (i.e., 10 x 4.3 = 43 mg/L).

Ammonia was added (at a Cl_2/N weight ratio of 3:1) to the chlorine residual (i.e., 2–11 mg/L) that remained after breakpoint chlorination. For this set of tests, the chloraminated water was held for ~1 day. The tests with a 50/50 blend of Weymouth DWTP influent and poorly nitrified EfOM were run in triplicate. The poorly nitrifried EfOM had a background level of NDMA (i.e., 27 ng/L) and NMOR (i.e., 18 ng/L). Figure 3 shows that the presence of NMOR in each sample was due to the background amount and was not due to any formation during the chlorination/chloramination. Although not shown, the same was found for mixtures of river water and well nitrified EfOM. In this analysis, it was presumed that there was little to no loss of the nitrosamines (e.g., due to photolysis) in a river downstream of the WWTP. However, that may not be the case, depending on the river depth and flow (*12*).

Alternatively, for NDMA, there was formation during chloramination over what was in the background EfOM sample. Although there was some variability in the triplicate samples for the 50/50 blend (Figure 4, top figure), there was an overall trend when the average values were used for those set of samples

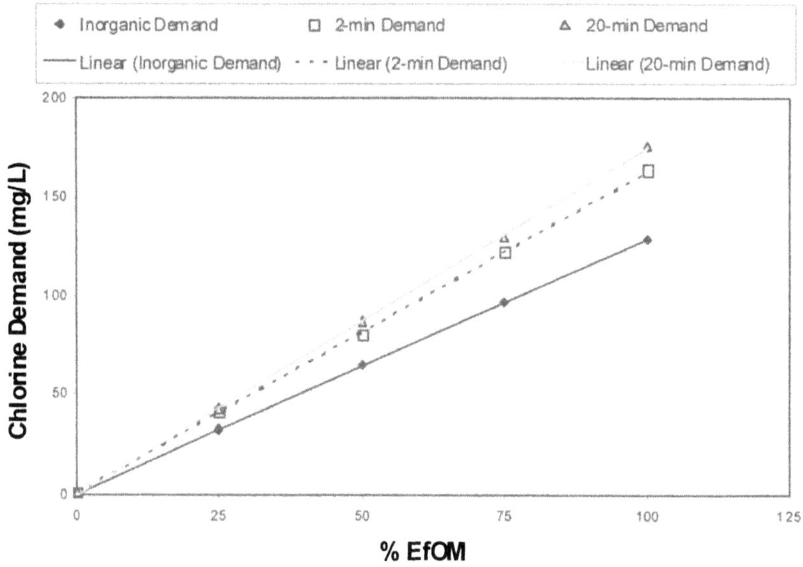

Figure 2. Chlorine demand during breakpoint chlorination of blends of Weymouth DWTP influent and poorly nitrified EfOM. (Reproduced with permission from reference 1. Copyright 2008 Awwa Research Foundation.)

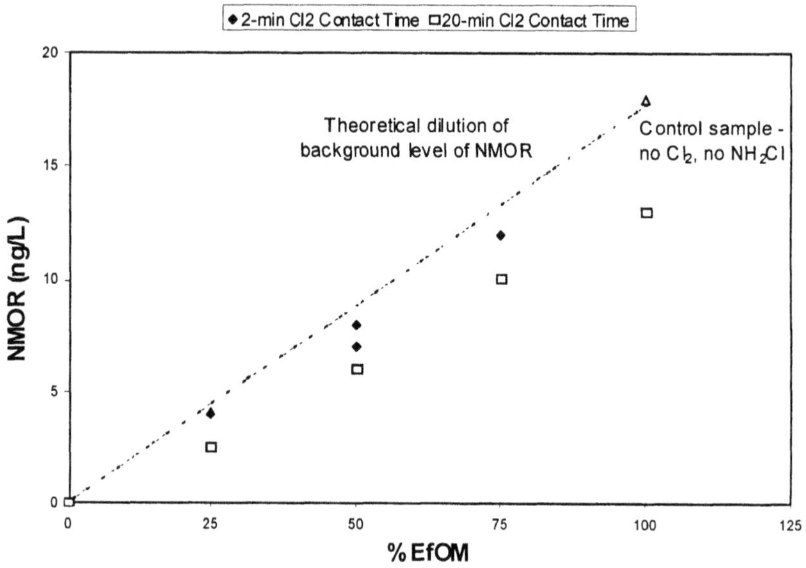

Figure 3. Occurrence of NMOR in blends of Weymouth DWTP influent and poorly nitrified EfOM after prechlorination/post-chloramination. (Reproduced with permission from reference 1. Copyright 2008 Awwa Research Foundation.)

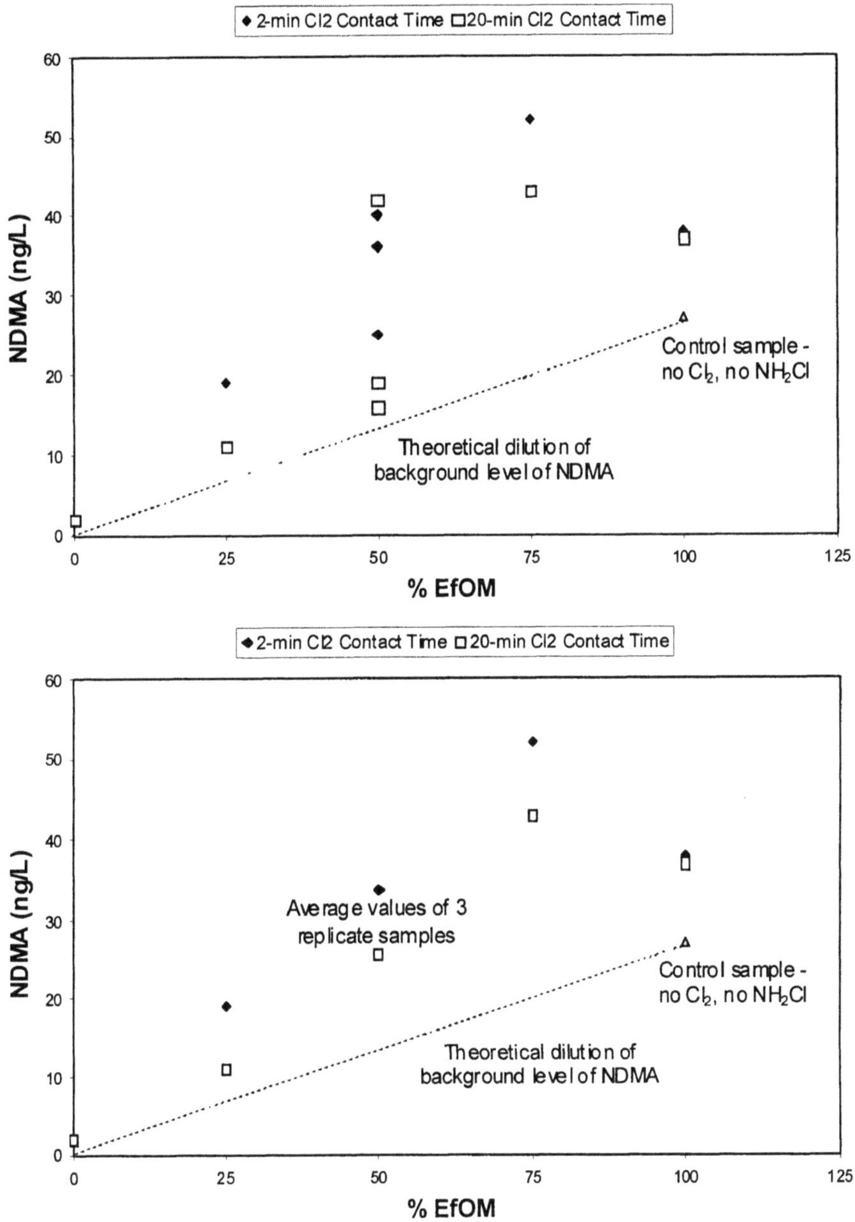

Figure 4. Impact of free-chlorine contact time on NDMA formation during post-chloramination: evaluation of blends of Weymouth DWTP influent and poorly nitrified EfOM (top figure: all data; bottom figure: average values of replicates). (Reproduced with permission from reference 1. Copyright 2008 Awwa Research Foundation.)

(Figure 4, bottom figure). A "delta" NDMA set of values was calculated, which represented the formation of NDMA over the background level of NDMA in the EfOM sample (Figure 5). Not including the 100-percent EfOM, there was a central tendency trend in delta NDMA formation going up with increasing EfOM blends. Moreover, the central tendency was for there to be less NDMA formation when a longer free-chlorine contact time was used.

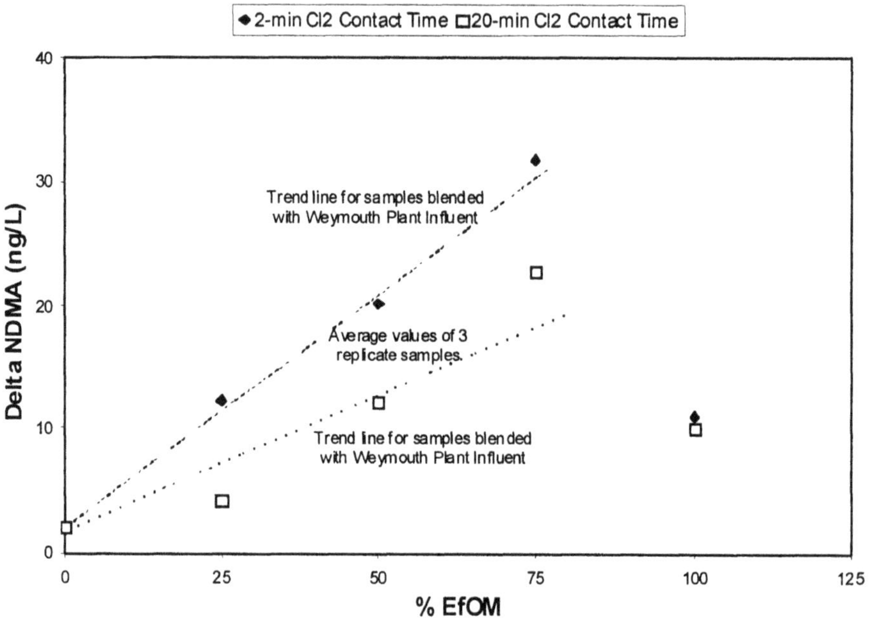

Figure 5. NDMA formation during post-chloramination over background amount of NDMA: evaluation of blends of Weymouth DWTP influent and poorly nitrified EfOM. (Reproduced from with permission reference 1. Copyright 2008 Awwa Research Foundation.)

Figure 6 shows the impact of free-chlorine contact time on NDMA formation during post-chloramination versus the formation of THMs (during chlorination/chloramination) for blends of river water and poorly nitrified EfOM. In these tests, post-chloramination was conducted at a Cl_2/N weight ratio of 3:1, and the samples were held for two days. When a free-chlorine contact time of 2 min was used, NDMA was not detected during post-chloramination in the river water and its formation increased from 16 to 315 ng/L with increasing EfOM content (i.e., from 25- to 100-percent EfOM). In this scenario, THM formation was below its MCL of 80 μg/L and was reliably below the MCL (with a 20-percent safety factor) for mixtures up to 75-percent EfOM. When a free-

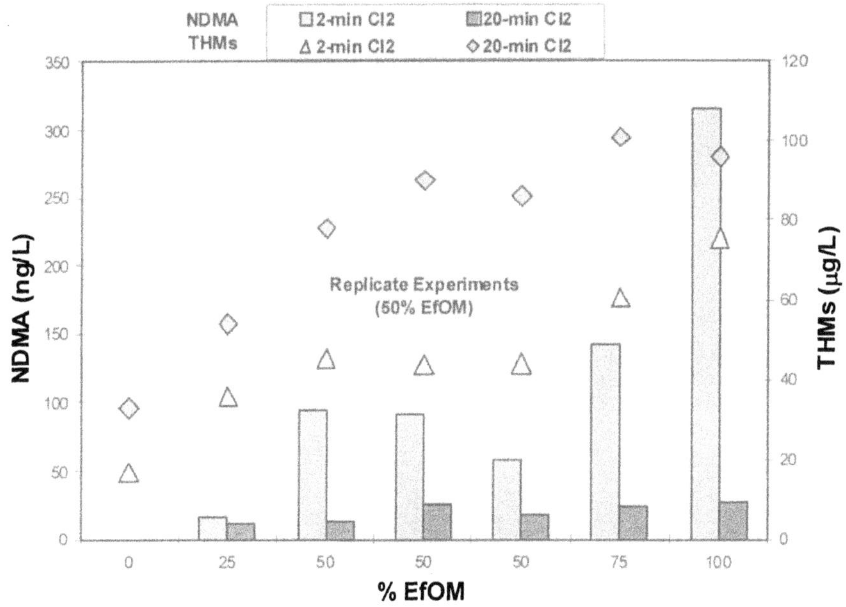

Figure 6. Impact of free-chlorine contact time on NDMA formation during post-chloramination versus the formation of THMs (during chlorination/ chloramination): evaluation of blends of river water and poorly nitrified EfOM. (Reproduced with permission from reference 1. Copyright 2008 Awwa Research Foundation.)

chlorine contact time of 20 min was used, NDMA was not detected during post-chloramination in the river water and its occurrence was 12 to 28 ng/L in river/EfOM mixtures. Note that 100-percent EfOM had a background level of 7 ng/L of NDMA. In this scenario, THM formation was often above its MCL (e.g., it was 78–101 µg/L for mixtures with 50- to 100-percent EfOM). However, in these tests, no coagulation was done, which would have removed a portion of the THM and NDMA precursors. Moreover, these tests were conducted at a summertime temperature. So these conditions represent a worst-case scenario. Nonetheless, these tests demonstrate that a balance needs to be achieved between the formation of regulated DBPs (e.g., THMs) and emerging DBPs of health concern (e.g., NDMA). In these tests, reducing the free-chlorine contact time from 20 to 2 min resulted in reducing THM formation by 21 to 51 percent (median = 40 percent). Alternatively, increasing the free-chlorine contact time from 2 to 20 min resulted in reducing NDMA formation during subsequent chloramination by 30 to 91 percent (median = 79 percent). A change in free-chlorine contact time had a greater impact on NDMA formation. Future research should evaluate intermediate free-chlorine contact times to find the most optimal conditions.

Figure 7 shows the impact of Cl_2/N ratio (i.e., 3:1, 4:1, 5:1 on a weight basis) on NDMA formation during post-chloramination (after a 2-min prechlorine contact time) for blends of river water and poorly nitrified EfOM. In general, there was no significant difference in NDMA formation with changes in this ratio.

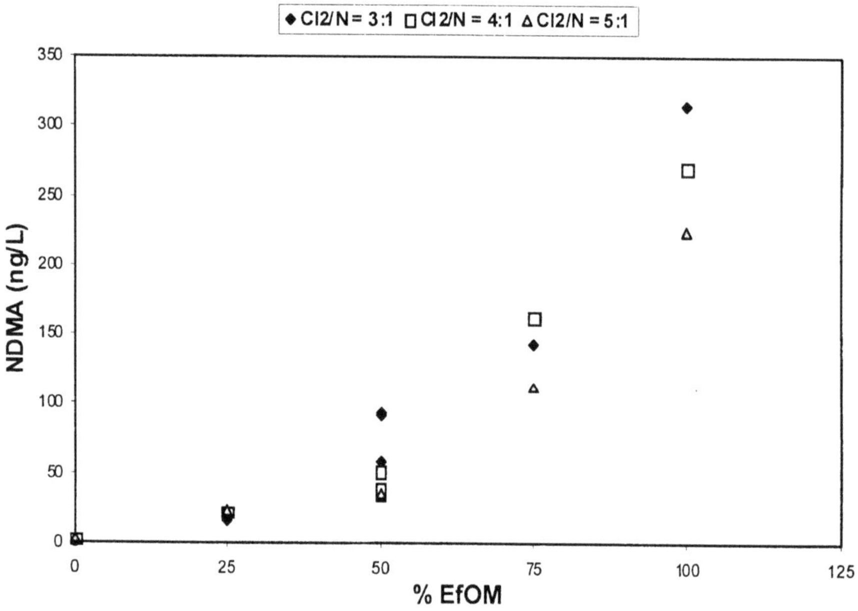

Figure 7. Impact of Cl_2/N ratio on NDMA formation during post-chloramination (after a 2-min prechlorine contact time): evaluation of blends of river water and poorly nitrified EfOM. (Reproduced with permission from reference 1. Copyright 2008 Awwa Research Foundation.)

Chlorination/Chloramination of Well Nitrified EfOM

Figure 8 shows the impact of free-chlorine contact time on NDMA formation during post-chloramination versus the formation of THMs (during chlorination/chloramination) for blends of river water and well nitrified EfOM. In these tests, post-chloramination was conducted at a Cl_2/N weight ratio of 3:1, and the samples were held for two days. When a free-chlorine contact time of 5 min was used, NDMA was <5 ng/L during post-chloramination in the river

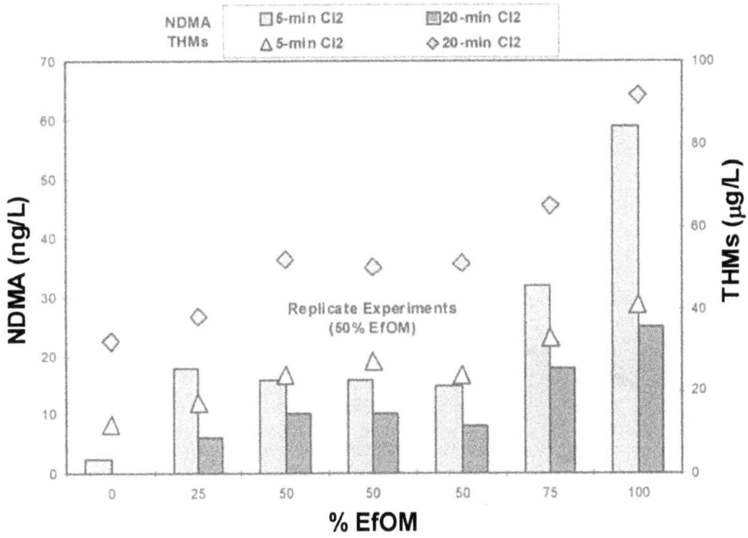

Figure 8. Impact of free-chlorine contact time on NDMA formation during post-chloramination versus the formation of THMs (during chlorination/ chloramination): evaluation of blends of river water and well nitrified EfOM (Reproduced with permission from reference 1. Copyright 2008 Awwa Research Foundation.)

water and its formation increased from 18 to 59 ng/L with increasing EfOM content. In this scenario, THM formation (i.e., 12–41 µg/L) was significantly below its MCL for all river/EfOM mixtures. When a free-chlorine contact time of 20 min was used, NDMA was not detected during post-chloramination in the river water and its occurrence was 6 to 25 ng/L in river/EfOM mixtures. Note that 100-percent EfOM had a background level of 10 ng/L of NDMA. In this scenario, THM formation was only above its MCL (i.e., it was 92 µg/L) for 100-percent EfOM (it was 32–65 µg/L for the river/EfOM mixtures). Figure 9 shows the impact of Cl_2/N ratio on NDMA formation during post-chloramination (after a 5-min prechlorine contact time) for blends of river water and well nitrified EfOM. In general, there was no significant difference in NDMA formation with changes in this ratio.

Conclusions

Although limited, these data suggest that NDMA and its precursors in a WWTP discharge may be problematic for a downstream DWTP that uses

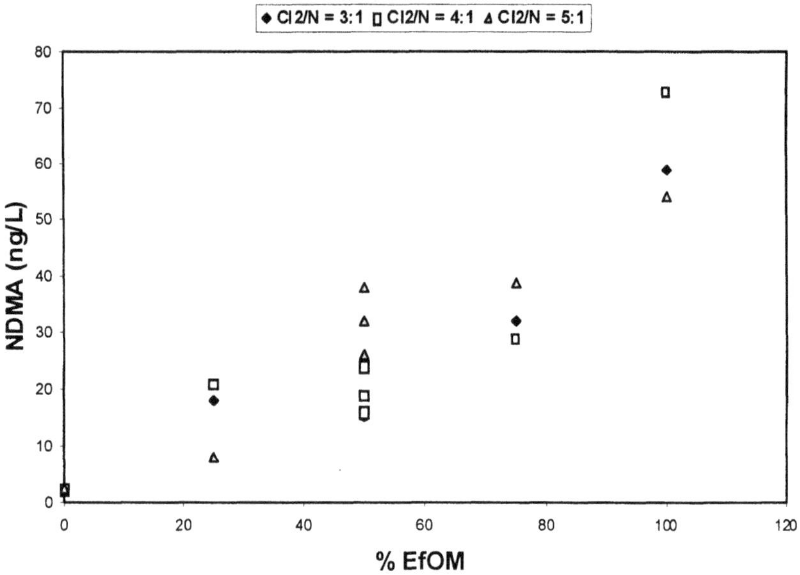

Figure 9. Impact of Cl_2/N ratio on NDMA formation during post-chloramination (after a 5-min prechlorine contact time): evaluation of blends of river water and well nitrified EfOM. (Reproduced from with permission reference 1. Copyright 2008 Awwa Research Foundation.)

chloramines. NDMA can be formed during chloramination at the DWTP; however, the amount formed depends on whether or not the water was prechlorinated and on the free-chlorine contact time. Prechlorination results in less formation of NDMA, but it also causes more formation of halogenated DBPs. Thus, a balance must be struck among the formation of halogenated DBPs, formation of NDMA, and adequate free-chlorine contact time to meet disinfection requirements. The choice of an optimal free-chlorine will need to be determined on a case-by-case basis (e.g., to meet regulatory requirements).

Because NDMA formation appears to be highly sensitive to the order of chlorine and ammonia addition and to mixing conditions, obtaining reproducible results in bench-scale tests may be challenging. Bench-scale tests may be useful for preliminary evaluations, but additional tests at pilot- and/or full-scale are probably required to optimize the treatment/disinfection process and to minimize NDMA formation. Despite some variability, these preliminary experiments demonstrate that the formation of NDMA—as well as background amount of NDMA (or NMOR)—can increase with increasing amounts of EfOM in the river. Therefore, a DWTP that uses a water supply that is highly impacted by EfOM may need additional treatment options (e.g., more precursor removal) in order to control the formation of NDMA.

Acknowledgments

The authors gratefully acknowledge that the Awwa Research Foundation is the joint owner of the technical information upon which this publication is based. The author thanks the Foundation and the U.S. government, through the Environmental Protection Agency (USEPA) for its financial, technical, and administrative assistance in funding and managing the project through which this information was discovered. The comments and views detailed herein may not necessarily reflect the views of the Awwa Research Foundation, its officers, directors, affiliates or agents, or the views of the U.S. Federal Government. The project manager is Alice Fulmer.

In addition, the participating utilities and agencies and their staff are acknowledged for their invaluable assistance and support.

References

1. Krasner, S. W.; Westerhoff, P.; Chen, B.; Amy, A.; Nam, S.-N.; Chowdhury, Z. K.; Sinha, S.; Rittmann, B. E. *Contribution of Wastewater to DBP Formation*; Awwa Research Foundation: Denver, Colo., 2008 (in press).
2. Mitch, W. A.; Sharp, J. O.; Trussell, R. R.; Valentine, R. L.; Alvarez-Cohen, L.; Sedlak, D. L. *N*-Nitrosodimethylamine as a drinking water contaminant: A review. *Environ. Eng. Sci.* **2003**, *20*(5), 389-404.
3. Choi, J. H.; Valentine, R. L. Formation of *N*-nitrosodimethylamine (NDMA) by reaction of monochloramine in a model water: A new disinfection by-product. *Wat. Res.*, **2002**, *36*(4), 817-824.
4. Mitch, W. A.; Sedlak, D. L. Formation of *N*-nitrosodimethylamine (NDMA) from dimethylamine during chlorination. *Environ. Sci. Technol.* **2002**, *36*(4), 588-595.
5. Taguchi, V.; Jenkins, S. D. W.; Wang, D. T.; Palmentier, J. P. F. P.; Reiner, E. J. Determination of *N*-nitrosodimethylamine by isotope dilution, high-resolution mass spectrometry. *Canadian Journal of Applied Spectroscopy* **1994**, *39*(3), 87-93.
6. Yoo, L. J.; Fitzsimmons, S.; Shen, Y. Determination of NDMA at part-per-trillion levels using positive chemical ionization from aqueous samples. In *Proc. of the 2000 Amer. Water Works Assoc. (AWWA) Water Quality Technology Conference*; AWWA: Denver, Colo., 2000.
7. Munch, D. J.; Hautman, D. P. Method 551.1. Determination of chlorination disinfection byproducts, chlorinated solvents, and halogenated pesticides/herbicides in drinking water by liquid-liquid extraction and gas chromatography with electron capture detection. In *Methods for the Determination of Organic Compounds in Drinking Water*; EPA-600/R-

95/131, Supplement III; USEPA Office of Research and Development, National Exposure Research Laboratory: Cincinnati, Ohio, 1995.

8. Schreiber, I. M.; Mitch, W. A. Influence of the order of reagent addition on NDMA formation during chloramination. *Environ. Sci. Technol.* **2005**, *39*(10), 3811-3818.
9. Charrois, J. W. A.; Hrudey, S. E. Breakpoint chlorination and free-chlorine contact time: Implications for drinking water *N*-nitrosodimethylamine concentrations. *Wat. Res.* **2007**, *41*(3), 672-682.
10. Chen, Z.; Valentine, R. L. The influence of pre-chlorination and pre-oxidation on the formation of N-nitrosodimethylamine (NDMA) from reaction of monochloramine with natural organic matter. **2007** (in press).
11. White, G. C. *Handbook of Chlorination and Alternative Disinfectants*, 4th ed.; John Wiley & Sons, Inc.: New York, NY, 1999.
12. Chen, B.; Westerhoff, P. K.; Krasner, S. W. Fate and transport of wasterwater-derived disinfection by-products in surface waters. In *Disinfection By-Products in Drinking Water: Occurrence, Formation, Health Effects, and Control;* Karanfil, T.; Krasner, S. W.; Westerhoff, P.; Xie, Y.; eds.; ACS Symposium Series 995. American Chemical Society: Washington, DC, 2008, Chapter 18 pp 257-273.

Chapter 22

Free Radical Chemistry of Advanced Oxidation Process Removal of Nitrosamines in Waters

Stephen P. Mezyk[1,*], Nicholas A. Landsman[1], Katy L. Swancutt[1], Christine N. Bradford[1], Casandra R. Cox[1], James J. Kiddle[2], Tim J. Clore[2]

[1]Department of Chemistry and Biochemistry, California State University at Long Beach, Long Beach, CA, 90840
[2]Department of Chemistry, Western Michigan University, Kalamazoo, MI, 49008

Carcinogenic nitrosamines, notably N-nitrosodimethylamine (NDMA), have been found in many water environments. Of major concern is the finding that NDMA can be formed under water disinfection conditions, from the reactions of chlorine or chloramines with dissolved dimethylamine, unsymmetrical 1,1-dimethylhydrazine or natural organic matter. The removal of nitrosamines from water by standard treatments is difficult, with neither activated carbon absoption or aeration stripping being efficient. To establish the feasibility of using free-radical-based advanced oxidation processes (AOPs) for the destructive removal of nitrosamines in waters, we have investigated their oxidative and reductive chemistry. Rate constants for the reactions of the hydroxyl radical and hydrated electron using electron pulse radiolysis, and removal efficiencies under continuous ^{60}Co irradiation, have been determined for some low-molecular weight alkyl/aryl substituted species. Mechanistic insight for the reaction of these two AOP radicals has been obtained from observed structure-reactivity relationships.

Introduction

Nitrosamines (R^1R^2N-NO) are mutagenic, teratogenic, and carcinogenic chemicals *(1-3)* that are ubiquitous by-products of various manufacturing, agricultural, and natural processes *(4)*. Nitrosoamines, particularly N-nitroso-dimethylamine (NDMA, $(CH_3)_2N$-NO) are present in foods and beverages that contain nitrite, or which have been exposed to nitrous oxides *(5-6)*. Recent studies have also shown that NDMA is formed under water disinfection conditions, from the reactions of monochloramine with dimethylamine *(7-8)*, unsymmetrical dimethylhydrazines *(8)* or natural organic matter *(9)*.

There are at least 300 known carcinogenic nitrosamines that have induced tumors in over thirty animal species *(10)*. The California Department of Health Services has presently set notification levels of 10 ng L^{-1} for NDMA in drinking water. To achieve such low nitrosamine concentrations is difficult; nitrosamines are not readily absorbed onto activated carbon, nor easily removed by air stripping. Mineral absorption using zeolites only removes trace amounts of nitrosamines *(11)*. Biodegradation is possible, with a reported degradation rate of 5 ng/mg-protein/min for *Pseudomonas mendocina (12)*.

Another approach is to use advanced oxidation processes (AOPs) to treat nitrosamine-contaminated waters. These technologies include ozone, UV/ozone, and UV/H_2O_2, which utilize oxidation via the hydroxyl radical ($^\bullet$OH), and heterogeneous catalysis by TiO_2, sonolysis, or the electron beam process, which produce a mixture of $^\bullet$OH radicals with reducing hydrated electrons (e^-_{aq}) and hydrogen atoms ($^\bullet$H). Currently the most widely used remediation method for NDMA is UV photolysis, but there are concerns of NDMA reformation after irradiation *(13)*. AOPs have been shown to be efficient at removing NDMA from water, and further reactions of hydroxyl radicals with degradation by-products (such as secondary amines) may help prevent reformation of nitrosamines.

To ensure that AOP treatment processes occur efficiently and quantitatively a full understanding of the kinetic and mechanistic chemistry involved under the conditions of use is necessary *(14)*. Therefore in this work we quantitatively established rate constants, reaction mechanisms, and overall degradation efficiencies in the free-radical-based destruction of 16 different nitrosamines in water. The structures of these nitrosamines are shown in Scheme 1.

Experimental

Nitrosamines were either commercially purchased (Sigma-Aldrich, > 99%) or synthesized from the corresponding amine using standard methodologies *(15)*. All sample solutions were prepared in water filtered through a Millipore Milli-Q system constantly illuminated by a UV lamp to maintain total organic carbon levels below 13 μg/L, as measured by an on-line analyzer.

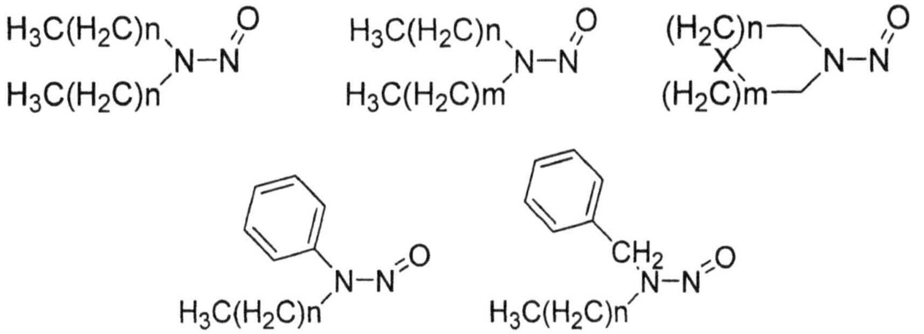

Scheme 1. Top, from left, structures of symmetric aliphatic (n = 0 – 3), asymmetric aliphatic (n, m = 0 – 3), ring (X = C, O; n,m = 0 - 3); bottom, from left, phenyl (n = 0 – 3) and benzyl (n = 0 – 2) containing nitrosamines.

The linear accelerator (LINAC) electron pulse radiolysis system at the Radiation Laboratory, University of Notre Dame, was used for the hydrated electron and hydroxyl radical rate constant determinations *(16)*. During the irradiation process, the solution vessels were bubbled with only the minimum amount of gas necessary to prevent air ingress, to minimize loss of the solute. Solution flow rates were adjusted so that each pulse irradiated a fresh sample.

The radiolysis of water by either fast electrons or gamma rays generates free radicals in solution according to *(17)*

$$H_2O \text{ -/\/\/}\!\!\rightarrow [0.28]^{\bullet}OH + [0.06]^{\bullet}H + [0.27]e_{aq}^{-} + [0.05]H_2 + [0.07]H_2O_2 + [0.27]\,H^{+} \quad (1)$$

where the numbers in brackets are the micromolar concentrations for each species produced for every joule of energy deposited into the solution (G-values).

LINAC dosimetry *(17)* was performed using N_2O-saturated 1.00 x 10^{-2} M KSCN solutions at λ = 475 nm ($G\varepsilon$ = 5.2 x 10^{-4} m^2 J^{-1}) with average doses of 3-5 Gy per 2-3 ns pulse. Total radical concentrations were between 2-4 μM/ pulse.

All kinetics experiments were performed at ambient temperature (20 ± 2°C) and in natural (pH~7.0) solution.

Degradation efficiency measurements were conducted using 20.0 mL of 1.0 mM nitrosamine aerated solutions subjected to a continuous, low-intensity (~100 Gy/min), ^{60}Co-irradiation source again at this laboratory. This irradiation also produces both oxidizing and reducing radicals according to Equation (1). For the steady-state irradiations, the radical production rate was typically about 20 μM/min for each species. Vials of each separate nitrosamine were irradiated for different doses up to 6.0 kGy, which ensured almost complete loss of the parent nitrosamine (final concentration values were typically 20-50 μM, significantly higher than the detection limit of < 1 μM). The capped irradiated solutions were then sealed and were kept at 4°C to prevent significant hydrolysis or further reaction of parents or products until analysis.

All samples were analyzed within several weeks. The extraction procedure used 5.00 mL of the irradiated nitrosamine sample, to which 100μg of NDEA or N-nitrosodibutylamine (NDBA) was added as a surrogate standard to measure extraction efficiency. Samples were extracted using three separate 3-4 mL aliquots of CH_2Cl_2. The extract was dried using anhydrous sodium sulfate (Fisher Scientific) and then quantitatively transferred into a 10.00 or 25.00 mL volumetric flask. A 1.00 mL aliquot of this latter sample was then transferred to an auto-sample vial, and had 20 μg of m-nitroxylene (Sigma-Aldrich) added as an internal standard. Standard calibration curves were also built on an internal basis for the surrogate standards. Extractions performed on the same solutions several weeks later gave the same results within experimental error (± 3%).

These samples and calibration standards were analyzed by an Agilent 6890N Gas Chromatography (GC) system in tandem with an Agilent 5973 mass selective detector (MSD). An Agilent 7683B auto-sampler was used to deliver sample injections to a JW Scientific DB-5ms column. GC parameters were as follows: Injection temp: 225°C, with pulsed splitless injection, pulse pressure: 30 psi, He as the carrier gas with a constant flow rate of 1.5 mL/min. Initial oven temperature was 45°C for 5 min, ramped at 15°C/min to 125°C, 5°C/min to 150°C, then 15°C/min to 250°C, and finally held for 5 min. MSD transfer line temperature was 225°C, MSD quad: 150°C, MSD source: 230°C. Electron ionization detection was used, with an ionizing energy of 70eV. MSD Chemstation software version D.01.02.16 was used for data analysis.

Results and Discussion

Hydroxyl Radical Reactions

For hydroxyl radical rate constant experiments, solutions were pre-saturated with nitrous oxide (N_2O), which removes dissolved oxygen and reacts quickly with hydrated electrons and hydrogen atoms to quantitatively convert them to oxidizing hydroxyl radicals *(17)*. For all the aliphatic nitrosamines in this study, no significant transient absorption was found in the range 250-800 nm upon oxidation, and therefore these rate constants were determined using SCN^- competition kinetics *(17)*, monitoring the change of absorption intensity of the produced $(SCN)_2^{\bullet-}$ transient at 475 nm.

Based on the hydroxyl radical competition by both SCN^- and the nitrosamine ($R^1R^2N\text{-}NO$) in these solutions:

$$^{\bullet}OH + R^1R^2N\text{-}NO \rightarrow H_2O + {}^{\bullet}R^1R^2N\text{-}NO \quad (2)$$

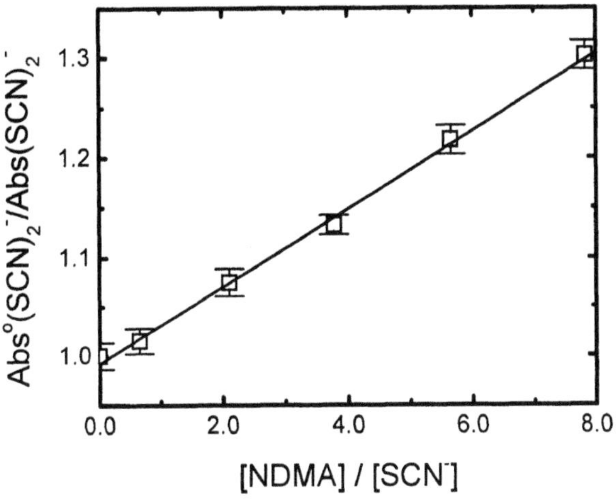

Figure 1. Thiocyanate competition kinetics plot for hydroxyl radical reaction with NDMA. Solid line is weighted linear fit, with slope of 0.0391 ± 0.0011. This corresponds to a second-order rate constant for NDMA reaction of $k_2 = (4.30 \pm 0.12) \times 10^8\ M^{-1}\ s^{-1}$.

$$^{\bullet}OH + SCN^- (+ SCN^-) \rightarrow OH^- + (SCN)_2^{\bullet-} \tag{3}$$

the following analytical expression for the change in yield of the product $(SCN)_2^{\bullet-}$ can be derived *(17)*:

$$\frac{Abs^o(SCN)_2^{\bullet-}}{Abs(SCN)_2^{\bullet-}} = 1 + \frac{k_2[R^1R^2N-NO]}{k_3[SCN^-]} \tag{4}$$

Therefore by plotting the ratio of the $(SCN)_2^{\bullet-}$ transient absorption intensity in the absence ($Abs^o(SCN)_2^{\bullet-}$) and presence ($Abs(SCN)_2^{\bullet-}$) of nitrosamine against the concentration ratio $[R^1R^2N\text{-}NO]/[SCN^-]$ one should get a straight line of slope k_2/k_3. A typical transformed plot obtained in this study for NDMA is shown in Figure 1. From the established value for hydroxyl radical reaction with thiocyanate ($k_3 = 1.05 \times 10^{10}\ M^{-1}\ s^{-1}$ *(17)*), the value of k_2 can be determined.

Similar measurements were made for all of the other aliphatic nitrosamines of interest in this study (see Table 1). In these data a consistent trend was observed, in which the reaction rate constant became faster as the nitrosamine complexity increased. This is to be expected, as previous electron paramagnetic spin trap measurements for NDMA have shown that this oxidation reaction consists of hydrogen atom abstraction from the molecule *(15,18)*. In the more complex nitrosamines, thermodynamic considerations suggest that the hydroxyl

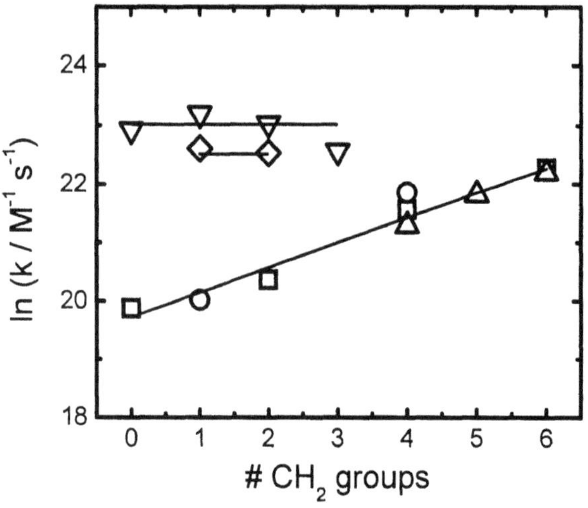

Figure 2. Correlation of measured hydroxyl radical reaction rate constants with number of methylene (-CH_2-) groups: (□, S) symmetric aliphatic, (○, A) asymmetric aliphatic, (△, R) aliphatic ring species, (▽, P) phenyl, and (◇, B) benzyl nitrosamines.

radical reaction would also be a hydrogen atom abstraction, but mainly from one of the methylene (-CH_2-) groups. This is readily seen in Figure 2, where we plot ln ($k/M^{-1}s^{-1}$) against the number of methylene groups in the aliphatic chains of these molecules. No specific correlation with the alkyl substituents being symmetric, asymmetric, or in a ring was observed for this oxidative reaction.

In contrast, the phenyl or benzyl-containing nitrosamines exhibited a transient UV absorption upon hydroxyl radical reaction, which allowed direct measurement of the kinetics of this oxidation. By varying the concentration of these nitrosamines rate constants for hydroxyl radical reaction were determined, and are also listed in Table 1 and shown in Figure 2. It can clearly be seen that these values are faster than for the aliphatic nitrosamines, and are effectively invariant with the number of methylene groups on the aliphatic ligand, indicating that the reaction is predominantly with the aromatic ring in these molecules.

Hydrated electron reactions

The reduction of nitrosamines by hydrated electron reaction also did not give any significant transient absorption over the UV/visible wavelength range so the rate constants for this reaction were determined by direct measurement of the rate of change of the hydrated electron absorption decay at 700nm (Figure 3a). These solutions also contained 0.10 M methanol to quantitatively scavenge hydroxyl radicals and hydrogen atoms *(17)*.

Table 1. Summary of hydroxyl radical reaction rate constants and hydroxyl radical degradation efficiencies, for nitrosamines of this study. (S) symmetric aliphatic, (A) asymmetric aliphatic, (R) aliphatic ring species, (P) phenyl, and (B) benzyl nitrosamines.

Species	# CH_2 groups	Type	10^{-8} $k_{\bullet OH}$ M^{-1} s^{-1}	% degradation efficiency
Aliphatic nitrosamines				
N-Nitrosodimethylamine[a)]	0	S	(4.30 ± 0.11)	80.0 ± 1.7
N-Nitrosomethylethylamine[b)]	1	A	(4.95 ± 0.21)	83.2 ± 4.8
N-Nitrosodiethylamine [b)]	2	S	(6.99 ± 0.28)	86.0 ± 3.3
N-Nitrosodipropylamine	4	S	(23.0 ± 0.6)	100.9 ± 2.3
N-Nitrosoethylbutylamine	4	A	(31.0 ± 1.3)	105.1 ± 4.7
N-Nitrosodibutylamine	6	S	(47.1 ± 1.9)	100.8 ± 1.7
N-Nitrosomorpholine	4	R	(17.5 ± 2.7)	92.9 ± 6.4
N-Nitrosopyrrolidine	4	R	(17.5 ± 0.7)	98.9 ± 6.0
N-Nitrosopiperidine	5	R	(29.8 ± 0.9)	98.0 ± 2.9
N-Nitrosohexamethyleneimine	6	R	(43.5 ± 1.4)	103.9 ± 1.7
Phenyl/Benzyl nitrosamines				
Methylphenyl nitrosamine	0	P	(90.1 ± 2.6)	-
Ethylphenyl nitrosamine	1	P	(118 ± 4)	-
Propylphenylnitrosamine	2	P	(100 ± 3)	-
Butylphenylnitrosamine	3	P	(63.1 ± 2.6)	-
Methylbenzyl nitrosamine	1	B	(66.1 ± 1.3)	-
Ethylbenzyl nitrosamine	2	B	(58.9 ± 1.7)	-

a) Kinetic data taken from Reference 18
b) Kinetic data taken from Reference 15

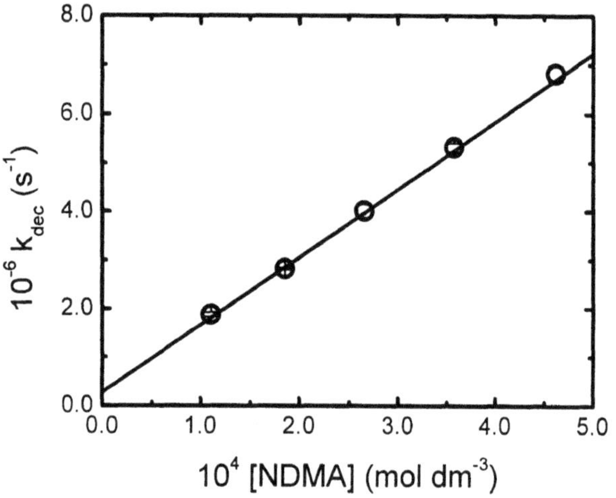

Figure 3. Second order rate constant plot for the reaction of hydrated electrons with NDMA. Fitted line corresponds to a reaction rate constant of $k = (1.41 \pm 0.02) \times 10^{10}\ M^{1}\ s^{-1}$.

For the pseudo-first-order decay kinetics obtained in the reaction:

$$e^{-}_{aq} + R^1R^2N\text{-}NO \rightarrow \text{products} \tag{5}$$

the bi-molecular reaction rate constants were obtained by plotting these values against the nitrosamine concentration, as shown in Figure 3b. All of the obtained reduction rate constants are summarized in Table 2.

Previous spin-trap measurements on the hydrated electron reduction of NDMA *(18)* showed that this reaction was only an addition to give a radical anion. The rate constants for hydrated electron reaction with all the nitrosamines studied are remarkably similar to NDMA (see Figure 4), indicating that a common reaction mechanism occurs. The electron adduct may be formed through interaction of the hydrated electron with the NO group, or form a type of conjugated structure with the π-orbitals of the molecule. However, these rate constants are significant in that although reduction of all these nitrosamines only appears to form adduct species, they may be capable of transferring their extra electron to another acceptor, such as dissolved organic matter or oxygen, to regenerate the parent nitrosoamine.

Degradation Efficiency measurements

Overall hydroxyl radical oxidative removal efficiencies for each aliphatic nitrosamine in aerated solution were also determined in this study, using a

Table 2. Summary of hydrated electron reaction rate constants for nitrosamines of this study. (S) symmetric aliphatic, (A) asymmetric aliphatic, (R) aliphatic ring species, (P) phenyl, and (B) benzyl nitrosamines.

Species	# CH_2 groups	Type	10^{-10} k_{e^-} M^{-1} s^{-1}
Aliphatic nitrosamines			
N-Nitrosodimethylamine[a)]	0	S	(1.41 ± 0.02)
N-Nitrosomethylethylamine[b)]	1	A	(1.67 ± 0.13)
N-Nitrosodiethylamine [b)]	2	S	(1.61 ± 0.12)
N-Nitrosodipropylamine	4	S	(1.53 ± 0.08)
N-Nitrosoethylbutylamine	4	A	(1.56 ± 0.07)
N-Nitrosodibutylamine	6	S	(1.55 ± 0.11)
N-Nitrosomorpholine	4	R	(2.18 ± 0.11)
N-Nitrosopyrrolidine	4	R	(1.90 ± 0.10)
N-Nitrosopiperidine	5	R	(1.74 ± 0.12)
N-Nitrosohexamethyleneimine	6	R	(1.54 ± 0.09)
Phenyl/Benzyl nitrosamines			
Methylphenyl nitrosamine	0	P	(2.27 ± 0.05)
Ethylphenyl nitrosamine	1	P	(2.21 ± 0.04)
Propylphenylnitrosamine	2	P	(2.22 ± 0.04)
Butylphenylnitrosamine	3	P	(1.37 ± 0.05)
Methylbenzyl nitrosamine	1	B	(1.81 ± 0.04)
Ethylbenzyl nitrosamine	2	B	(1.83 ± 0.03)

a) Kinetic data taken from Reference 18
b) Kinetic data taken from Reference 15

steady-state/low intensity (0.0202 kGy/min) ^{60}Co-irradiation source at the Radiation Laboratory, University of Notre Dame. Values for the phenyl and benzyl-alkyl species were not possible to obtain by this methodology. The loss of nitrosamine in irradiated aerated solution was by first-order kinetics. To provide quantitative removal data that can be used under all conditions, we plot the absolute change of nitrosamine concentration with dose, as seen in Figure 5a. The initial slopes of these plots are linear (values summarized in Table 1), and correspond to total G-values for destruction of these nitrosamines in aerated water. At higher doses, there is significant deviation from linearity, indicating that competition from formed products becomes a factor.

The efficiency of free-radical-induced removal of the aliphatic nitrosamines from water can be readily converted to a percentage basis, using Equation (1).

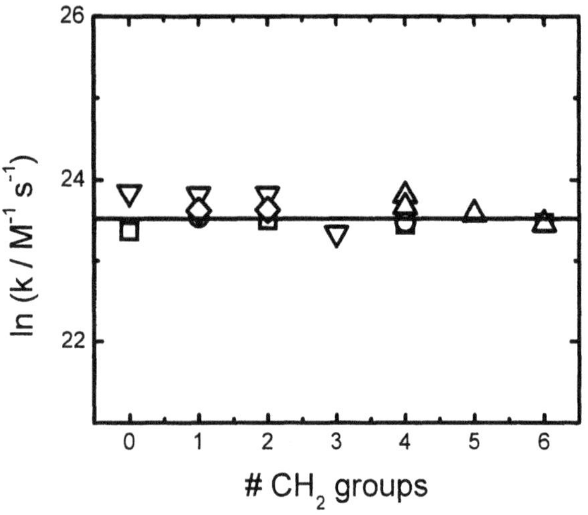

Figure 4. Correlation of measured hydrated electron reaction rate constants with number of methylene ($-CH_2-$) groups in nitrosamines (□, S) symmetric aliphatic, (O, A) asymmetric aliphatic, (Δ, R) aliphatic ring species, (∇, P) phenyl, and (◇, B) benzyl nitrosamines.

Although three radicals are produced in the radiolysis of water, as the reduction of all these nitrosamines by the hydrated electron only results in an electron-adduct species we can ignore the contributions of this radical in overall nitrosamine degradation. The hydrogen atom has also been shown to react with NDMA *(18)*, with a rate constant of $k = (2.01 \pm 0.03) \times 10^8\ M^{-1}\ s^{-1}$, however, this reaction mechanism was again postulated to be only addition to the nitroso group in these molecules. Moreover, in aerated solutions, the reaction of both hydrated electrons ($k = 1.9 \times 10^{10}\ M^{-1}\ s^{-1}$ *(17)*) and hydrogen atoms ($k = 1.2 \times 10^{10}\ M^{-1}\ s^{-1}$ *(17)*) with dissolved oxygen, ($[O_2] = 2.5 \times 10^{-4}$ M) will be the dominant reaction for these species. Therefore, only the reactions of the hydroxyl radical will result in nitrosamine degradation under our experimental conditions. Therefore, by dividing our initial slopes by $G_{\bullet OH} = 0.28\ \mu mol/J^{-1}$, we can calculate percentage efficiencies for nitrosamine removal. These values are also given in Table 1, and shown in Figure 5b.

It is immediately clear that for the higher molecular weight nitrosamines, defined as those which contain more than three methylene groups, the hydroxyl radical reaction is 100% efficient, which means that *every* hydroxyl radical reaction results in *complete* destruction of the nitrosamine. However, for the three smaller aliphatic species of this study, the degradation efficiency is significantly less than unity, with a lower limit value for NDMA of 80% found.

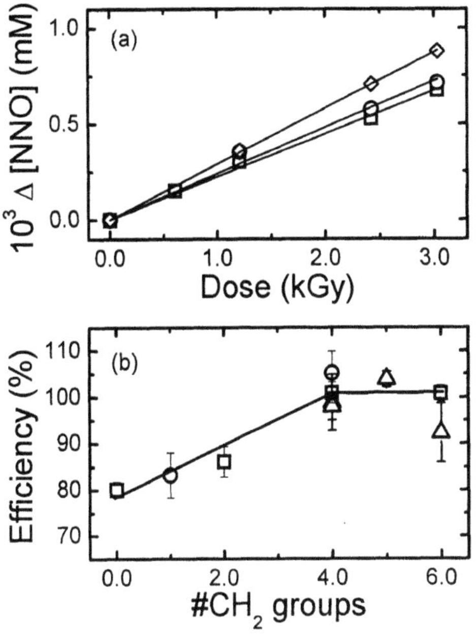

Figure 5. a) Steady-state ^{60}Co irradiation removal of nitrosamines in aerated, unbuffered high quality water at natural pH (~7.0). Data plotted as absolute change in nitrosamine concentration for nitrosodimethylamine (□), nitrosodiethylamine (O) and nitrosohexamethyleneimine (◇). Solid lines correspond to linear fits to initial data, with slopes determined as (2.241 ± 0.047) x 10^{-4}, (2.410 ± 0.092) x 10^{-4}, and (2.908 ± 0.018) x 10^{-4} kGy^{-1}, respectively. b) Percentage efficiency for hydroxyl-radical-induced reaction with nitrosamines in water, plotted against number of methylene groups in each nitrosamine for symmetric aliphatic (□), unsymmetric aliphatic (O), and aliphatic ring (Δ) structures. Error bars correspond to one standard deviation.

This finding has important implications for AOP applications and computer models, as it appears that excess hydroxyl radicals will be required to achieve quantitative removal of low molecular weight nitrosamines from waters.

At this time we have not established the reason for the lower degradation efficiencies measured for these three nitrosamines. In these oxidations (Equation (2)), it would be expected that the initially formed methylene carbon-centered radical would react with dissolved oxygen to form a peroxyl radical:

$$^{\bullet}R^1R^2N\text{-}NO + O_2 \rightarrow {}^{\bullet}O_2R^1R^2N\text{-}NO \quad (6)$$

These peroxyl radicals are relatively unreactive, and usually dimerize to form intermediate tetroxides *(20),* which then decompose to give a variety of smaller, oxygen-containing, products. This peroxyl formation reaction for NDMA has been determined previously *(18)*, with a rate constant of $k = (5.3 \pm 0.6) \times 10^6\ M^{-1}\ s^{-1}$ measured. This rate constant is several orders of magnitude slower than usually observed for carbon-centered radical reaction with dissolved oxygen *(21)*, which was attributed (*18*) to considerable spin-density delocalization from the original $^{\bullet}CH_2(CH_3)NNO$ species formed into the N-NO bond(s), thereby giving a much less reactive radical species. For these more complex nitrosamines the formation of carbon-centered radicals not adjacent to the N-NO group may prevent this electron delocalization, and therefore it is possible that under our experimental conditions some nitrosamine repair reactions of the form

$$^{\bullet}R^1R^2N\text{-}NO + {}^{\bullet}H \rightarrow R^1R^2N\text{-}NO \quad (7)$$

$$^{\bullet}R^1R^2N\text{-}NO + e_{aq}^- (+ H^+) \rightarrow R^1R^2N\text{-}NO \quad (8)$$

might compete with peroxyl radical formation, especially as the concentration of dissolved oxygen is depleted. These repair reactions may be slower in the larger nitrosamines, due to steric and resonance stabilization factors, thus increasing their degradation efficiency.

Product studies

No major products were detected in these irradiations using our GC/MS methodology. However, preliminary investigations of the production of free nitric oxide (NO) has been performed using electron paramagnetic resonance measurements with a Fe(II)-MGD spin-trap. For NDMA, experiments were performed under a variety of redox conditions, as seen in Figure 6. Under redox neutral conditions (N_2-saturated to remove dissolved oxygen) almost quantitative yield (~88 %) of NO production is seen, however, under only oxidative (< 1%), only reductive (< 1%), and aerated (~3 %)conditions, very little of this free radical is produced.

Implications

It is also important to note that these efficiency measurements were made in very high quality water, which limits the subsequent chemistry of the formed carbon-centered radicals. However, these kinetic data also provide a quantitative foundation for the evaluation of AOP efficiency in removing nitrosamines from waters. These oxidation rate constants are relatively slow, which suggests that in waters containing high levels of dissolved natural organic matter (NOM) the hydroxyl radicals produced by AOPs might not have any significant impact on nitrosamine removal. For example, in typical wastewaters containing 7 ppm NOM (580 μM NOM assuming 12 g C per mole C), and 10 ppb NDMA (135 nM) at pH 8.0 and a typical alkalinity of 100 mg/L (as $CaCO_3$, corresponding to ~1.0 mM HCO_3^-), the hydroxyl radical reaction will be partitioned according to:

$$^{\bullet}OH + NOM \rightarrow H_2O + NOM^{\bullet} \qquad k_{NOM} = 2.25 \times 10^8\ M^{-1}\ s^{-1} \tag{9}$$

$$^{\bullet}OH + NDMA \rightarrow H_2O + NDMA^{\bullet} \qquad k_{eff} = 0.8 \times 4.30 \times 10^8 = 3.44 \times 10^8\ M^{-1}\ s^{-1} \tag{10}$$

$$^{\bullet}OH + HCO_3^- \rightarrow OH^- + CO_3^{-\bullet} \qquad k_{HCO3^-} = 8.5 \times 10^6\ M^{-1}\ s^{-1}\ \ \textit{(17)} \tag{11}$$

Based on the relative rates of these three reactions, the fraction following reaction (10) would only be 0.033%. This fraction would remain constant under these conditions, independent of the rate of hydroxyl radical production or the AOP used.

Acknowledgments

Work described herein at the Radiation Laboratory, University of Notre Dame, which is supported by the Office of Basic Energy Sciences of the U.S. Department of Energy. This work was performed under Research Corporation Grant #CC6469, with partial support also from Howard Hughes Medical Institute (HHMI) Grant #52002663 and California State University, Long Beach Women and Philanthropy. TJC would like to acknowledge his REU program #DBI0552517 at Western Michigan University. We would also like to thank the Institute for Integrated Research in Materials, Environments, and Society (IIRMES) at CSULB for use of their GC/MS facilities.

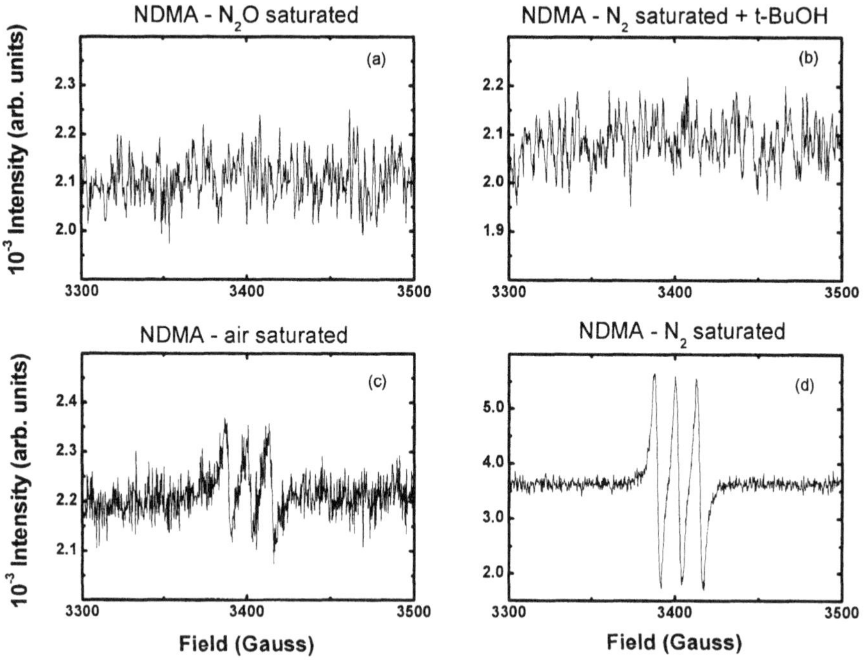

Figure 6. EPR MGD-NO spin-trap data for 1.0 x 10^{-3} M NDMA solutions subjected to ^{60}Co irradiation (~150-200 Gy) under different redox conditions (see text). Vertical scales are arbitrary units, but scaled equivalently.

References

1. U.S.E.P.A. CASRN 62-75-9, N-Nitrosodimethylamine, IRIS Substance File, 1997.
2. Magee P. N.; Barnes, J. M. Carcinogenic nitroso compounds. *Adv. Cancer Res.* **1967**, *10*, 163-246
3. Magee, P. N., Toxicity of nitrosamines: Their possible human health hazards. *Food Cosmet. Toxicol.* **1971**, *9*, 207-218.
4. Tricker, A.R.; Spiegelhalder, B.; Preussmann, R., Environmental exposure to preformed nitroso compounds. *Cancer Surv.*, **1989**, *8*, 207.
5. Mirvish, S.S., Role of N-nitroso compounds and N-nitrosation in etiology of gastric, esophageal, nasopharyngeal and bladder cancer and contribution to cancer of known exposures to NOC. *Cancer Lett.*, **1995**, *93*, 17-48.
6. Lijinsky, W., N-Nitroso compounds in the diet. *Mutat. Res.*, **1999**, *443*, 129.
7. Sen, N.P.; Seaman, S.W.; Bergeron, C.; Brousseau, R., Trends in the levels of N-nitrosodimethylamine in Canadian and imported beers. *J. Agric. Food Chem.*, **1996**, *44*, 1498-1501.

8. Mitch, W.A.; Sedlak, D.L., Formation of N-Nitrosodimethylamine from dimethylamine during chlorination. *Env. Sci. Technol.*, **2002**, *36*, 588-595.
9. Choi, J.; Valentine, R.L., N-Nitrosodimethylamine formation by free-chlorine-enhanced nitrosation of dimethylamine. *Water Res.*, **2002**, *36*, 817.
10. Hecht, S.S. Approaches to cancer prevention based on understanding of N-nitrosamine carcinogenesis. *Proc. Soc. Exp. Biol. Med.*, **1997**, *216*, 181-191.
11. Zhu, J.H.; Yan D.; Xai, J.R.; Ma, L.L.; Shen, B., Attempt to adsorb N-nitrosamines in solution by use of zeolites. *Chemosphere*, **2001**, *44*, 949-956.
12. Sharp, J.O.; Wood, T.K.; Alvarez-Cohen, L., Aerobic biodegradation of N-nitrosodimethylamine (NDMA) by axenic bacterial strains. *Biotech. Bioeng.*, **2005**, *89*, 608-618.
13. Liang S.; Min J.H.; Davis M.K.; Green J.F.; Remer D.S., Use of pulsed-UV processes to destroy NDMA. *J. Am. Water Works Assoc.* **2003**, *95*, 121-131.
14. Crittenden, J.C.; Hu, S.; Hand, D.W.; Green, S.A., A kinetic model for H_2O_2 /UV process in a completely mixed batch reactor. *Wat. Res.*, **1999**, *33*, 2315.
15. Mezyk, S.P.; Ewing, D.; Kiddle, J.J.; Cooper, W.J.; Madden, K.P., Kinetics and mechanisms of the reactions of hydroxyl radicals and hydrated electrons with nitrosamines and nitramines in water. *J. Phys. Chem. A.*, **2006**, *110*, 4732-4737, and also Supplemental Material.
16. Whitman, K.; Lyons, S.; Miller, R.; Nett, D.; Treas, P; Zante, A.; Fessenden, R.W.; Thomas, M.D.; Wang, Y., Linear accelerator for radiation chemistry research at Notre Dame 1995. In *Proceedings of the '95 Particle Accelerator Conference and International Conference on High Energy Accelerators*, Texas, USA, **1999**.
17. Buxton, G. V.; Greenstock, C. L.; Helman, W. P.; Ross, A. B., Critical review of rate constants for reactions of hydrated electrons, hydrogen atoms and hydroxyl radicals ($OH/O^{\cdot}$) in aqueous solution. *J. Phys. Chem. Ref. Data,* **1988**, *17*, 513-887.
18. Mezyk, S.P.; Cooper, W.J.; Madden, K. P.; Bartels D. M., Free radical destruction of N-Nitrosodimethylamine in water. *Environ. Sci. Technol.*, **2004**, *38*, 3161-3167.
19. Mezyk, S.P.; Neubauer, T.; Cooper, W.J.; Peller, J., Free-radical-induced oxidative and reductive degradation of sulfa drugs in water: Absolute kinetics and efficiencies of hydroxyl radical and hydrated electron reaction. *J. Phys. Chem. A.*, 2007, *111*, 9019-9024.
20. von Sonntag, C.; Schuchmann, H.-P. Peroxyl Radicals in Aqueous Solution. In *Peroxyl Radicals*, Ed. Alfassi, Z.B. Chichester: Wiley, **1997**, 173-234.
21. Neta, P.; Huie, R.E.; Ross, A.B., Rate constants for reactions of peroxyl radicals in fluid solutions. *J. Phys. Chem. Ref. Data*, **1990**, *19*, 413-513.

Chapter 23

Degradation of Halogenated Disinfection Byproducts in Water Distribution Systems

Raymond M. Hozalski, William A. Arnold, Chanlan Chun, Timothy M. LaPara, Jeong-Yub Lee, Carrie R. Pearson, and Ping Zhang

Department of Civil Engineering, University of Minnesota, Minneapolis, MN 55455-0116

Water distribution systems are complex environments frequently containing corroded iron pipes and biofilms. To thoroughly understand the fate of halogenated disinfection byproducts (DBPs) in these systems, two degradation processes were investigated: abiotic degradation (i.e. hydrolysis and reductive dehalogenation) and biodegradation. DBPs were selected from 6 different compound classes representing both regulated DBPs (i.e. trihalomethanes or THMs, and haloacetic acids or HAAs) and non-regulated or "emerging" DBPs. Batch experiments were conducted to investigate the pathways and kinetics of DBP degradation. As expected, the relative importance of hydrolysis, abiotic reductive dehalogenation, and biodegradation depends on the DBP structure and on the environmental conditions (i.e. pH, temperature, dissolved oxygen, Fe minerals present, bacteria present, etc.). From our results, chloropicrin (i.e. trichloronitromethane) and most brominated DBPs are highly susceptible to abiotic reductive dehalogenation, trichloroacetonitrile and trichloropropanone are the most susceptible to hydrolysis, and HAAs are readily biodegraded under aerobic conditions. Knowledge of DBP degradation mechanisms and rates in distribution systems is important for selecting DBP monitoring locations, modeling DBP fate, and for predicting exposure to these compounds. Such information could also be useful for developing treatment systems for DBP removal.

The interface between the pipe wall and the flowing water in a drinking water distribution system is a complicated environment. Distribution system pipes are commonly comprised of unlined cast or ductile iron, a powerful reductant, and the water contains both dissolved oxygen and free or combined chlorine, which are potent oxidizing agents. The interaction of oxidant and reductant leads to corrosion of the iron pipe and formation of a variety of partially to fully oxidized iron mineral phases. The pipe surface also harbors biofilms that contain viable bacteria (*1*, *2*). The iron metal surface, corrosion products, and bacteria in distribution systems are capable of degrading halogenated disinfection byproducts (DBPs).

Several full-scale monitoring campaigns have demonstrated that trihalomethane (THM) levels generally increase with increasing residence time along the distribution system in the presence of a free chlorine residual. Conversely, haloacetic acids (HAAs) often decrease along the distribution system when chlorine levels are low or absent (*e.g.*, *3*, *4*). Although decreases, increases and no change in DBP levels with increasing residence time in distribution systems have been observed, several parameters such as chlorine residual, type of disinfectant, retention time, natural organic matter (NOM), pH, and temperature are believed to be important factors in DBP fate. Chlorine residual is considered a crucial parameter controlling DBP concentrations with its dual impact of increasing formation rates and decreasing biological activity. When DBPs were observed to increase, the chlorine residuals (free chlorine) tended to be high (> 2 mg/L) (*5*, *6*, *7*) while lower chlorine residuals (< 1 mg/L) were associated with decreases in the concentrations of some DBPs (*e.g.*, dichloroacetic acid) (*3*, *8*).

Additionally, Krasner et al. (*5*, *9*) documented the occurrence of a wide variety of non-regulated or "emerging" DBPs in treated drinking waters including haloacetonitriles, haloketones, haloacetaldehydes, and halonitromethanes. The dominant compound in each of these DBP classes and the reported concentration ranges were as follows: dichloroacetonitrile (DCAN, 0.1-5 μg/L), 1,1,1-trichloropropanone (1,1,1-TCP, 0-5 μg/L), dichloroacetaldehyde (DCAh, 0.4-14 μg/L), and trichloronitromethane (TCNM, 0-2 μg/L) (*9*). The fates of these compounds in water distribution systems have not been well studied.

Iron Pipe Corrosion and Abiotic DBP degradation

The interior surface of an unlined cast iron water distribution pipe is typically exposed to an oxidizing environment that causes corrosion and the buildup of corrosion products. Due to the variability in treated water chemistry, hydraulic conditions, and pipe age, the presence and distribution of different minerals varies from system to system and even within individual systems.

Goethite (75.6%), magnetite (21.5%), and lepidocrocite (2.9%) were the dominant corrosion products found on the inner wall of a steel water pipe in Melbourne, Australia (*10*). Investigations of iron pipes from Champaign, IL and Boston, MA also found that goethite and magnetite were the dominant corrosion products on the pipe surfaces (*11*). Finally, in a water distribution pipe from Columbus, Ohio, green rust was reported to be a major iron corrosion product (*12*).

Many DBPs are small, halogenated organic compounds that can be transformed via hydrolytic and reductive pathways. Reduction at the pipe wall could be mediated by zero-valent iron or by the ferrous iron contained in or sorbed to iron oxide corrosion products. Numerous halogenated organic compounds are susceptible to reduction by these materials, including the solvents tetrachlorethene and trichloroethene. Work in our laboratory has shown that many DBPs including HAAs (*13*, *14*) and TCNM (*15*) are susceptible to reduction by zero-valent iron. Certainly the buildup of corrosion products on the pipe walls and the use of corrosion inhibitors would tend to limit the availability of Fe(0) surfaces. Corroding iron pipes, however, also create high surface area iron oxides that can also catalyze reductive dehalogenation reactions.

Hydrolysis may also lead to DBP losses in a distribution system (*16-18*). For example, haloacetonitriles can undergo hydrolysis to form haloacetamides, and in basic solutions haloacetamides further hydrolyze to form haloacetic acids (*18*). Chloroform (TCM) can also be formed upon hydrolysis of 1,1,1-TCP (*19*) or chloral hydrate/trichloroacetaldehyde (TCAh).

Biological degradation of DBPs

Many researchers have observed the biodegradation of haloacetic acids in soil and aquatic environments (*e.g., 20-25*) with most of the work in soils focused on the herbicide trichloracetic acid (TCAA). There are two potential mechanisms for the biodegradation of HAAs: hydrolysis-oxidation and reductive dechlorination. Because the effluent from drinking water treatment plants usually contains 8 to 10 mg/L of dissolved oxygen, suspended bacteria and the biofilm on the pipe walls typically are exposed to aerobic conditions. Thus, the hydrolysis-oxidation pathway is more likely to occur. Because THMs are more oxidized than oxygen, their direct biodegradation is not thermodynamically favorable under aerobic conditions (*26*).

Previous research has shown that bacteria are able to aerobically degrade HAAs either cometabolically (*23*) or as a sole carbon and energy source (*21*). Nevertheless, because the concern of most of these studies was cleanup of areas contaminated with spills of HAAs (i.e. TCAA) used as herbicides, HAA concentrations were orders of magnitude higher than those observed in surface waters and in drinking water distribution systems. A small number of studies have investigated the degradation of HAAs in the environment and suggested that biodegradation is the dominant loss mechanism (*27-29*). HAA loss in water distribution systems is typically attributed to biodegradation (*4*, *5*, *30-32*). To our knowledge, Williams and colleagues (*33*) are the only group to have demonstrated that bacteria collected from a water distribution system were capable of degradation of dichloroacetic acid (DCAA) and other HAAs. Nevertheless, the degradation kinetics and organisms involved in the degradation were not well characterized.

The first two complete gene sequences encoding for 2-haloacid dehalogenases were reported by Schneider et al. (*34*) in P*seudomonas* sp. strain CBS3. Genes coding for dehalogenases (referred to in the following as *deh*) able to degrade 2-haloacids have been found in a number of organisms (*e.g.*, *35-38*). Sequence alignments of known *deh* genes suggested the existence of two groups of dehalogenases, which were related to each other but did not show obvious homology. Separate alignment of the two groups allowed the identification of conserved regions and the design of degenerate primers (*39*) specific for each group. It turned out that with very few exceptions all described 2-haloacid dehalogenases were encoded by genes that can be assigned to either group I or group II. The *deh* genes provide potential target sites for enumerating HAA degraders in environmental samples using quantitative PCR (qPCR).

Experimental Approach

Batch experiments were used to evaluate the pathways and kinetics of abiotic DBP degradation in the presence of iron metal, synthetic iron minerals (goethite, magnetite, green rust), and iron corrosion products obtained from distribution systems. Selected compounds from the following DBP classes were investigated: THMs, HAAs, halonitromethanes, haloacetonitriles, halopropanones, and haloacetaldehydes. For the biodegradation work, HAA degraders were enriched from wastewater activated sludge (positive control) and water distribution systems and the HAA biodegradation kinetics of the enrichment cultures and selected isolates were investigated. In addition, the bacteria responsible for HAA biodegradation were identified and characterized. Ultimately, we plan to develop a qPCR method for enumerating HAA degraders in water distribution system samples so that we can extrapolate the batch kinetic data to these systems. Experimental details are provided below.

Abiotic Batch Experiments

Details of experimental protocols and analytical methods have been reported previously (*15*, *40*). Briefly batch reactors contained deoxygenated, buffered water (*e.g.*, 25-50 mM MOPS) and iron metal, a synthetic iron oxide mineral, or corrosion solids collected from pipes extracted from distribution systems at a pre-determined loading (0.8 – 3 g/L). For experiments with synthetic goethite and magnetite, 1 mM Fe(II) was also added. Experiments were initiated by spiking the target DBP into the reactor. To facilitate monitoring the decay of the parent compound and formation of daughter products, initial DBP concentrations (typically 20 – 100 μM) were 10-100 fold higher than levels typically found in distribution systems. All reactors were completely filled (no headspace). Reactors were mixed on a rotator to overcome any mass transfer limitations. Reactions were monitored for 1-2 half lives of the parent compound. Blanks (no solid present) were used to account for hydrolysis. The loss of the parent compound via reduction/hydrolysis and the formation of daughter products were fit to a pseudo-first order kinetic model.

For all the target DBPs except HAAs, samples were extracted using methyl *t*-butyl ether and analyzed using gas chromatography (GC) with electron capture detection. Selected reaction products were detected using headspace GC with flame ionization detection or high pressure liquid chromatography (*15*, *40*). HAAs were quantified without derivitization using capillary electrophoresis (*14*). Ferrous iron concentrations were determined via the Ferrozine method (*41*) and characterization of the synthetic minerals and the pipe solids was performed using X-ray diffraction.

Establishing HAA-Degrading Enrichment Cultures

To enrich for HAA degraders, a medium was prepared containing inorganic minerals supplemented with either monochloroacetic acid (MCAA), DCAA, or TCAA as sole carbon and energy source. The inorganic mineral medium contained (per liter DI H_2O): 0.03 g $MgSO_4$, 1.96 g $Na_2HPO_4{\cdot}7H_2O$, 0.37 g KH_2PO_4, 0.50 g NH_4Cl, 0.0006 g $CaCl_2$, and 0.1 mL of SL7 trace mineral solution (0.75 g $FeCl_2.4H_2O$, 0.03 g H_3BO_3, 0.05 g $MnSO_4$, 0.06 g $Co(NO_3)_2{\cdot}6H_2O$, 0.066 g $ZnSO_4{\cdot}7H_2O$, 0.0125 g $NiCl_2{\cdot}6H_2O$, 0.007 g $CuCl_2$, 0.0125 g $Na_2MoO_4{\cdot}2H_2O$, and 4.4 mL of 37% HCl per 500 mL DI H_2O). After autoclaving and cooling, the mineral medium was aseptically amended with 1 mM MCAA, DCAA, or TCAA (i.e. 94.5 mg/L MCAA, 129 mg/L DCAA, or 163.5 mg/L TCAA) by spiking with the respective HAA stocks (filter-sterilized).

The batch reactors were inoculated with activated sludge biomass (positive control), tap water bacteria collected on nylon membrane filters (47 mm diameter, 0.2 μm pore size), or biofilm scraped from the inside of cast iron water mains. For water distribution system biomass sampling, water and pipe samples were collected from four systems in the United States and three in the United Kingdom covering a range of water qualities (chlorine type, residual concentration, heterotrophic plate counts, HAA levels, etc.). The hydraulic residence times of the sample sites are not known. The batch reactors (500 mL Erlenmeyer flasks containing 200 mL of culture media) were incubated in the dark at room temperature (23 ± 1°C) with shaking at 100 rpm. Bacterial growth (optical density at 600 nm, OD_{600}) and HAA concentrations were monitored over time by a UV/Vis spectrophotometer and by capillary electrophoresis, respectively. HAAs were respiked to ~1 mM when depleted, and pH was maintained at ~7.2 by periodic addition of 2 M NaOH.

The HAA enrichment cultures were harvested by centrifuging at 10,000*g* for 25 minutes, washed once with fresh medium, and resuspended in medium supplemented with 1 mM of the respective HAA and 15% glycerol. The cell suspension was then dispensed into cryo-vials (1 mL/vial) and stored at -70°C for subsequent biodegradation kinetics experiments. To preserve a sample of an enrichment culture for bacterial community analysis, 0.5 mL of the cell suspension was centrifuged at 16,000*g* for 5 minutes. The cell pellet was resuspended in 0.5 mL lysis buffer (120 mM sodium phosphate, 5% sodium dodecyl sulfate) and stored at -20°C for subsequent genomic DNA extraction.

Isolation and Identification of HAA Degraders

Agar plates were prepared containing the same mineral medium as used for enrichment, 10 mM of HAA (MCAA, DCAA, or TCAA), and 1.4% washed agar. The phosphate buffer concentration in the mineral medium was increased to 25 mM to provide improved buffer capacity and the medium pH was adjusted to ~7.2 with 1 M NaOH. A pH indicator (bromocresol purple) was added at 5 mg per liter medium, which yields a purple color at pH ≥ 6.8 and yellow at pH ≤ 5.7 (*21*). The HAA was added after the medium was autoclaved and cooled to ~45 °C. Strains turning the purple color to yellow, due to release of hydrochloric acid during HAA degradation, were considered to be HAA degraders.

HAA degraders were isolated using the spread plate method and purified using the streak plate method. Aqueous samples from the enrichment cultures were spread onto their respective HAA plates (i.e. the MCAA enriched bacteria were spread on MCAA plates). The plates were incubated at room temperature for 7 days. Single colonies were selected and streaked onto fresh plates. The isolates were re-streaked up to 3 times to ensure that the strains were pure.

Results and Discussion

DBP reduction by Fe(0)

A summary of the abiotic reactivity of the DBPs studied is provided in Table I. TCNM, trichloroacetonitrile (TCAN), 1,1,1-TCP, TCAA, and TCM were all reduced by Fe(0) (*14*, *15*, *42*). The dichloro- and monochloro- species were also reduced (except those produced from chloroform), albeit at slower rates. Experiments using bromo- or bromochloroacetic acids demonstrated that brominated species react more rapidly, with bromine being preferentially removed over chlorine (*13*, *14*). In general, reaction proceeded via hydrogenolysis (replacement of a halogen by hydrogen) (*13*, *42*). The exception was TCNM, which reacted via a combination of hydrogenolysis and reductive α-elimination (*15*). Reaction kinetics varied over orders of magnitude ranging from 0.22 min^{-1} for TCNM and 0.34 min^{-1} for tribromoacetic acid (TBAA) to 1.4×10^{-5} min^{-1} for MCAA at an iron loading of 2.4 – 3 g/L. In fact, the most reactive species (TCNM, TCAN, TBAA) were mass transfer limited as is the reduction of oxygen by Fe(0) (*42*, *43*). Thus, the mass transfer limited compounds are most likely to be amenable to reduction at walls comprised of uncorroded (i.e. new) iron pipes (*42*), for they react as quickly as the competing oxidants (oxygen, disinfectant) present in drinking water.

DBP reduction mediated by iron minerals

Carbonate green rust, Fe(II)/magnetite, and Fe(II)/goethite were capable of reducing most of the DBPs studied (*40*, *44*). Chlorinated acetic acids were only reduced by green rust, and TCM was unreactive with all of the minerals studied. Reductions occurred via hydrogenolysis. The exception was for TCNM with Fe(II)/magnetite or Fe(II)/goethite—in these systems, TCNM reacted via a combination of hydrogenolysis and reductive α-elimination. Green rust was the most potent reductant of the minerals studied. After normalizing for the surface density of reactive sites, it was determined that Fe(II)/magnetite and Fe(II)/goethite degraded DBPs at similar rates. Reduction of some DBPs did occur in the presence of magnetite alone and aqueous Fe(II) alone, but at rates much slower than those observed for sorbed Fe(II) and green rust. For DBPs degraded via both reduction and hydrolysis (*e.g.*, 1,1,1-TCP; Figure 1), the reduction and hydrolysis rate constants were obtained by fitting the loss of parent DBP and formation of reduction and hydrolysis products to pseudo-first order kinetic models.

Among the trichlorinated DBPs, the reactivity trend in the presence of Fe minerals was TCNM > TCAN > 1,1,1-TCP ~ TCAh >> TCAA >> TCM. As shown in Figure 2, the rate constants for Fe(II)/magnetite and Fe(II)/goethite normalized by the surface density of sorbed Fe(II) were similar and correlated with the one-electron reduction potential of the DBPs. A correlation between these rate constants and the electronegativity of the R group in R-CCl_3 was also found (*40*). This suggests that the rates of abiotic DBP reduction in distribution systems due to reaction with these commonly occurring minerals can be

Table I. Summary of abiotic degradation rates for selected DBPs

DBP	*Hydrolysis*	*Reduction by Fe(0) (2.4-3 g/L)*	*Reduction by Fe minerals (0.8 g/L magnetite, 2.4 g/L carbon green rust)*
$Cl_3C(=O)OH$	0	++	0, ++
$Cl_2HC(=O)OH$	0	+	0, 0
$ClH_2C(=O)OH$	0	0	0, 0
$Br_3C(=O)OH$	0	+++	NM, +++
$Br_2HC(=O)OH$	0	+++	NM, ++
$BrH_2C(=O)OH$	0	++	0, 0
Cl_3CH	0	0	0, 0
Cl_3CNO_2	0	+++	+++, +++
Cl_2HCNO_2	0	+++	+++, +++
Cl_3CCN	+	+++	+, +++
$Cl_3CC(=O)CH_3$	+	+++	0, +
$Cl_3CC(=O)H$	0	NM	0, 0

NOTE: Reaction conditions: pH 7.5, temperature = 20°C, deoxygenated MOPS buffer; Ranking: 0, t1/2 > 1 week; +, t1/2 = 1 day to 1 week; ++, t1/2 = 1 hour to 1 day; +++, t1/2 < 1 hour; NM, not measured; Magnetite: Buffer solution initially contained 1 mM aqueous Fe(II).

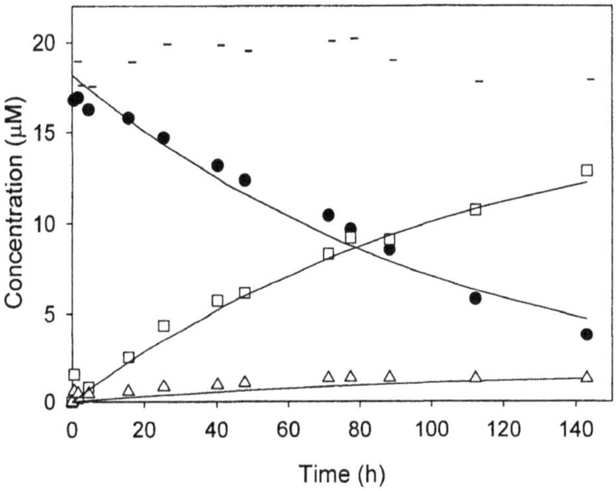

Figure 1. Degradation of 1,1,1-TCP (●) by Fe(II)/magnetite. Products are 1,1-DCP (△) and TCM (□). The dash (–) represents the observed carbon mass balance, and solid lines are model fits. Experimental conditions: pH 7.5, room temperature, mineral loading of 0.8 g/L and 1 mM total Fe(II). (Reproduced from reference 40. Copyright 2005 American Chemical Society.)

predicted if the surface areas of the minerals present and the Fe(II) release rates/availability are known.

To this end, additional experiments investigating the reduction of TCNM by iron corrosion products collected from water distribution system pipes were conducted. As shown in Figure 3, the pseudo-first order rate constant for TCNM reduction is a function of the water soluble Fe(II) in the solids. The water soluble Fe(II) was measured by equilibrating a sample of the pipe solid in buffer solution for 24 hours in a separate experiment. The solids producing higher levels of Fe(II) likely have more sorbed Fe(II) on the solids resulting in faster DBP degradation. Additional experiments were performed to investigate the potential for dissolved oxygen (DO) to compete with TCNM for reactive sites on the iron corrosion products. The rates of TCNM reduction in the presence of DO were still significant but decreased by ~10-fold, indicating that only those species that react rapidly enough to compete with other oxidants (such as TCNM, TCAN, and brominated DBPs) are likely to be susceptible to reduction at corroded iron pipe walls in drinking water distribution systems.

Hydrolysis

Hydrolysis of some DBPs was also observed. Base-catalyzed hydrolysis of 1,1,1-TCP and TCAh led to the production of TCM and acetic or formic acid, respectively. TCAN hydrolysis resulted in formation of trichloracetamide. Over time scales longer than those of these experiments (maximum of 150 hours), trichloroacetamide will hydrolyze to form TCAA (18). Although hydrolysis leads to the removal of these non-regulated DBPs species, the products (chloroform, TCAA) are regulated DBPs that are obviously of concern. Hydrolysis (pH 7.5, 20-25°C) of the HAAs, TCM, and the halonitromethanes was not observed for experiments lasting up to 2 weeks.

Biodegradation of DBPs

Aerobic HAA-degrading bacteria were successfully enriched from wastewater activated sludge using MCAA, DCAA, or TCAA as sole carbon and energy source. Enrichment of HAA degraders from water distribution systems proved difficult, especially for chloraminated systems. The likelihood of a successful enrichment from tap water increased for systems with relatively high HAA concentrations, low residual chlorine, and high heterotrophic plate counts. MCAA degraders have also been enriched from pipe wall biofilms.

Fifteen HAA-degrading isolates were obtained from the wastewater enrichment cultures and a preliminary identification was performed by PCR amplification and sequencing of partial 16S rRNA gene sequence (matches $\geq$

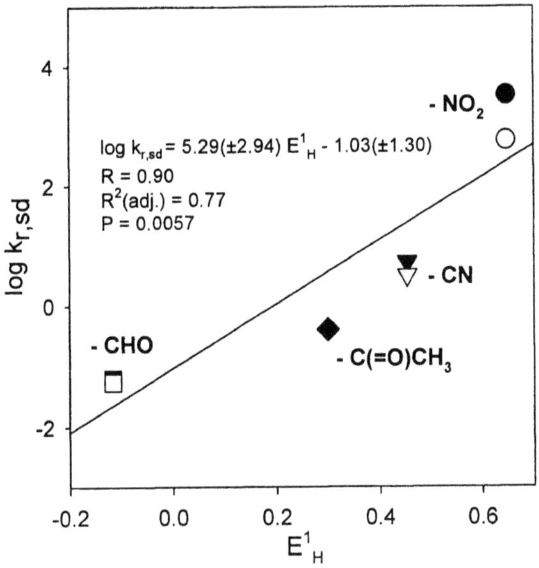

Figure 2. Correlation of abiotic reductive dechlorination rate constant at pH 7.5 and room temperature normalized by surface density of sorbed Fe(II) ($k_{r,sd}$, $h^{-1}mmol^{-1}m^2$) with one electron reduction potential (E^1_H) for DBPs of the form Cl_3CR. Open symbols: Fe(II)/goethite, solid symbols: Fe(II)/magnetite. (Reproduced from reference 40. Copyright 2005 American Chemical Society.)

99%). Eight MCAA degraders (M1-M7) and seven DCAA degraders (D1-D7) were selected based on differences in colony morphology. Some of the isolates were highly similar at the 16S rRNA gene sequence level. Nucleotide sequence analysis of their 16S rRNA genes revealed that 3 unique MCAA isolates and 4 unique DCAA isolates were obtained. All of the MCAA-degrading isolates (*Pseudomonas* sp. BCNU171, *Ultramicrobacterium* str. ND5, and *Pseudomonas plecoglossicida*) and DCAA-degrading isolates (*Delftia tsuruhatensis*, *Xanthobacter autotrophicus*, *Afipia felis*, and *Ralstonia* sp. PA1-3) from the wastewater enrichment were members of the α, β, or γ-*Proteobacteria* groups. These organisms are largely consistent with other bacterial strains isolated from previous researchers from other environments (*e.g.*, contaminated soils). Only two HAA-degrading isolates (both DCAA degraders) were obtained from tap water, *Methylobacterium* sp. and *Afipia felis*. The discovery of a HAA-degrading methylobacterium is novel as it extends the known range of species capable of growth on DCAA and also because methylobacteria were previously thought to be obligate for growth on methane or methanol.

The biodegradation of MCAA and DCAA was rapid by cultures enriched from wastewater. These cultures were able to repeatedly degrade spikes of

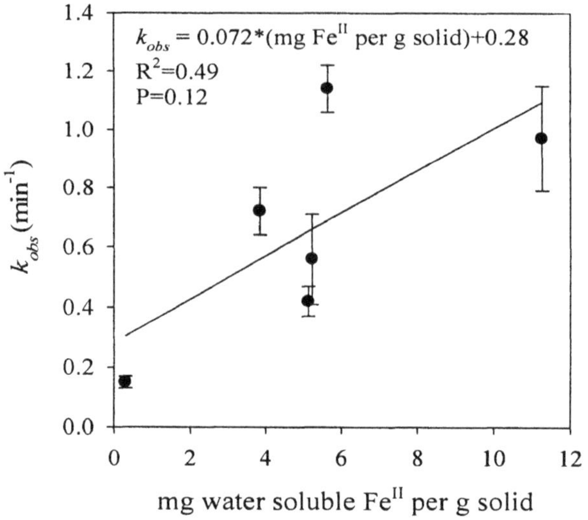

Figure 3. Correlation of TCNM degradation rate constant (k_{obs}) at pH 7.5 and room temperature with water soluble total iron concentration in samples of pipe solids obtained from water distribution systems.

MCAA and DCAA within one day. In contrast, the enrichment of TCAA-degrading cultures was much slower – degradation of spikes required ~20 days. Maximum HAA degradation rates for the wastewater enrichment cultures ranged from 6.6 μg HAA/μg protein/d (for TCAA) to 47.0 μg HAA/μg protein/d (for DCAA) for initial concentrations ranging from 10 to 370 μg/L. A tap water enrichment exhibited slower DCAA degradation kinetics (9.4 μg HAA/μg protein/d).

With regard to substrate range, all HAA degrading-isolates obtained in this research are capable of degrading more than one HAA. For example, all organisms capable of degrading MCAA could also degrade monobromo- and monoiodoacetic acid. All organisms capable of degrading dichloroacetic acid could also degrade the monohalogenated HAAs, but the opposite was not always true. Few isolates were able to degrade TCAA.

Conclusions

Overall, the results of this research have important implications for understanding DBP fate in distribution systems, selecting monitoring locations, considering new DBPs to monitor, and estimating exposures. For example, THMs are stable but could form in distribution systems; thus, monitoring at the maximum residence time location is most conservative. DBPs that may degrade rapidly, such as halonitromethanes (*e.g.*, TCNM) due to reduction in iron pipe, 1,1,1-TCP due to hydrolysis, or HAAs due to biodegradation, should be monitored at the plant effluent and perhaps at selected locations throughout the distribution system. Finally, the results of this work could lead to the development of new treatment systems for removing DBPs from chlorinated water or wastewater.

References

1. LeChevallier, M.W.; Babcock, T.M.; Lee, R.G. Examination and characterization of distribution system biofilms. *Appl. Environ. Microbiol.* **1987**, *53*, 2714-2724.
2. Zhang, M.; Semmens, M.J.; Schuler, D.; Hozalski, R.M. Evaluation of biostability and microbiological quality in a chloraminated distribution system. *J. AWWA* **2002**, *94*, (9): 112-122.
3. LeBel, G.L.; Benoit, F.M.; Williams, D.T. A one-year survey of halogenated disinfection by-products in the distribution system of treatment plants using three different disinfection processes. *Chemosphere* **1997**, *34*, 2301-2317.
4. Williams, D.T.; Lebel, G.L.; Benoit, F.M. Disinfection By-Products in Canadian Drinking Water. *Chemosphere* **1997**, *34*, 299-316.
5. Krasner, S.W.; McGuire, M.J.; Jacangelo, J.G.; Patania, N.L.; Reagan, K.M.; Aieta, E.M. The occurrence of disinfection by-products in US drinking water. *J. AWWA* **1989**, *81*, (8): 41-53.
6. Nieminski, C.E.; Chaudhuri, S.; Lamoreaux ,T. The occurrence of DBPs in Utah drinking waters. *J. AWWA* **1993**, *85*, (9): 98-105.
7. Singer, P.C.; Obolensky, A.; Greiner, A. DBPs in chlorinated North Carolina drinking waters. *J. AWWA* **1995**, *87*, (10): 83-92.
8. Chen, W.J.; Weisel, C.P. Halogenated DBP concentrations in a distribution system. *J. AWWA* **1998**, *90*, (4): 151-163.
9. Krasner, S.W.; Weinberg, H.S.; Richardson, S.D.; Pastor, S.J.; Chinn, R.; Sclimenti, M.J., Onstad, G.D.; Thruston, A.D., Jr. Occurrence of a New Generation of Disinfection Byproducts. *Environ. Sci. Technol.* **2006**, *40*, 7175-7185.
10. Lin, J.; Ellaway, M.; Adrien, R. Study of corrosion material accumulated on the inner wall of steel water pipe. *Corrosion Science* **2001**, *43*, 2065-2081.
11. Sarin, P.; Snoeyink, V.L.; Bebee, J.; Kriven, W.M.; Clement, J.A. Physico-chemical characteristics of corrosion scales in old iron pipes. *Wat. Res.* **2001**, *35*, 2961-2969.
12. Tuovinen, O.H.; Button, K.S.; Vuorinen, A; Carlson, L.; Mair, D.M.; Yut, L.A. Bacterial, chemical, and mineralogical characteristics of tubercles in distribution pipelines. *J. AWWA* **1980**, *72*, (11): 626-635.

13. Hozalski, R.M.; Zhang, L.; Arnold, W.A. Reduction of Haloacetic Acids by Fe^0: Implications for Treatment and Fate. *Environ. Sci. Technol.* **2001**, *35*, 2258-2263.
14. Zhang, L.; Arnold, W.A.; Hozalski, R.M. Kinetics of haloacetic acid reactions with Fe(0). *Environ. Sci. Technol.* **2004**, *38*, 6881-6889.
15. Pearson, C.R.; Hozalski, R.M.; Arnold, W.A. Degradation of chloropicrin in the presence of Fe(0). *Environ. Toxicol. Chem.* **2005**, *24*, (12):48-53.
16. Urbansky, E.T. The fate of the haloacetates in drinking water--chemical kinetics in aqueous solution. *Chem. Rev.* **2001**, *101*,: 3233-3243.
17. Zhang, X.; Minear, R.A. Decomposition of trihaloacetic acids and formation of the corresponding trihalomethanes in water. *Wat. Res.* **2002**, *36*, 3665-3673.
18. Glezer, V.; Harris, B.; Tal, N.; Iosefzon, B.; Lev, O. Hydrolysis of haloacetonitriles: Linear free energy relationship, kinetics and products. *Wat. Res.* **1999**, *33*, 1938-1948.
19. Obolensky, A.; Davis, B.J.; Narangajavana, K.; Ceci, L.; Oalickal, G.; Eyring, A. Disinfection byproducts in three water treatment plant distribution systems. *Proc. AWWA Water Qual. Technol. Conf.*, Tampa, FL, 1999.
20. Lode, O. Microbial decomposition of trichloroacetic acid. *Acta Agric. Scand.* **1967**, *17*, 140-148.
21. Yu, P.; Welander, T. Growth of an aerobic bacterium with trichloroacetic acid as the sole source of energy and carbon. *Appl. Microbiol. Biotech.* **1995**, *42*, 769-774.
22. Landmeyer, J.E.; Bradley, P.M.; Thomas, J.M. Biodegradation of disinfection byproducts as a potential removal process during aquifer storage recovery. *J. Am. Water Resour. Assoc.* **2000**, *36*, 861-867.
23. Weightman, A.L.; Weightman, A.J.; Slater, J.H. Microbial dehalogenation of trichloroacetic acid. *World J. Micro. Biot.* **1992**, *8*, 512-518.
24. DeWever, H.; Cole, J.R.; Fettig, M.R.; Hogan, D.A.; and Tiedje, J.M. Reductive dehalogenation of trichloroacetic acid by *Trichlorobacter thiogenes* gen. nov., sp. nov. *Appl. Environ. Microb.* **2000**, *66*, 2297-2301.
25. McRae, B.M.; LaPara, T.M.; Hozalski, R.M. Biodegradation of haloacetic acids by bacterial enrichment cultures. *Chemosphere* **2003**, *55*, 915-925.
26. Vogel, T.M. Natural bioremediation of chlorinated solvents. *Handbook of Bioremediation*. Lewis Publishers, 1993; p. 201-225.
27. Boethling, R.S.; Alexander, M. Effect of concentration of organic chemicals on their biodegradation by natural microbial communities. *Appl. Environ. Microbiol.* **1979**, *37*, 1211-1216.
28. Ellis, D.A.; Hanson, M.L.; Sibley, P.K.; Shahid, T.; Fineberg, N.A.; Solomon, K.R.; Muir, D.C.G.; Mabury, S.A. The fate and persistence of trifluoroacetic and chloroacetic acids in pond waters. *Chemosphere* **2001**, *42*, 309-318.

29. Hashimoto, S.; Tadashi, A.; Otsuki, A. Distribution, Sources, and Stability of Haloacetic Acids in Tokyo Bay, Japan. *Environ. Toxicol. and Chem.* **1998**, *17*, 798-805.
30. Baribeau, H.; Krasner, S.W.; Chinn, R.; and Singer, P.C. Impact of biomass on the stability of haloacetic acids and trihalomethanes in a simulated distribution system. *Proc. AWWA Water Quality Technology Conf.*, Salt Lake City, UT, 2000.
31. Rodriguez, M.J.; Serodes, J-B.; Levallois, P. Behavior of trihalomethanes and haloacetic acids in a drinking water distribution system. *Wat. Res.* **2004**, *38*, 4367-4382.
32. Williams, S.L.; Rindfleisch, D.F.; Williams, R.L. Deadend on haloacetic acids (HAA). *Proc. AWWA Water Quality Technology Conf.*, San Francisco, CA, 1994.
33. Williams, S.L.; Williams, R.L.; Gordon, A.S. The impact of bacterial degradation of haloacetic acids (HAA) in the distribution system. *Proc. AWWA Water Quality Technology Conf.*, Boston, MA, 1996.
34. Schneider, B.; Muller, R.; Frank, R.; Lingens, F.. Complete nucleotide sequences and comparison of the structural genes of two 2-haloalkanoic acid dehalogenases from *Pseudomonas* sp. strain CBS3. *J. Bacteriol.* **1991**, *173*, 1530-1535.
35. Van der Ploeg, J.; Willemsen, M.; van Hall, G.; Hanssen, D.B. Adaptation of *Xanthobacter autotrophicus* GJ10 to bromoacetate due to activation and mobilization of the haloacetate gene by insertion element IS1247. *J. Bacteriology* **1995**, *177*, 1348-1356.
36. Nardi-Dei, V.; Kurihara, T.; Okamura, T.; Liu, J.Q.; Koshikawa, H.; Ozaki, H.; Terashima, Y.; Esaki, N.; Soda, K. Comparative studies of genes encoding thermostable L-2-halo acid dehalogenase from *Pseudomonas sp.* strain YL, other dehalogenases, and two related hypothetical proteins from *Escherichia coli. Appl. Environ. Microbiol.* **1994**, *60*, 3375-3380.
37. Nardi-Dei, V.; Kurihara, T.; Park, C.; Esaki, N.; Soda, K. Bacterial DL-2-haloacid dehalogenase from *Pseudomonas sp.* strain 113: gene cloning and structural comparison with D- and L-2-haloacid dehalogenases. *J. Bacteriol.* **1997**, *179*, 4232-4238.
38. Sota, M.; Endo, M.; Keiji, N.; Kawasaki, H.; Tsuda, M. Characterization of a class II defective transposon carrying two haloacetate dehalogenase genes from *Delftia acidovorans* plasmid pUO1. *Appl. Environ. Microbiol.* **2002**, *68*, 2307-2315.
39. Hill, K.E.; Marchesi, J.R.; Weightman, A.J. Investigation of two evolutionary unrelated halocarboxylic acid dehalogenase gene families. *J. Bacteriol.* **1999**, *181*, 2535-2547.
40. Chun, C.; Hozalski, R.M.; Arnold, W.A. Degradation of disinfection byproducts by synthetic goethite and magnetite. *Environ. Sci. Technol.* **2005**, *39*, 8525-8532.

41. Stookey, L.L. Ferrozine- a new spectrophotometric reagent for iron. *Anal. Chem.* **1970**, *42*, 779-781.
42. Lee, J.; Hozalski, R.M.; Arnold, W.A. Effects of dissolved oxygen and iron aging on the reduction of trichloronitromethane, trichloracetonitrile, and trichloropropanone. *Chemosphere* **2007**, *66*, 2127-2135.
43. Scherer, M.M.; Westall, J.C.; Ziomek-Moroz, M.; Tratnyek, P.G. Kinetics of carbon tetrachloride reduction at an oxide-free iron electrode. *Environ. Sci. Technol.* **1997**, *31*, 2385-2391.
44. Chun, C.; Hozalski, R.M.; Arnold, W.A. Degradation of disinfection byproducts by carbonate green rust. *Environ. Sci. Technol.* **2007**, *41*, 1615-1621.

Indexes

Author Index

Ando, Ryo, 65
Andrews, Robert, 289
Arnold, William A., 334
Blatchley, Ernest R., III, 172
Bougeard, Cynthia M. M., 95
Boyd, Jessica M., 289
Bradford, Christine N., 319
Bull, Richard J., 51
Chang, Hyun-Shik, 158
Chen, Baiyang, 257, 274
Chiu, Chao-An, 140
Chow, Christopher W. K., 214
Chun, Chanlan, 334
Clore, Tim J., 319
Cox, Casandra R., 319
Dixon, Mike, 227
Dotson, Aaron D., 274
Drikas, Mary, 214, 227
Echigo, Shinya, 65
Fabris, Rolando, 214
Goslan, Emma H., 95
Hong, Ying, 123
Hozalski, Raymond M., 334
Hrudey, Steve, 289
Hsu, Kang-Mei, 36
Hua, Guanghui, 108
Itoh, Sadahiko, 65
Janmohamed, Imran H. S., 95
Jefferson, Bruce, 95
Karanfil, Tanju, 2, 123, 242
Kiddle, James J., 319
Kilduff, James E., 242
Korshin, Gregory V., 158, 198
Krasner, Stuart W., 2, 257, 304
Landsman, Nicholas A., 319
LaPara, Timothy M., 334
Lee, Chih Fen Tiffany, 304
Lee, Jeong-Yub, 334
Lee, Wontae, 274
Li, Jing, 172
Li, Xing-Fang, 289
Makdissy, Gladys, 80
Mezyk, Stephen P., 319
Morgan, Geraint H., 95
Morran, Jim Y., 227
Muellner, Mark G., 36
Parsons, Simon A., 95
Pearson, Carrie R., 334
Plewa, Michael J., 36
Pope, Phillip G., 182
Reckhow, David A., 51, 80, 108
Rees, Paula S., 80
Richardson, Susan D., 36
Roberson, J. Alan, 22
Roccaro, Paolo, 198
Schramm, Jessica, 304
Sclimenti, Michael J., 304
Song, Hocheol, 123
Speitel, Gerald E., Jr., 182
Swancutt, Katy L., 319
Tan, Yongrui, 242
Vagliasindi, Federico G. A., 198
Wagner, Elizabeth D., 36

Wang, Gen-Shuh, 140
Watson, Jonathan S., 95
Westerhoff, Paul, 2, 257, 274
Woodbeck, Matthew, 289
Xie, Yuefeng, 2
Zhang, Ping, 334
Zhao, Yuan Yuan, 289

Subject Index

A

Abiotic batch experiments, disinfection by-products degradation, 337 338
Abiotic disinfection by-products degradation and iron pipe corrosion, 335 336
Activated carbon. *See* Granular activated carbon; Powdered activated carbon
Adelaide, Australia, Myponga Reservoir source water for trihalomethane precursor reduction treatments, 214 226
Adsorbent technologies with coagulation, combined treatments for reduction of trihalomethane precursors, 220 224
Adsorption processes for disinfection by-products loss, 261
Adsorption significance for disinfection by-products fate, 267, 269*f*
Advanced oxidation process
  effect on nitrosamine formation, 299 301
  free radical chemistry, nitrosamine removal, 319 333
Algae. *See Microcystic aeruginosa, Oscillatoria prolifera*
Algal growth promotion of dissolved organic nitrogen, 280
Algal suspensions, trihalomethanes formation, 146 148*f*
Alum coagulation and activated carbon adsorption, DPBs formation potentials, 243
Amino acids and amino sugars, dissolved organic nitrogen components, 277, 280 284*t*
Analytical devices for haloacetic acid measurements, 99 100
Ancipa reservoir source water for spectroscopic indices, comparisons, halogenation, natural organic matter, quantification, 198 212
Anion exchange efficiencies and activated carbon, dissolved organic matter removal, comparisons, 244 245
Anion exchange resin processes for removing dissolved organic matter, 87, 220 224*f*, 227 241, 242 256
  See also MIEX resin entries; Resin entries
Arginine,
  dissolved organic nitrogen component, 282 284*t*
  in recreational water, volatile DBP formation, 173 174*f*, 179 180
Arizona, Saguaro Lake, nitrogen-rich waste water, 284
Australia
  Mt Pleasant Water Treatment Plant, MIEX DOC processes, 228 238
  Myponga Reservoir source water for trihalomethane precursor reduction, 214 226
  regulations, DBPs in drinking water, 14
AwwaRF project, Organic Nitrogen in Drinking Water and Reclaimed Wastewater, 279

B

Big Sioux Aquifer, waters, spectroscopic studies, chromophores in natural organic matter chlorination and disinfection by-product formation, 158 171
Biodegradation, disinfection by-products, 336 337, 342 344
Biodegradation rate constants for DBPs loss processes, 262 265
Biological wastewater treatments, effect on dissolved organic nitrogen, 282 284*f*
Bladder cancer risk, 11 12, 52
Bromamine reactivity, 183 184, 187 191*f*
Bromide effect on formation
  dihalogenated haloacetic acids, and speciation during chloramination, 129, 133 136*f*
  trihalomethanes and haloacetic acids, 102, 104, 105*f*
Bromide in drinking water sources, 243 244
Bromide removal by M43 resin, 249 250*f*
Bromide spiked natural waters, chlorination, 111, 114 117*f*
Bromine incorporation factor, 253
Bromine-substituted haloamines, reactivity in forming haloacetic acids, 182 197
Bromine substitution factor, 119 120
Bromine substitution in trihalomethanes, 252 254*f*
Bromine total organic, relation to brominated trihalomethanes and haloacetic acids during drinking water chlorination, 109 123
Bromoacetic acids, mutagenicity, 66
Bromochloramine reactivity, 184, 190, 192 195*f*
Bromoiodoacetamide, 46 47*f*
  genotoxicity, 41 42*f*

C

California Department of Health Services, 320
California State Project water, 306
Cambridge, Massachusetts, chlorination, natural waters, 112 113*t*
Canada, Repentigny, Quebec, chlorination, natural waters, 112 113*t*
Carbonaceous disinfection by-products, toxicity, 44 46*f*
CCLs. *See* Containment Candidate Lists
China, regulations, disinfection by-products in drinking water, 13 14
Chinese hamster ovary cell, cytotoxicity and genotoxicity assays, 37 43*f*
Chloramination
  haloacetic acid formation and speciation, 124 140
  nitrosamine formation, 298 299, 300*f*
  nitrosodimethylamine formation, 307 308
  reverse osmosis isolates, natural waters, 126 127
Chlorination
  bromide spiked natural waters, 114 117*f*
  combination with chloramination, well nitrified effluent organic matter, 314 315
  natural organic matter fractions, 117 120*t*
  natural waters, 120 121*f*
Chlorine demand, effect of MIEX treatment, 232*f*–233
3-Chloro-4-(chloromethyl)-5-hydroxy-2-(5H)furanone, 57
3-Chloro-4-(dichloromethyl)-2-furanone, 59

Chloroform
fate-and-transport modeling, receiving water, wastewater-derived DPBs, 267, 270 271*f*
formation, pre-chlorination effects, 146 148, 151, 153 154*f*
Chromophores, spectroscopic studies
natural organic matter chlorination and disinfection by-products formation, 158 171
natural organic matter chlorination, differential absorbance, 166, 168*f*–169
Chromosomal aberration test for organic bromine contribution to genotoxicity, chlorinated water, 68, 69, 72 75
Cincinnati, Ohio, waters, spectroscopic studies, chromophores in natural organic matter chlorination and disinfection by-products formation, 158 171
Coagulation with adsorbent technologies for reduction of trihalomethane precursors, 220 224*f*
Colorado River water, 306
Comet assay. *See* Single cell gel electrophoresis
Compliance data, Disinfection By-products Rule, 26 28
Connecticut, various locations, raw and treated water collections, disinfection by-products precursor content, 84
Containment Candidate Lists (CCLs), USEPA, 29
Creatinine in recreational water, volatile disinfection by-products formation during chlorination, 173 174*f*, 176 178, 179 180
Cryptosporidium, 10, 31
Cyanogen chloride, volatile disinfection by-products formation during chlorination, recreational water, 174*f*, 175, 178 179
Cytotoxicity and genotoxicity indices, disinfection by-products, 44 46*f*
Cytotoxicity assay, chronic, disinfection by-products, 37 40

D

Dallas, Texas, chlorination, natural waters, 112 113*t*
Data adjustment for disinfection by-products, precursor density, power function modeling, 82 83
DBP. *See* Disinfection by-products
DBPR. *See* Disinfection By-Products Rule
Degradation efficiency measurements, hydroxyl radical oxidative removal efficiencies, nitrosamines, 327 330
Developmental toxicants, disinfection by-products, 61*t*–62
Dialysis pretreatment in measurement of dissolved organic nitrogen, 275 276, 280
Dichloroacetic acid, fate-and-transport modeling in receiving waters, wastewater-derived disinfection by-products, 269 271*f*
Dichloroacetonitrile, 57 58*t*
fate-and-transport modeling in receiving waters, wastewater-derived DBPs, 269 271*f*
volatile DBPs formation during chlorination, recreational water, 174*f*, 175, 178 180
2,6-Dichloro-1,4-dibenzoquinone 4-(N-chloro) imine, formation, 59 60*f*
Dichloromethylamine, volatile disinfection by-products formation during chlorination, recreational water, 174*f*, 175, 176 178*f*
Differential absorbance
chromophores in natural organic matter chlorination and DBPs formation, 158 171

relation to disinfection by-product formation, 203 206*f*
See also UV-absorbance
Differential fluorescence index, natural organic matter, 203
Dihaloacetic acid precursors, 89
correlation between specific UV absorbance, 84 86*f*
Dihalogenated haloacetic acids (DXAAs), formation and speciation during chloramination, 127 136*f*
Disinfection by-products (DBPs) degradation, halogenated, water distribution systems, 334 348
Disinfection by-products formation and natural dissolved organic matter removal, ion exchange and powdered activated carbon comparison, 242 256
Disinfection by-products formation and natural organic matter chlorination, chromophore roles, spectroscopic studies, 158 171
Disinfection by-products from wastewaters in surface waters, 257 273
Disinfection by-products in drinking water
formation, occurrence, control, health effects, and regulation (overview), 2 19
mammalian cell toxicity, 36 50
novel, identification, possible toxicological concern, 51 64
regulations, 12 14, 22 35
Disinfection by-products in recreational water, formation from organic-N precursor chlorination, 172 181
Disinfection by-products loss rate parameters, estimation, and modeling framework, 258 262
Disinfection byproducts precursor content, natural organic matter extracts, 80 94
natural organic matter fractions, internally consistent sets, 84 87*f*
Disinfection by-products reduction
by iron (0), 339 340
mediated by iron minerals, 340 342, 343*f*
Disinfection By-Products Rule (DBPR), USEPA, 25 26, 124 125
compliance data, 26 28
Disinfection by-products speciation and bromine incorporation, 252 254*f*
Dissolved organic carbon, molecular weight distribution, removed by MIEX pre-treatment with GAC filtration, 235 238*f*
Dissolved organic carbon removal, MIEX pre-treatment effect on GAC performance, 230 232*f*
Dissolved organic matter (DOM)
dihalogenated haloacetic acids, formation and speciation, chloramination, 135, 137
from nitrogen-rich waters, 284 287
heterogeneity, 243
organic precursors, disinfection by-products, 6 7
removal efficiencies, powdered activated carbon with alum, 218 220*f*
reverse osmosis isolation, 125 126*t*
uptake, effects on molecular weight and SUVA by weak base resin and GAC, 247 249*f*
Dissolved organic nitrogen occurrence and characterization, 274 288
DOM. *See* Dissolved organic matter
Drinking water and disinfection by-products (overview), 2 19
Drinking water chlorination, brominated, trihalomethanes, haloacetic acids, and total organic bromine relationships, 109 123

Drinking water treatment, organic nitrogen, 279 284*f*
Drinking waters
effluent-impacted, nitrosamine formation, 304 318
treated in UK, haloacetic acid and trihalomethane, 95 108
DXAAs. *See* Dihalogenated haloacetic acids

E

Effluent-impacted drinking waters, nitrosamine formation, 304 318
Effluent organic matter
poorly nitrified, breakpoint chlorination, 309 314*f*
well nitrified, chlorination/chloramination, 314 315
Elgin, Illinois, chlorination, natural waters, 112 113*t*
Enhanced Surface Water Treatment Rules (ESWTRs), USEPA, 24 26, 31
Environmental Protection Agency (US). *See USEPA entries*
European Union regulations, disinfection by-products in drinking water, 12 13*t*
Extracellular organic matter, trihalomethanes formation, 148 150

F

Fate-and-transport modeling, 258 259
in receiving waters, wastewater-derived disinfection by-products, 267, 269 271*f*
Florence, Nebraska, waters, spectroscopic studies, chromophores in natural organic matter chlorination and disinfection by-product formation, 158 171
Florida, Hillsborough River waters, spectroscopic studies, chromophores in natural organic matter chlorination and disinfection by-products formation, 158 171
Florida, Tampa, chlorination, natural organic matter fractions, 111 112, 117 120*t*
Fluorescence in natural organic matter, 199 200
chlorination, 201, 203, 204*f*, 207 210*f*
Fluorescence indices, halogenation and DBP release, 207 210*f*
Free-chlorine contact time impact on NDMA formation, 307 315
Fulvic acids in analysis of published data on hydrophobic-based natural organic matter fractions, 87 92

G

GAC. *See* Granular activated carbon
GC-ECD (gas chromatography-electron capture detection), 99 100, 113
GC-MS for haloacetic acid measurements, 99 100
Genotoxicity
acute, disinfection by-products, 40 43*f*
and cytotoxicity indices, disinfection by-products, 44 46*f*
chlorinated water, organic bromine contribution, 65 77
Giardia lamblia, 24
Granular activated carbon (GAC)
effect on trihalomethanes formation, 250 252*f*
filters, 10 11

with ion exchange, natural dissolved organic matter removal, DBP formation, 242 256
with MIEX resin, trihalomethane precursor removal, 227 241
Greenville, South Carolina, drinking water treatment plant, dissolved organic matter concentration from influent, 125 126*t*
Ground Water Rule (GWR), 26
Growth condition trends, algae, chlorophyll a concentration measurement, 144 145*f*

H

HAA. *See* Haloacetic acids
Haloacetaldehydes, disinfection by-products in drinking water, 5
Haloacetic acids (HAA)
degrading enrichment cultures, establishment, 338 339
precursors, correlation with specific UV absorbance, 84 86*f*
reduction by activated carbon and ion exchange resin, 250 252*f*
regulation standard (US), 30
Haloacetic acids and trihalomethanes, brominated, relation to total organic bromine during drinking water chlorination, 109 123
Haloacetic acids formation
and speciation during chloramination, 124 140
by bromine-substituted haloamines, 182 197
parameters, 102 107
Haloamine reactivities, 185, 186*t*–196
forming haloacetic acids, 182 197
Halogenated DBP degradation in water distribution systems, 334 348
Halogenated quinones, formation, 59 60*f*
Halonitriles, 57 58*t*
potential developmental toxicants, 61*t*–62
Haloquinone derivatives, 57
Health effects, disinfection by-products in drinking water, 11 12
Hillsborough River, Florida, waters, spectroscopic studies, chromophores in natural organic matter chlorination and disinfection by-products formation, 158 171
Histidine in recreational water, volatile disinfection by-products formation during chlorination, 173 174*f*, 178 180
HPLC-MS/MS analysis, N-nitrosamines, 291 295*f*
Human cell toxicogenomic analysis, disinfection by-products, 46 48*f*
Hydrated electron reactions and rate constants, nitrosamines, 324, 326 327*t*
Hydrolysis, disinfection by-products, 261, 342
Hydrolysis rate constants for disinfection by-products loss processes, 264*t*, 266, 267*f*
Hydrophilic-based natural organic matter fractions. *See* Hydrophobic-based natural organic matter fractions
Hydrophobic-based natural organic matter fractions, analysis of published data, 87 92*f*
Hydroxyl radical oxidative removal efficiencies, degradation efficiency measurements, nitrosamines, 327 330
Hydroxyl radical reactions and hydroxyl radical degradation efficiencies, nitrosamines, 322 325*t*
Hypoiodous acid. *See* Iodinated acids

I

IARC, 290
Illinois, Elgin, chlorination, natural waters, 112 113*t*
Incubation temperature, effect on algal cell characteristics, 151 155
Indices, cytotoxicity and genotoxicity, disinfection by-products classes, 44 46*f*
International Agency for Research on Cancer, 290
Iodinated acids, disinfection by-products in drinking water, 3
Ion exchange and powdered activated carbon comparison, natural dissolved organic matter removal and disinfection byproduct formation, 242 256
Iron elemental in disinfection by-products reduction, 339 340
Iron minerals in disinfection by-products reduction mediation, 340 342, 343*f*
Iron pipe corrosion and abiotic disinfection by-products degradation, 335 336
Italy, Ancipa reservoir source water for spectroscopic indices, comparisons, halogenation, natural organic matter, quantification, 198 212

J

Japan, Lake Biwa, water source, 67
Japan regulations, disinfection by-products in drinking water, 13

K

Kjeldahl nitrogen for measurement of dissolved organic nitrogen, 275 276

L

Lake Austin source water, haloamine reactivities, 187 189*f*, 192 195*f*
Lake Biwa, Japan, water source, 67
Lake Washington, Washington, waters, spectroscopic studies, chromophores in natural organic matter chlorination and disinfection by-products formation, 158 171
Lead Copper Rule, USEPA, violations, 27 28*f*

M

M43, anion-exchange resin
  bromide removal, 249 250*f*
  natural organic matter removal, 245
Mammalian cell toxicity, disinfection by-products in drinking water, 36 50
Manitoba, Winnipeg, bromide spike natural waters, chlorination, 111, 114 116
Massachusetts
  Cambridge, chlorination, natural waters, 112 113*t*
  Springfield, chlorination, natural organic matter fractions, 111 112
  various locations, raw and treated water collections, DBP precursor content, 84
Maximum contaminant levels (MCL), 25 26
Maximum Residual Disinfectant Levels (MRDL), 25
Membrane Introduction Mass Spectrometry, 185 186*f*
Metedeconk River source water, haloamine reactivities, 188, 190 191*f*, 194 196
Microcystic aeruginosa, pre-chlorination induced DOC and

disinfection by-product formation in treatment processes, 141 156
MIEX resin, combined treatment, adsorbent technologies with coagulation, for reduction of trihalomethane precursors, 220 224*f*
MIEX resin, dissolved organic carbon process with granular activated carbon filtration, trihalomethane precursor removal, 227 241
See also Anion exchange entries; *Resin entries*
Milwaukee outbreak, 31
Missouri River water, spectroscopic studies, chromophores in natural organic matter chlorination and disinfection by-product formation, 158 171
Molecular weight distribution
dissolved organic carbon removed by MIEX pre-treatment, GAC filtration, 235 238*f*
dissolved organic matter solutions, measurements, 245
MRDL. *See* Maximum Residual Disinfectant Levels
Mt Pleasant Water Treatment Plant, South Australia, MIEX DOC processes, 228 238
Multi-barrier approach to drinking water treatment, 215
Myponga Reservoir source water for trihalomethane precursor reduction treatments, 214 226
Myrtle Beach, South Carolina, drinking water treatment plant, dissolved organic matter concentration from influent, 125 126*t*

N

National Interim Primary Drinking Water Regulations, 23
Natural dissolved organic matter removal and DBP formation, ion exchange and powdered activated carbon comparison, 242 256
Natural organic matter (NOM)
chlorination and DBP formation, chromophore roles, spectroscopic studies, 158 171
chlorination, UV-absorbance and fluorescence spectra, 201, 203, 204*f*
extracts, disinfection byproduct precursor content, 80 94
fractions, chlorination, 117 120*t*
fractions, DBP precursor content in internally consistent sets, 84 87*f*
halogenation quantification, spectroscopic indices, comparisons, 198 212
Natural waters, chlorination, 112 113*t*, 120 121*f*
NDMA. *See* N-Nitrosodimethylamine
Nebraska, Florence, waters, spectroscopic studies, chromophores in natural organic matter chlorination and disinfection by-product formation, 158 171
Negotiated rulemaking, USEPA, 24 28
New York, Tomhannock Reservoir, natural water for dissolved organic matter removal and subsequent disinfection byproduct formation, 245 255
Newport News, Virginia, chlorination, natural waters, 112 113*t*
Nitric oxide production in free-radical-based advanced oxidation processes, removal of nitrosamines, 330 331
Nitrogen-containing disinfection by-products, toxicity, 44 45*f*
Nitrogen-rich waters, composition of dissolved organic matter, 284 286*f*

Nitrogenous disinfection by-products in drinking water, 4 5
Nitrosamine formation in effluent-impacted drinking waters, 304 318
Nitrosamines, 31, 291 295*f*
hydrated electron reactions, 324, 326 327*t*
hydroxyl radical oxidative removal, degradation efficiency measurements, 327 330
hydroxyl radical reactions and hydroxyl radical degradation efficiencies, 322 325*t*
removal, free radical chemistry, 319 333
source waters treated with disinfection processes, 289 303
SPE enrichment and HPLC separation in HPLC-MS/MS analysis, 291 293*f*, 297
N-Nitrosodimethylamine (NDMA), 4, 290, 293, 295*f*, 297 301, 305
fate-and-transport modeling in receiving waters, wastewater-derived DBPs, 269 271*f*
formation during breakpoint chlorination, 309 314*f*
formation during chloramination, 307 308
formation, free-chlorine contact time impact on, 307 315
hydroxyl radical reactions rate constants and degradation efficiencies, 322 325*t*
N-Nitrosodiphenylamine identification as a disinfection by-product, 293 295*f*
NOM. *See* Natural organic matter
Novel disinfection by-products, identification, possible toxicological concern, 51 64

O

Ohio, Cincinnati, waters, spectroscopic studies, chromophores in natural organic matter chlorination and disinfection by-products formation, 158 171
Oklahoma, Tulsa, bromide spiked natural waters, chlorination, 111, 114 117*f*
Ontario Ministry of the Environment, drinking water quality standard for NDMA, 290
Organic bromine contribution, genotoxicity, chlorinated water, 65 77
Organic-N precursor chlorination, formation of disinfection by-products in recreational water, 172 181
Organic Nitrogen in Drinking Water and Reclaimed Wastewater, AwwaRF project, 279
Oscillatoria prolifera, 284
Oxidation/disinfection strategies (overview), 9 10
Ozone water treatment, effect on nitrosamine formation, 299, 300*f*

P

pH effects
dihalogenated haloacetic acids, formation and speciation, chloramination, 127, 129 136
trihalomethanes and haloacetic acids, formation, 102, 103*f*, 105*f*
Photolysis rate constants for disinfection by-products loss, 260
Pipe wall and flowing water interface in drinking water distribution systems, 335

Population attributable risks, chlorinated drinking water, USEPA, 52 53
Potential carcinogens, screening, 51 62
Powdered activated carbon
compared to ion exchange, NOM removal and DBP formation, 242 256
evaluation protocol, 216 217
performance characterization, 217 220*f*
with coagulation, combined treatment for trihalomethane precursor reduction, 220 224*f*
Power function modeling, data adjustment for DBPs, precursor density, 82 83
Pre-chlorination effects
for algae cultivated at different temperature, 151 155
on nonpurgeable dissolved organic carbon concentration, bulk algal suspensions, 144
on trihalomethane and haloacetic acid species distribution, 150 152*f*
Pre-chlorination induced dissolved organic matter and disinfection by-product formation from *Microcystic aeruginosa*, 141 156
Precursor content, disinfection byproducts, natural organic matter extracts, 80 94
Prioritization scheme, identification, potential toxic novel DBPs, 53 56
Pseudomonas, 320, 337

Q

QSTR analysis, potential disinfection by-products, 60 62
QSTR program for identification, potential toxic novel DBPs, TOPKAT, 53 56
Quantitative structure toxicity relationship program. *See* QSTR program
Quebec, Repentigny, chlorination, natural waters, 112 113*t*
Quenching agents, effect on haloacetic acids formation routes during chloramination, 137 138
Quinones, halogenated, formation, 59 60*f*

R

Rate constant estimation for disinfection by-products loss processes, 262 269*f*
Recreational water, volatile disinfection by-products formation from organic-N precursor chlorination, 172 181
Reg-Neg. *See* Negotiated rulemaking
Regulations, disinfection by-products in drinking water, 12 14, 22 35
Repentigny, Quebec, chlorination, natural waters, 112 113*t*
Reproductive outcomes, chlorinated drinking water, 52
Resin isolation, nitrogen rich natural organic matter, 277 279*t*
Resin sorption effects, trihalomethanes formation, 250 252*f*
See also Anion exchange entries; MIEX resin entries
Reverse osmosis isolation, dissolved organic matter, 125 126*t*

S

Safe Drinking Water Act (SDWA), 23, 29 30
Safe Drinking Water Information System (SDWIS/FED) for tracking compliance, 26 28
Saguaro Lake, Arizona, nitrogen-rich waste water, 284
SDWA. *See* Safe Drinking Water Act
Seattle, Washington, Lake Washington waters, spectroscopic studies, chromophores in natural organic matter chlorination and DPB formation, 158 171
Sicily, Italy, Ancipa reservoir source water for spectroscopic indices, comparisons, halogenation, natural organic matter, quantification, 198 212
Single cell gel electrophoresis, genotoxicity, 40 43*f*
Sioux Falls, South Dakota, waters, spectroscopic studies, chromophores in natural organic matter chlorination and disinfection by-product formation, 158 171
South Australia, Mt Pleasant Water Treatment Plant, MIEX DOC processes, 228 238
South Carolina
  Greenville, drinking water treatment plant, DOM from influent, 125 126*t*
  Myrtle Beach, drinking water treatment plant, DOM from influent, 125 126*t*
South Dakota, Sioux Falls, waters, spectroscopic studies, chromophores in NOM chlorination and DBP formation, 158 171
SPE enrichment and HPLC separation with tandem MS and multiple reaction monitoring in HPLC-MS/MS analysis, nitrosamines, 291 293*f*, 297
Specific ultra violet absorbance (SUVA), 243
Spectroscopic indices for halogenation NOM quantification, 198 212
Spectroscopic studies, chromophores in NOM chlorination and DBP formation, 158 171
Springfield, Massachusetts, chlorination, NOM fractions, 111 112, 120*t*
Straw man rule, USEPA, 24
Surface Water Treatment Rule, 24
Surface waters, fate and transport of wastewater-derived DBPs, 257 273
SUVA. *See* Specific ultra violet absorbance
Suwannee River, nitrogenous disinfection by-products, 4
Swimming pool water. *See* Recreational water

T

Tampa, Florida, chlorination, natural organic matter fractions, 111 112
Temperature effects
  on algae cultivation, pre-chlorination, 151 155
  on dissolved organic nitrogen, 280 281*f*
  on formation of trihalomethanes and haloacetic acids, 104, 106
  on natural organic matter halogenation, 202*f*, 203, 205
Texas, Dallas, chlorination, natural waters, 112 113*t*
Texas, Waco, chlorination, natural waters, 112 113*t*
THAAs. *See* Trihaloacetic acids
THMs. *See* Trihalomethanes

Timeframe, nitrosamine formation in water distribution systems, 293 295*f*
TOBr. *See* Total organic bromine
TOCl. *See* Total organic chlorine
TOG, coal-based activated carbon for adsorption and anion exchange efficiencies, 245
Tomhannock Reservoir, natural water for dissolved organic matter removal and subsequent disinfection byproduct formation, 245 255
TOPKAT, QSTR program for identification, potential toxic novel DBPs, 53 56, 59
Total Coliform Rule, USEPA, 27 28*f*, 29 30
Total organic bromine (TOBr)
  activity inducing chromosomal aberrations, 72 75
  formation, effect of bromide ion concentration and chlorine dose, 69 72*f*
  relationship with brominated disinfection byproducts, 109 123
Total organic chlorine (TOCl), formation, effect of bromide ion concentration and chlorine dose, 69 70*f*
Total organic halogen (TOX) formation, effect of bromide ion concentration and chlorine dose, 69 72*f*
Total Trihalomethane Rule, USEPA, 23 24
TOX. *See* Total organic halogen
Toxicity to mammalian cells
  carbonaceous disinfection by-products, 44 46*f*
  disinfection by-products in drinking water, 36 50
  nitrogen-containing disinfection by-products, 44 45*f*
Toxicological concern, identification, novel disinfection by-products, 51 64
Trichloramine, volatile DBPs formation during chlorination, recreational water, 174*f*, 175, 176 178*f*, 179 180
Trihaloacetic acids, correlation with specific UV absorbance, 84, 86*f*
Trihalomethane precursors
  analysis of published data on hydrophobic-based NOM fractions, 88 89*f*
  combined treatments for enhanced reduction, 214 226
  removal using MIEX DOC process with granular activated carbon, 227 241
Trihalomethanes and haloacetic acids brominated, relation to total organic bromine during drinking water chlorination, 109 123
Trihalomethanes, bromine substitution, 252 254*f*
Trihalomethanes, formation
  from algal suspensions, 146 148*f*
  from extracellular organic matter, 148 150
  MIEX treatment effects, 233 235
  parameters, 102 107
  resin sorption effects, 250 252*f*
Trihalomethanes, iodinated, disinfection by-products in drinking water, 3
Troy, New York, Tomhannock Reservoir, natural water for dissolved organic matter removal and subsequent disinfection byproduct formation, 245 255
Tulsa, Oklahoma, bromide spiked natural waters, chlorination, 111, 114 117*f*

U

UK treated drinking waters, haloacetic acid and trihalomethane concentrations, 95 108
Uniform Formation Conditions protocol for chlorination, 246
Unregulated by US, disinfection by-products, 31 32
Unregulated Contaminant Monitoring Rule, 31, 290
Urea in recreational water, volatile disinfection by-products formation during chlorination, 173 174*f*, 179 180
USEPA, population attributable risks, chlorinated drinking water, 52 53
USEPA regulations, disinfection by-products in drinking water, 12, 13*t*, 22 35
USEPA rulemaking, 22 49
UV-absorbance
  and fluorescence spectra in natural organic matter chlorination, 201, 203, 204*f*
  dihaloacetic acid precursors, 84 86*f*
  removal with alum and powdered activated carbon, 217 218*f*
  See also Differential absorbance
UV water disinfection, effect on nitrosamine formation, 299, 300*f*

V

Variations, spatial and temporal, disinfection by-products in drinking water, 5 6
Virginia, Newport News, chlorination, natural waters, 112 113*t*
Volatilization controlling factors for disinfection by-products, 261
Volatilization rate constants for disinfection by-products loss processes, 266, 268*f*

W

Waco, Texas, chlorination, natural waters, 112 113*t*
Washington, Lake Washington waters, spectroscopic studies, chromophores in natural organic matter chlorination and disinfection by-product formation, 158 171
Wastewater-derived disinfection by-products in surface waters, 257 273
Wastewater treatments, effect on dissolved organic nitrogen, 282 284*f*
Water characterization (UK water), 101 102
Water disinfection treatments, nitrosamine formation, 296 301
Water distribution systems
  halogenated disinfection byproduct degradation, 334 348
  nitrosamine formation timeframes, 293 295*f*
Water quality parameters, coagulation with adsorbent technologies, reduction of trihalomethane precursors, 220 224*f*
Winnipeg, Manitoba, bromide spiked natural waters, chlorination, 111, 114 116
World Health Organization, guidelines, disinfection by-products in drinking water, 12

X

XAD-8 resin in removal, hydrophobic acids, 87

www.ingramcontent.com/pod-product-compliance
Ingram Content Group UK Ltd.
Pitfield, Milton Keynes, MK11 3LW, UK
UKHW020616190926
15216UKWH00014B/3

9 780841 269507